T0207352

Communications in Computer and Information Science 1931

Rationale
The CCIS series is devoted to the publication of proceedings of computer science conferences. Its aim is to efficiently disseminate original research results in informatics in printed and electronic form. While the focus is on publication of peer-reviewed full papers presenting mature work, inclusion of reviewed short papers reporting on work in progress is welcome, too. Besides globally relevant meetings with internationally representative program committees guaranteeing a strict peer-reviewing and paper selection process, conferences run by societies or of high regional or national relevance are also considered for publication.

Topics
The topical scope of CCIS spans the entire spectrum of informatics ranging from foundational topics in the theory of computing to information and communications science and technology and a broad variety of interdisciplinary application fields.

Information for Volume Editors and Authors
Publication in CCIS is free of charge. No royalties are paid, however, we offer registered conference participants temporary free access to the online version of the conference proceedings on SpringerLink (http://link.springer.com) by means of an http referrer from the conference website and/or a number of complimentary printed copies, as specified in the official acceptance email of the event.

CCIS proceedings can be published in time for distribution at conferences or as post-proceedings, and delivered in the form of printed books and/or electronically as USBs and/or e-content licenses for accessing proceedings at SpringerLink. Furthermore, CCIS proceedings are included in the CCIS electronic book series hosted in the SpringerLink digital library at http://link.springer.com/bookseries/7899. Conferences publishing in CCIS are allowed to use Online Conference Service (OCS) for managing the whole proceedings lifecycle (from submission and reviewing to preparing for publication) free of charge.

Publication process
The language of publication is exclusively English. Authors publishing in CCIS have to sign the Springer CCIS copyright transfer form, however, they are free to use their material published in CCIS for substantially changed, more elaborate subsequent publications elsewhere. For the preparation of the camera-ready papers/files, authors have to strictly adhere to the Springer CCIS Authors' Instructions and are strongly encouraged to use the CCIS LaTeX style files or templates.

Abstracting/Indexing
CCIS is abstracted/indexed in DBLP, Google Scholar, EI-Compendex, Mathematical Reviews, SCImago, Scopus. CCIS volumes are also submitted for the inclusion in ISI Proceedings.

How to start
To start the evaluation of your proposal for inclusion in the CCIS series, please send an e-mail to ccis@springer.com.

Bin Xin · Naoyuki Kubota · Kewei Chen ·
Fangyan Dong
Editors

Advanced Computational Intelligence and Intelligent Informatics

8th International Workshop, IWACIII 2023
Beijing, China, November 3–5, 2023
Proceedings, Part I

Editors
Bin Xin
Beijing Institute of Technology
Beijing, China

Naoyuki Kubota
Tokyo Metropolitan University
Tokyo, Japan

Kewei Chen
Ningbo University
Ningbo, China

Fangyan Dong
Ningbo University
Ningbo, China

ISSN 1865-0929 ISSN 1865-0937 (electronic)
Communications in Computer and Information Science
ISBN 978-981-99-7589-1 ISBN 978-981-99-7590-7 (eBook)
https://doi.org/10.1007/978-981-99-7590-7

This Springer imprint is published by the registered company Springer Nature Singapore Pte Ltd.
The registered company address is: 152 Beach Road, #21-01/04 Gateway East, Singapore 189721, Singapore

Preface

This volume contains the papers from the 8th International Workshop on Advanced Computational Intelligence and Intelligent Informatics (IWACIII 2023).

IWACIII is an international symposium funded in 2009 by Kaoru Hirota, a professor from the School of Automation, Beijing Institute of Technology, and is held every two years. Unremittingly, IWACIII welcomed its 8th grand event in 2023. IWACIII 2023 was jointly organized by Beijing Institute of Technology and Beijing Association of Automation, Beijing, P. R. China. It provided a forum for scientists and engineers from all over the world to present their theoretical results and techniques in the field of computational intelligence and intelligent informatics.

The topics included in this edition of the event covered the following fields connected to computational intelligence and intelligent informatics: Intelligent information processing, Pattern recognition and computer vision, Intelligent optimization and decision-making, Advanced control, Multi-agent systems, Robotics, and various applications of computational intelligence methods such as neural networks, fuzzy reasoning, evolutionary computing, machine learning, and deep learning. IWACIII 2023 received in total 118 initial submissions from China, Japan and Russia. Finally, 56 papers were accepted. All the accepted papers were peer reviewed by two qualified reviewers, in a single-blind process.

The proceedings editors wish to thank the dedicated scientific committee members and all the other reviewers for their contributions. We also thank the professional editors from Springer for their trust and for publishing the proceedings of IWACIII 2023.

November 2023

Bin Xin
Naoyuki Kubota
Kewei Chen
Fangyan Dong

Organization

Scientific Committee

Program Committee Chairs

Bin Xin	Beijing Institute of Technology, China
Naoyuki Kubota	Tokyo Metropolitan University, Japan
Kewei Chen	Ningbo University, China
Fangyan Dong	Ningbo University, China

Program Committee Members

Yaping Dai	Beijing Institute of Technology, China
Jie Chen	Beijing Institute of Technology, China
Luefeng Chen	China University of Geosciences (Wuhan), China
Xin Chen	China University of Geosciences (Wuhan), China
Elmer P. Dadios	De La Salle University, Philippines
Haobin Dong	China University of Geosciences (Wuhan), China
Hao Fang	Beijing Institute of Technology, China
Toshio Fukuda	Nagoya University, Japan
Kenji Fujimoto	University of Tsukuba, Japan
Edwardo F. Fukushima	Tokyo University of Technology, Japan
Tomomi Hashimoto	University of Tokyo, Japan
Yutaka Hatakeyama	Kochi University, Japan
Yong He	China University of Geosciences (Wuhan), China
Yukio Horiguchi	Kyoto University, Japan
Yukinobu Hoshino	Kochi University of Technology, Japan
Norikazu Ikoma	Nippon Institute of Technology, Japan
Abdullah M. Iliyasu	Prince Sattam Bin Abdulaziz University, Kingdom of Saudi Arabia
Masahiro Inuiguchi	Osaka Metropolitan University, Japan
Hisao Ishibuchi	Osaka Metropolitan University, Japan
Hitoshi Iyatomi	Hosei University, Japan
Janusz Kacprzyk	Polish Academy of Sciences, Poland
Kazuhiko Kawamoto	Chiba University, Japan
Seiichi Kawata	Advanced Institute of Industrial Technology, Japan

Donggyun Kim	Mokpo National Maritime University, South Korea
Syoji Kobashi	Hyogo University, Japan
Ichiro Kobayashi	Ochanomizu University, Japan
László T. Kóczy	Széchenyi István University of Györ, Hungary
Kentarou Kurashige	Muroran Institute of Technology, Japan
Ru Lai	Beijing Institute of Technology, China
Changhe Li	China University of Geosciences (Wuhan), China
Zhihua Li	China University of Geosciences (Wuhan), China
Xiaozhong Liao	Beijing Institute of Technology, China
Guoping Liu	University of South Wales, UK
Xiangdong Liu	Beijing Institute of Technology, China
Zhentao Liu	China University of Geosciences (Wuhan), China
Hongbin Ma	Beijing Institute of Technology, China
Yutaka Matsuo	Tokyo University of Technology, Japan
Masahiro Moniwa	Tokyo University of Technology, Japan
Yuki Nakagawa	RTI Inc., Japan
Yosuke Nakanishi	Waseda University, Japan
Hajime Nobuhara	University of Tsukuba, Japan
Yusuke Nojima	Osaka Metropolitan University, Japan
Clement N. Nyirenda	University of the Western Cape, South Africa
Tomomasa Ohkubo	Tokyo University of Technology, Japan
Kouhei Ohnishi	Keio University, Japan
Sumio Ohno	Tokyo University of Technology, Japan
Kazushi Okamoto	University of Electro-Communications, Japan
Isao Ono	Tokyo Institute of Technology, Japan
Quan Pan	Northwestern Polytechnical University, China
Gyei-Kark Park	Mokpo National Maritime University, South Korea
Witold Pedrycz	University of Alberta, Canada
Nguyen Hoang Phuong	Thang Long University, Vietnam
Anca L. Ralescu	University of Cincinnati, USA
Dan A. Ralescu	University of Cincinnati, USA
Imre J. Rudas	Óbuda University, Hungary
Hidenori Sakaniwa	Hitachi Ltd., Japan
Hirosato Seki	Osaka University, Japan
Jinhua She	Tokyo University of Technology, Japan
Dawei Shi	Beijing Institute of Technology, China
Atsushi Shibata	Advanced Institute of Industrial Technology, Japan
Takanori Shibata	AIST Information Technology Research Institute, Japan

Eri Sato-Shimokawara	Tokyo Metropolitan University, Japan
Zhuoyue Song	Beijing Institute of Technology, China
Joe Spencer	University of Liverpool, UK
Wei Su	Changchun University of Science and Technology, China
Jian Sun	Beijing Institute of Technology, China
Takao Terano	Tokyo Institute of Technology, Japan
Kiyohiko Uehara	Ibaraki University, Japan
Yuki Ueno	Tokyo University of Technology, Japan
Junzheng Wang	Beijing Institute of Technology, China
Qinglin Wang	Beijing Institute of Technology, China
Kok Wai Wong	Murdoch University, Australia
Min Wu	China University of Geosciences (Wuhan), China
Qinghe Wu	Beijing Institute of Technology, China
Yuanqing Xia	Beijing Institute of Technology, China
Yonghua Xiong	China University of Geosciences (Wuhan), China
Li Xu	Okayama Prefactural University, Japan
Toru Yamaguchi	Tokyo Metropolitan University, Japan
Takahiro Yamanoi	Hokkai Gakuen University, Japan
Yamazaki Yoichi	Kanagawa Institute of Technology, Japan
Fei Yan	Changchun University of Science and Technology, China
Jianqiang Yi	Chinese Academy of Sciences, China
Ryuichi Yokoyama	Waseda University, Japan
Shinichi Yoshida	Kochi University of Technology, Japan
Tomohiro Yoshikawa	Suzuki University of Medical Science, Japan
Li Yu	Zhejiang University of Technology, China
Chuanke Zhang	China University of Geosciences (Wuhan), China
Guohun Zhu	University of Queensland, Australia

Organizing Committee

Hongbin Ma	Beijing Institute of Technology, China
Jinhua She	Tokyo University of Technology, Japan
Liqun Han	Chinese Society of Educational Development Strategy, China & Beijing Technology and Business University, China
Bin Xin	Beijing Institute of Technology, China
Naoyuki Kubota	Tokyo Metropolitan University, Japan
Kewei Chen	Ningbo University, China
Fangyan Dong	Ningbo University, China

Yukinobu Hoshino	Kochi University of Technology, Japan
Eri Sato-Shimokawara	Tokyo Metropolitan University, Japan
Xiangyuan Zeng	Beijing Institute of Technology, China
Shinichi Yoshida	Kochi University of Technology, Japan
Zhiyang Jia	Beijing Institute of Technology, China
Sijie Yin	Beijing Institute of Technology, China
Takenori Obo	Tokyo Metropolitan University, Japan
Qing Wang	Beijing Institute of Technology, China
Shuai Shao	Beijing Institute of Technology, China
Aulia S. Azhar	Tokyo Metropolitan University, Japan
Hong Huang	Beijing Institute of Technology, China
Junji Nishino	University of Electro-Communications, Japan
Minling Zhu	Beijing Information Science & Technology University, China
Rongli Li	Beijing Institute of Technology, China

Local Committee

Dawei Shi	Beijing Institute of Technology, China
Yuan Li	Beijing Institute of Technology, China

Contents – Part I

Intelligent Information Processing

3D Point Cloud-Based Lithium Battery Surface Defects Detection Using Region Growing Proposal Approach 3
Zia Ur Rehman, Xin Wang, Abdulrahman Abdo Ali Alsumeri, Malak Abid Ali Khan, and Hongbin Ma

Reducing Communication Consumption in Collaborative Visual SLAM with Map Point Selection and Efficient Data Compression 15
Weiqiang Zhang, Lan Cheng, Xinying Xu, and Zhimin Hu

Optimal Information Fusion Descriptor Fractional Order Kalman Filter 24
Xiao Liang, Guangming Yan, Yanfeng Zhu, Tianyi Li, and Xiaojun Sun

Multi-sensor Data Fusion Algorithm for Indoor Fire Detection Based on Ensemble Learning 37
Lei Wang and Jia Zhang

Research on Water Surface Environment Perception Method Based on Visual and Positional Information Fusion 50
Qin Na, Zhe Zuo, Ning Xu, ZhenYu Zhang, and Yi Lu

Novel Fault Diagnosis Method Integrating D-L2-FDA and AdaBoost 63
Yang Zhao, Wei Ke, Wei Zhang, Yi Luo, Qun-Xiong Zhu, Yan-Lin He, Yang Zhang, Ming-Qing Zhang, and Yuan Xu

Structural Health Monitoring of Similar Gantry Crane Based on Federated Learning Algorithm 75
Zexuan Peng, Zhaohui Zhang, Xiaoyan Zhao, Tianyao Zhang, and Qi Wu

Accelerated Lifetime Experiment of Maximum Current Ratio Based on Charge and Discharge Capacity Confinement 89
Baoji Wang, Boyan Li, Qixuan Wang, and Lei Dong

Adaptive Design of Uni-Variate Alarm Systems Based on Statistical Distance Measures 101
Mohsen Asaadi, Koorosh Aslansefat, Iman Izadi, and Fan Yang

Correlation Analysis Between Insomnia Severity and Depressive Symptoms of College Students Based on Pseudo-Siamese Network 116
Ya-fei Wang, Yan-ling Zhu, Peng Wu, Meng Liu, and Hui Gao

Construction and Research of Pediatric Pulmonary Disease Diagnosis and Treatment Experience Knowledge Graph Based on Professor Wang Lie's Experience 128
Qingyu Xie and Wei Su

A Novel SEIAISRD Model to Evaluate Pandemic Spreading 139
Hui Wei and Chunyan Zhang

Keyword-based Research Field Discovery with External Knowledge Aware Hierarchical Co-clustering 153
Kai Sugahara and Kazushi Okamoto

An End-to-End Intent Recognition Method for Combat Drone Swarm 167
Hui He, Zhihong Peng, Peiqiao Shang, Wenjie Wang, and Xiaoshuai Pei

An Attention Detection System Based on Gaze Estimation Using Self-supervised Learning 178
Xiang-Yu Zeng, Bo-Yang Zhang, and Zhen-Tao Liu

Effects of Pseudo Labels in Pose Estimation Models Using Semi-supervised Learning 189
Harunobu Ariga and Yuki Shinomiya

Sequential Masking Imitation Learning for Handling Causal Confusion in Autonomous Driving 200
Huanghui Zhang and Zhi Zheng

Proposal of Timestamp-Based Dynamic Context Features for Music Recommendation 215
Yasufumi Takama, Lin Qian, and Hiroki Shibata

Method to Control Embedded Representation of Piece of Music in Playlists 226
Hiroki Shibata, Kenta Ebine, and Yasufumi Takama

Design and Implementation of ANFIS on FPGA and Verification with Class Classification Problem 241
Moegi Utami, Yukinobu Hoshino, and Namal Rathnayake

Intelligent Optimization and Decision-Making

Beacon Localization Method Based on Flower Pollination-Fireworks Algorithm 255
Zhaofeng Du, He Huang, and Bin Xin

Parameter Identification for Fictitious Play Algorithm in Repeated Games 270
Hongcheng Dong and Yifen Mu

An Improved Hypervolume-Based Evolutionary Algorithm for Many-Objective Optimization 283
Chengxin Wen, Lihua Li, and Hongbin Ma

Reinforcement Learning-Based Policy Selection of Multi-sensor Cyber Physical Systems Under DoS Attacks 298
Zengwang Jin, Qian Li, Huixiang Zhang, and Changyin Sun

A UAV Penetration Method Based on the Improved A* Algorithm 310
Shitong Zhang, Qing Wang, Bin Xin, and Yujue Wang

Hybrid D-DEPSO for Multi-objective Task Assignment in Hospital Inspection 324
Chun Mei Zhang, Xin Yao Ma, and Bin Zhai

An Analysis of the Generalized Tit-for-Tat Strategy Within the Framework of Memory-One Strategies 338
Yunhao Ding, Jianlei Zhang, and Chunyan Zhang

Stochastic Resource Allocation with Time Windows 348
Yang Li and Bin Xin

Author Index 359

Contents – Part II

Pattern Recognition and Computer Vision

Pipe Alignment with the Image Based Visual Servo Control ... 3
Ivan Kholodilin, Nikita Savosteenko, Nikita Maksimov, Dmitry Khriukin, and Maksim Grigorev

A System for Estimating the Importance of Speech Based on Acoustic Features ... 11
Jiating Liu and Sumio Ohno

Zero-Shot Action Recognition with ChatGPT-Based Instruction ... 18
Nan Wu, Hiroshi Kera, and Kazuhiko Kawamoto

Algorithm for Human Abnormal Behavior Recognition Based on Improved Spatial Temporal Graph Convolutional Networks ... 29
Qi Wu, Xiaoyan Zhao, Zhaohui Zhang, Tianyao Zhang, and Zexuan Peng

Helmet Detection Algorithm of Electric Bicycle Riders Based on YOLOv5 with CBAM Attention Mechanism Integration ... 43
Si-Yue Fu, Dong Wei, and Liu-Ying Zhou

Plane Defect Detection Based on 3D Point Cloud ... 57
Mingsong Bai, Shuang Wu, Hongbin Ma, and Ying Jin

An Improved TrICP Point Cloud Registration Method Based on Automatically Trimming Overlap Regions ... 70
Pengcheng Jiang and Yuan Li

Research on Estimation of Kyphosis Degree Based on Monocular Camera for Achieving Furniture's Adaptive Height Adjustment ... 81
Qingwei Song, Naoyuki Kubota, and Yuqi Zhang

Exploring Whether CNN-Based Segmentation Models Should Extract Features in Earlier or Later Stages for MRI Images ... 93
Hibiki Umeda and Yuki Shinomiya

Cognitive Impairment Detection System based on Image Segmentation and Artificial Intelligence Art ... 105
Yuqi Zhang, Qingwei Song, Takenori Obo, and Naoyuki Kubota

Developing a Searching Sheep Application Using Machine Learning 117
Chengyuan Dong and Yihsin Ho

Using Non-deep Learning to Recognize High and Low Valence Emotions on Young Adults by HRV 129
Yidi Jing and Eri Sato-Shimokawara

Simulation for Development of Microcomputer Car with White Line Following Controller 141
Junichi Sasagawa, Michio Watamori, and Yukinobu Hoshino

Validation of Contour Extraction Using YOLACT for Analysis of NK Cell Chemotaxis 150
Reiji Okawa, Yukinobu Hoshino, Shoya Kusunose, Shinpei Yamamoto, Takashi Ushiwaka, and Nagamasa Maeda

Improving the Efficiency of Image Recognition for Yuzu Fruit Counting Using Object Recognition Models 156
Takahiro Sugiyama and Shinichi Yoshida

A Study on Explainability of Deep Learning Model for Image Classification Using CycleGAN 167
Taiga Nakajima and Shinichi Yoshida

Research on Algorithms of Lateral Face Recognition Based on Data Generation 182
Zimin Zhang, Zhaohui Zhang, Xiaoyan Zhao, and Tianyao Zhang

Advanced Control

Design and Operation Control of an Indoor Storage Crane 197
Rahman Mizanur, Yiming Duan, Malak Abid Ali Khan, Zia Ur Rehman, and Hongbin Ma

Design of a Rotating Inverted Pendulum Control System Based on Qube-Servo2 209
Haoran Wang, Qing Wang, and Yujue Wang

Dual-Loop Control Based on Tube-Based MPC for UAVs with Disturbance 223
Bowen Hong, Zhiwei Chen, Yongming Han, and Zhiqiang Geng

Design of Intelligent Twin-Screw Extruder Control System Based on Improved PSO-BP Neural Network 237
Xuanhao Yang, Hongzhan Zhang, and Wei Xiao

Finite-Time Stabilization-Based Neural Control for the Synchronous Generator 250
Honghong Wang, Bing Chen, Chong Lin, and Gang Xu

A Constant Air Flow Controller Based on Interval Type-2 Fuzzy PID Controller 262
Bojin Shang, Xiaohan Wang, Shuai Shao, and Yaping Dai

Multi-agent Systems

Neural Network Control of Distributed Cooperative Formation of Multi-agent System 283
Si Kheang Moeurn and Bin Xin

Moving-Target Enclosing Control for Multiple Nonholonomic Mobile Agents Under Input Disturbances 293
Yaning Jin, Shuang Ju, and Jing Wang

Characteristics Verification of the Luggage Transportation Problem Using Relative Vectors in Multi-agent Reinforcement Learning 304
Daisuke Hashimoto and Yukinobu Hoshino

Robotics

Variable Photo-Model Stereo Vision Pose and Size Detection for Home Service Robot 319
Hongzhi Tian and Jirong Wang

Motion Capture Modeling of Dexterous Hand for Intelligent Sensing 329
Xiaoyan Zhao, Siyi Cui, Zhaohui Zhang, Qi Cao, Yuan Yuan, Xianhao Wu, and Shaowen Zheng

Design of a Left-Right-Independent Pedaling Machine for Lower-Limb Rehabilitation 343
Shigeki Kuroda, Jinhua She, Rennong Wang, Daisuke Chugo, Keio Ishiguro, Hiromi Sakai, and Hiroshi Hashimoto

Author Index 351

Intelligent Information Processing

3D Point Cloud-Based Lithium Battery Surface Defects Detection Using Region Growing Proposal Approach

Zia Ur Rehman, Xin Wang, Abdulrahman Abdo Ali Alsumeri, Malak Abid Ali Khan, and Hongbin Ma(✉)

State Key Laboratory of Intelligent Control and Decision of Complex Systems, School of Automation, Beijing Institute of Technology, Beijing 100081, China
mathmhb@bit.edu.cn

Abstract. Detecting the lithium battery surface defects is a difficult task due to the illumination reflection from the surface. To overcome the issue related to labeling and training big data by using 2D techniques, a 3D point cloud-based technique has been proposed in this paper. The 3D point cloud-based defect detection of lithium batteries used feature-based techniques to downscale the point clouds to reduce the computational cost, extracting the normals of the points and calculating their differences to detect the defects of the battery which assure the quality of the product. This paper offers a novel strategy using 3D point clouds to get beyond the labeling and training challenges involved with conventional 2D approaches. This 3D point cloud-based approach for lithium battery fault identification makes use of feature-based methods to improve the point cloud data and lessen the computing burden. In our work, the experiments show that the feature-based technique precisely detects the affected surface of the battery.

Keywords: 3D point cloud · Defects detection · Region growing proposal

1 Introduction

Lithium-ion batteries have become widely used energy storage batteries due to their high energy density, low self-discharge rate, absence of memory effects, and relatively low production cost. They are lightweight and powerful, but are also prone to leakage and catching fire, making monitoring of the production process and early detection of electrode defects crucial. Defective lithium batteries can greatly impact the battery qualification rate in industrial production. However, detecting defects in lithium batteries with aluminum/steel shells is challenging due to the reflective surface and limitations of 2D computer vision detection methods [1]. To overcome issues with deformation and occlusion characteristics, literature [2] has devised a method that uses adversarial networks and spatial dropout networks. The objective is to make fine differences between specific object classes to enhance object detection. Their approach improves features of the regional proposal networks created by Faster Region-based Convolutional Neural Network (R-CNN) and creates a useful mechanism for locating smaller regions using a

B. Xin et al. (Eds.): IWACIII 2023, CCIS 1931, pp. 3–14, 2024.
https://doi.org/10.1007/978-981-99-7590-7_1

single high-level feature map. The ability to detect small objects will especially benefit from this development. They also use an offline deep neural network for a variety of detection tasks and an action-driven approach for object tracking. However, the main goal of their suggested strategy is to apply learning-based detection and tracking techniques to distinguish between appearance models. Deep learning is used to train on large datasets, ensuring a strong and reliable relationship between detection and tracking. Furthermore, there are not enough defective lithium battery samples available to effectively train deep neural networks. Manual defect detection is a slow and tedious process, which lowers production automation and efficiency. Even skilled inspectors can only detect around 75% of defects, and small defects are often missed. When inspections become fatigued, their detection accuracy decreases, introducing instability to the quality inspection process. To overcome these issues, researchers have suggested using machine vision technology for accurate and reliable defect detection. Currently, the most popular methods for defect detection are time-domain, frequency-domain, and deep learning approaches.

3D object detection refers to the process of detecting objects within a 3D space, such as a scene captured by a 3D sensor, or a point cloud generated from a 2D image. This task is more challenging than 2D object detection because the objects may be oriented in different directions and can be occluded by other objects in the scene. 3D object detection can be used in various applications such as autonomous vehicles, robotics, and augmented reality. Techniques used for 3D object detection include 3D CNNs, LiDAR point cloud processing, and stereo vision [3]. A 3D visual measurement system is a promising solution for detecting surface defects based on their roughness and height. This paper proposes an integrated approach to address the problem of lithium battery surface defect detection based on region growing proposal algorithm.

2 Previous Work

Current methods for object detection and computer vision mainly rely on deep learning and neural networks. Applications of this active study area can be found in many fields, such as autonomous driving, aerial photography, security, and surveillance. Modern object identification techniques frequently use rectangular bounding boxes to pinpoint an object's exact location. However, these cutting-edge object identification algorithms encounter several difficulties, such as problems with lighting, occlusion, viewing angle, camera rotation, and financial limitations [4]. The implementation of deep learning-based object recognition holds the promise of greatly increasing recognition speed and strengthening these systems' resistance to outside influence considering these difficulties. The field of computer vision and its applications have advanced thanks to this development in a significant way.

This section provides an overview of recent research in two important areas related to the paper. Surface defect detection and 3D point data processing, these topics have received significant attention in recent years, and the focus has been on developing more efficient and accurate methods for these tasks. The most popular techniques for defect identification in the field of machine vision based on the temporal point cloud of the frame structure are registration alignment based on contour point cloud and point cloud contour fitting. Ordered and disordered 3D point clouds are the two categories. Strict coordinate intervals are used to elegantly arrange and confine the ordered point cloud. No point can be precisely determined in the section due to the disorganized point cloud's dispersed spatial points. The use of neighboring data points is necessary for the section's creation. To produce the projection trajectory of the point on the given section, Khameneifar and Feng [5] first fit a local quadric surface to the point neighborhood. Creation of a dispersed point cloud. Dandage and Harshad K developed the LSDD approach, which is an effective method for quickly detecting defects using a limited number of training images. Their experiments demonstrate that with only 26 source images, the LSDD method can create two augmented multi-scale datasets consisting of 19,309 and 6,889 image patches for training and testing, respectively, achieving an accuracy of 93.67% in detecting defective image patches in the first stage, and attaining a mean precision rate of 90.78% and a mean recall rate of 93.89% for identifying the type of defect in the second stage. The proposed two-stage classification approach demonstrates a higher sensitivity for defect detection with an accuracy of 69%, compared to a more intuitive one-stage classification approach.

Furthermore, the two-stage approach outperforms the one-stage scheme in identifying specific types of defects. When compared with the YOLOv3 detector, the approach of Dandage and Harshad shows fewer misdetections of defects. While several object detection methods have been proposed with various solutions, they have mainly focused on qualitative detection rather than quantitative detection [5]. Due to the need for a significant amount of data and a powerful computer to train and test the 2D detection model, the process is slow and time-consuming. As a result, researchers are now turning to 3D detection techniques to overcome these challenges.

Ke Wu [6] proposed a few-shot learning technique for detecting 3D defects in lithium batteries. This method involves using a multi-exposure-based structured light method to create a 3D representation of the battery's shape, the defective part of the 3D point cloud is then transformed into 2D images using the height-gray transformation. In another approach, Zong developed a detection system that consists of two image capture modules and a turntable. These modules are binocular stereo vision systems with monochrome cameras, a color camera, and a speckle projector. The speckle projector reconstructs the 3D point clouds of the object's surface using stereo digital image correlation (stereo-DIC) technology. The 3D point cloud of the defective area is obtained by segmenting the defect using image and point cloud segmentation algorithms based on the point-image mapping relationship. Finally, the 3D characteristic parameters of defects are calculated using the corresponding 3D point cloud of the defective area. Xu and Changlu [7] have suggested a technique to detect and classify surface defects in lithium battery pole pieces. The method combines multiple image features and a partial swarm optimization-support vector machine (PSO-SVM) to accomplish this. This approach entails using image subtraction and contrast adjustment to preprocess the defected image, which reduces the influence of non-defective regions while enhancing the defect features. The defected area of the image is then extracted using the Canny algorithm and the AND logical operation. The texture, edge, and HOG [8] features are combined to extract the features of the defected area in the image. Finally, a support vector machine optimized using particle swarm optimization is used to identify and categorize the defects in the images automatically.

G. Lanza developed a Single-Point-Analysis approach using optical measurement techniques to establish inline quality assurance during the production of cell foils. The method involves a filtering algorithm that monitors the inter-point distance of each point in the designated range. Additionally, a mean distance for each single measurement point is calculated by analyzing the distances to its direct neighbors. The main objective is to detect particles and defects, such as scratches, and dents on the cell surface. The analysis developed by Lanza is capable of clearly identifying these common cell failures [9]. In their work, M. Himmelsbach presented a perception system for ground robot mobility that is based on LIDAR technology. The system includes 3D object detection, classification, and tracking, and was tested on the autonomous ground vehicle MuCAR-3. The successful implementation of this system enabled the vehicle to navigate safely in various scenarios, including urban traffic and off-road convoy situations [10].

Nasrollahi utilized PointNet in their work to detect surface defects from point cloud datasets obtained by scanning bridge surfaces. PointNet was chosen for its robustness in handling missing or corrupted data. The datasets were collected from concrete bridges and manually annotated, resulting in five datasets with a total of 3,572 annotated segments. The segments were classified into two classes, defect, and non-defect, with around 15% of the points being labeled as defects [11]. Xinhua Liu has proposed a process for detecting defects using an improved K-Nearest Neighbor (KNN) algorithm and Euclidean clustering segmentation. The process begins by implementing an improved Voxel density strategy for KNN, which enhances the point filtering process to increase efficiency. Next, an improved clustering segmentation strategy is applied to distinguish point clouds with defect features. The geometric features of each surface defect are determined using an outline fitting algorithm based on the least square method. The experimental results demonstrate that the proposed method for detecting surface defects has achieved an accuracy of 99.2% with an average time consumption of 35.3ms for data processing using a testing dataset. Despite the small size of the training dataset, the initial results are promising. Future work will focus on increasing the accuracy of the model by using a larger dataset for training [12].

3 Methodology

Region growing is a popular algorithm in image processing and computer vision for segmentation and defect detection. The basic idea of region growing is to group pixels or points that share common characteristics, such as similar intensity values or surface normals. In this proposal, we will focus on using region growth for normals estimation and defect detection. Here is an outline of the proposed algorithm as illustrated in the Fig. 1. A-point cloud P representing a 3D surface Estimate surface normals at each point using a method such as principal component analysis (PCA)-based normal estimation or a weighted average of neighboring normal. Initialize a set S of seed points, which are used to start the region-growing process. Seed points can be chosen randomly or based on some heuristic, such as high curvature or low normal consistency. For each seed point s in S, grow a region by iteratively adding neighboring points that have similar surface normal. The similarity threshold can be set based on the standard deviation of the surface normal in the point cloud or some other criterion. A common approach is to use a region growing criterion that compares the angle between the normal at the current point and the normal at the seed point. If the angle is below a certain threshold, the point is added to the region. Once a region is grown, perform some analysis on the region to detect defects or other features of interest. For example, if the region is small or has a high curvature, it may be considered a defect. Other features such as edges or corners can also be detected by analyzing the shape of the region. Repeat these steps for all seed points in S. Output the set of regions that have been detected as defects or features of interest.

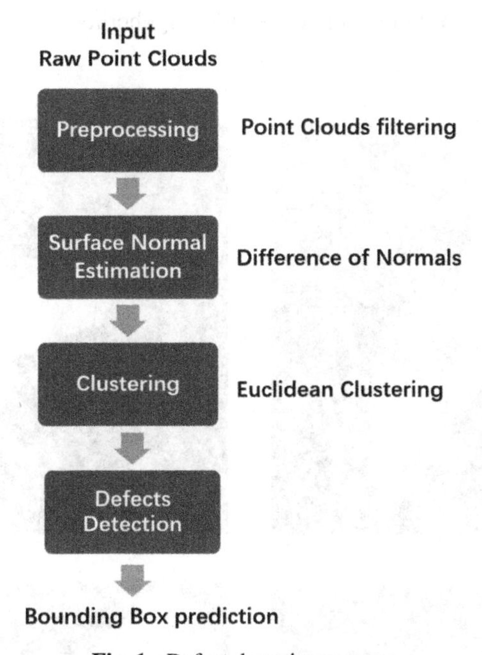

Fig. 1. Defect detection process

There are many variations of the region-growing algorithm, and the specific implementation may vary depending on the application and the point cloud data type. Some parameters that can be tuned include the size and shape of the neighborhood used to estimate the surface normals, the similarity threshold used for region growth, and the criteria used to detect defects or features of interest.

3.1 Data Acquisition and Preprocessing

Point cloud data acquisition for lithium battery defects involves using specialized equipment and software to capture and analyze the surface characteristics of a battery. The process typically involves scanning the surface of the battery using a 3D scanner or other similar device to collect data on the battery's surface geometry and topology. This data is then used to create a 3D point cloud, which is a digital representation of the battery's surface made up of numerous individual points that collectively form a 3D model. Our dataset includes information on different types of defects, such as scratches and dents, observed under varying lighting and weather conditions.

Figure 2 displays some examples from our collected data. In the proposed work we used a 3D depth-sensing camera to capture the point clouds of the battery surface, which contain different types of defects and the number of points for analyzing the defects of the battery is different for each scan from 27313461 to 2737032 which are very high for processing. To overcome this issue, we used Moving Least Squares (MLS) algorithms to down-sample and smoothen the point clouds. The MLS is a technique for reconstructing a surface from a set of unorganized data points by using higher-order polynomial interpolations within a fixed point's neighborhood.

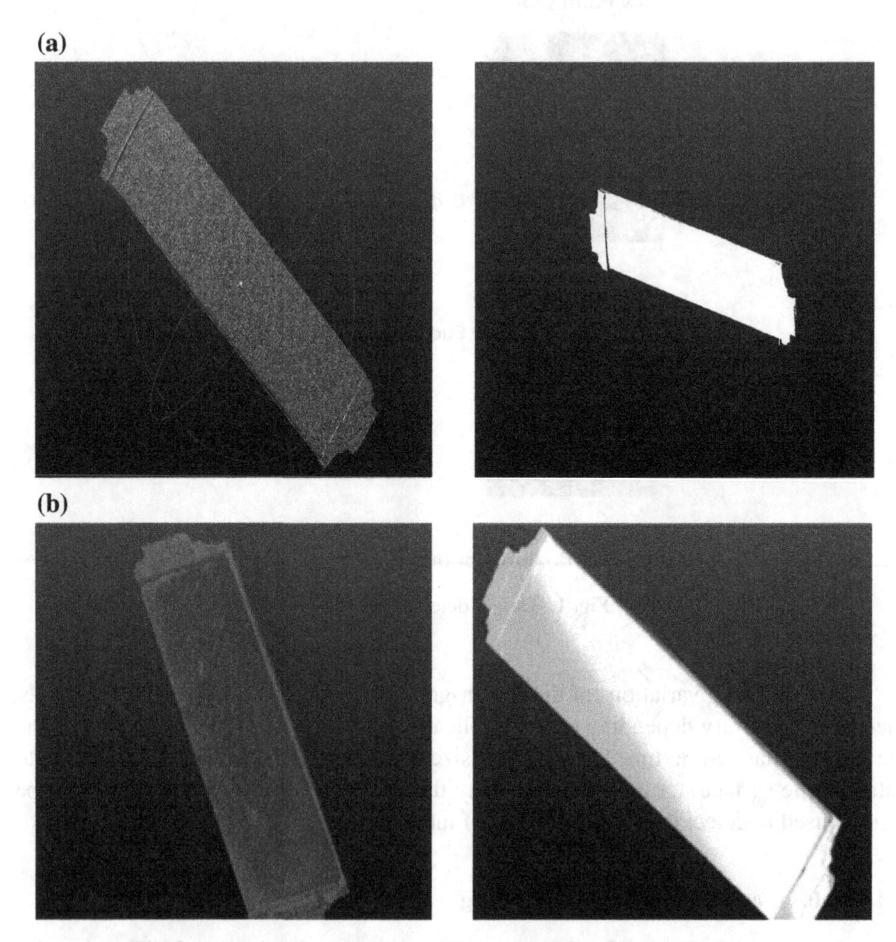

Fig. 2. (a) Point clouds without defects. (b) Point clouds with defects

This approach was originally proposed by Lancaster and Salkauskas in 1981, [13, 14] and further developed by Levin. In our specific application, we are utilizing a second-degree polynomial in R^n to approximate the cloud surface, as this type of surface most closely resembles to battery's surface.

The mathematical model of the MLS algorithm can be described as follows. Given a function $f : R^n \rightarrow R$ and a set S of points $= \{a_j, f_j | f(a_j) = f_j\}$ where a j is a point in R^n and f_j is a scalar value such that $f(a_j) = f_j$, the MLS approximation of the point a_j is determined by minimizing the error functional [15].

$$f_{MLS}(a_j) = \sum_j (\|f(a_j) - f_j\|)^2 \theta \|a - a_j\| \tag{1}$$

3.2 Normals Estimation and Segmentation

Surface normals are vectors perpendicular to a geometric surface at a specific point, and they provide important information about the local properties of the surface [16]. There are several methods for estimating surface normals, one of which involves estimating the normal of a plane tangent to the surface. This method involves using a least-square plane fitting estimation algorithm on a point cloud P_N and a query point P_b. Neighboring points P_T are considered to determine a tangent plane S, which is represented by a point a and a normal vector N_X. By defining the distance from each point P_i in P_K to the plane S, it is possible to estimate surface normals accurately. To identify the damaged area on a battery surface, it is necessary to segment the 3D point cloud data into regions that exhibit similar surface characteristics, such as normal vector angles and curvature differences. This process involves dividing the original point cloud into two distinct components: damaged regions and non-damaged regions. The aim is to partition the point cloud into sub-clouds based on information about the normals and curvatures of the surface [17, 18]. The surface normals illustration can be seen in the Fig. 3.

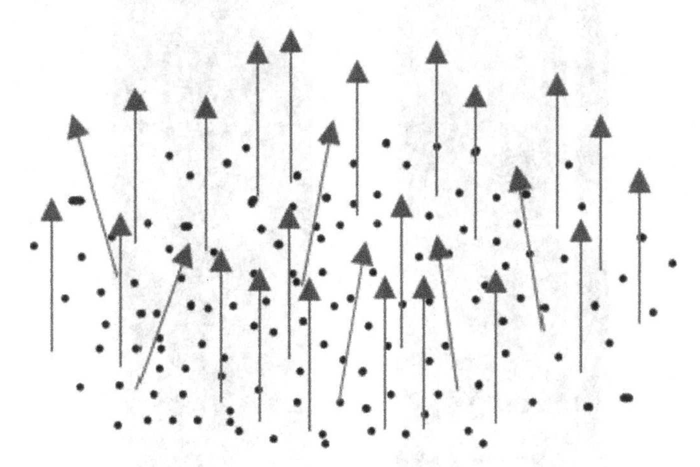

Fig. 3. Surface normals illustration

4 Experiments and Discussion

The proposed method has been tested on different point clouds for both defective and non-defective surface of the battery which successfully detect the surface, the range of the input clouds are between 27313461 to 2737032 which are then down sampled and smoothed by MLS algorithms after filtering the points clouds are between 382953 to 419252. After the difference of normals filtering and thresholding steps, the segmented areas of the point cloud data that have a significant change in surface normals can be further analyzed using clustering algorithms such as k-means or hierarchical clustering. Clustering is a technique used for grouping similar data points into clusters based on their features. In the case of defect detection in point cloud data of lithium batteries, the features used for clustering can include the location, size, shape, and type of defects. Once the clustering algorithm has identified the different clusters of defects, each cluster can be visualized as a different color or shape in the 3D point cloud model of the battery. Figure 4 displays the concatenated cluster of all individual clusters for visualizing the defects as point clouds in yellow color. This method enables the differentiation between different types of defects based on their size and shape, providing a clear visual representation of the battery's internal structure and any present defects. Clustering can also be utilized to generate statistics and metrics for each cluster of defects, such as the number of defects, average size and location of defects, and other quantitative measures. These metrics are useful for evaluating the severity and extent of the defects and for developing strategies to enhance the performance and reliability of the battery. It provides an overview of all parameters utilized in this experiment, as well as their corresponding values or ranges. This approach enables the precise analysis and quantification of

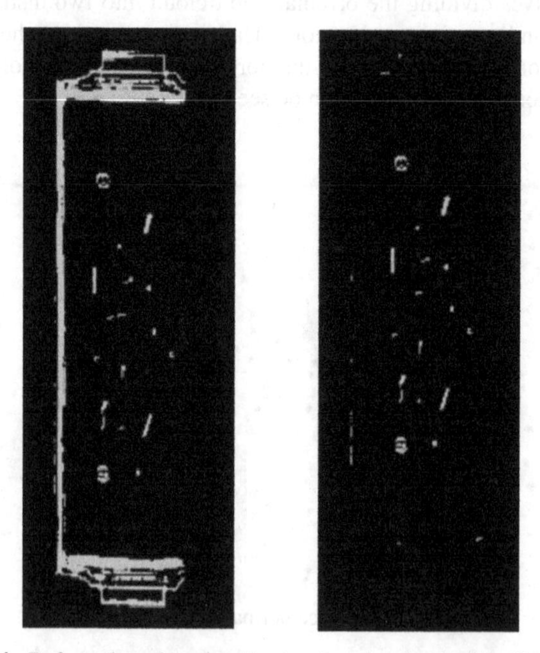

Fig. 4. Defects detection visualization in point cloud using filtering.

defects in lithium batteries, which is crucial for the development of improved battery technologies.

The difference of normal magnitude is 0.05 and number of filter points are very depending on the size of the input point cloud, and the number of clusters is proportional to the number of defects in the surface as shown in Fig. 4 and the detailed information of the parameters are shown in Table 1. The processing time varies from 6 to 10 s which is acceptable as the battery surface inspection needs to be done thoroughly.

Table 1. Experimental parameters details

Parameters	Range/Value
Point clouds before filtering	27313461–2737032
Point clouds after filtering	382953–404885
Difference of normal	0.05
Inlier points number	373845–404885
Outlier points number	12389–14367
Filtered point clouds	39128–41237
Number of clusters	17–27
Minimum cluster size	10
Maximum cluster size	400

These values are obtained after testing on optimized code on a PC with the specification of 2.3 GHz intel Core i-7 32 GB RAM on Microsoft Visual Studio 2021 with support of Point cloud library using Ubuntu Platform.

Defects by bounding box prediction in point cloud data of lithium batteries are another way to identify and analyze the different types of defects present in the battery. This technique involves using a bounding box to enclose each defect in the point cloud data, based on its size and shape. By bounding defects with boxes, it becomes easier to identify and analyze different types of defects and to understand their location and size within the battery. This can provide valuable insights into the battery's internal structure and help to develop strategies for improving its performance and reliability. Figure 5 showcases the auto-bounding information around the defects, by bounding box for the corresponding defects in the point cloud data. This approach allows for the precise localization and identification of defects within the battery structure. The bounding box encapsulates the dimensions and orientation of the defect, providing a detailed representation of its spatial characteristics. This labeling information can be utilized to generate quantitative metrics for each defect, such as its volume, surface area, and location within the battery structure. Additionally, this information can be used to compare defects across different battery samples, allowing for the identification of trends and patterns in the occurrence and distribution of defects. The use of bounding boxes is a valuable technique for the characterization and analysis of defects in lithium batteries and can provide insights for the development of enhanced battery technologies.

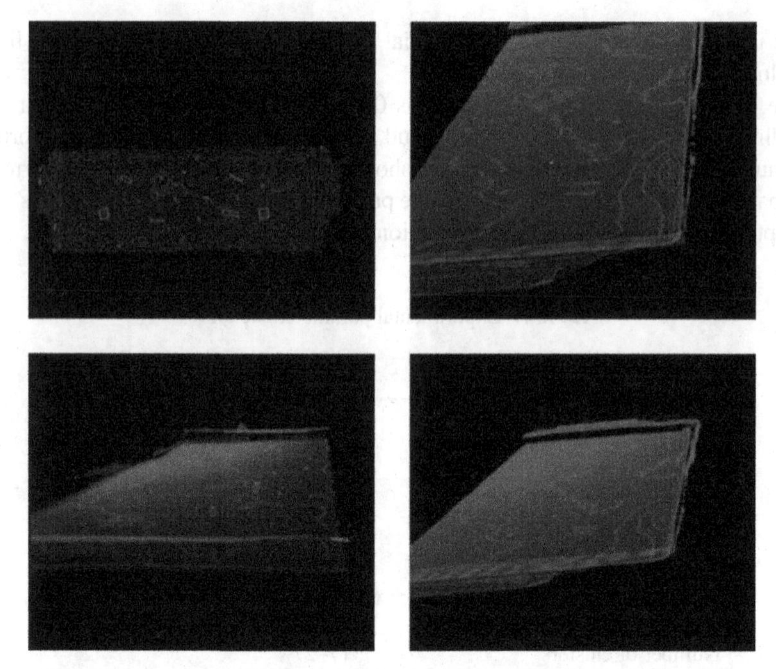

Fig. 5. Defects detection visualization in point cloud using bounding box.

5 Conclusion and Future Work

In this work, we presented a framework for defect detection on lithium battery surfaces based on the characterization of the point cloud data. The proposed methodology consists of two primary stages. The first stage is dedicated to detecting defects in the Point Cloud. This involves segmenting the Point Cloud to differentiate between defect regions and non-defect regions. To achieve this, a computer vision algorithm has been developed based on the Region-Growing method. The algorithm utilizes local surface information such as points normal and curvature to identify various undesired deformations on the airplane surface.

Defects by bounding boxes in 3D point cloud defect data are important for several reasons. By bounding defects with boxes, it becomes easier to identify and analyze different types of defects and to understand their location and size within the battery. This can help to improve the accuracy of defect identification and characterization. Bounding the defects allows for quantitative analysis of the size and shape of each defect, which can provide valuable insights into the severity and extent of the defects. By bounding the defects, it becomes easier to visualize the different types of defects in the 3D point cloud model of the battery. This can provide a clear representation of the battery's internal structure and any defects that may be present. Defects can be useful for generating detailed reports on the condition of the battery, which can be used to communicate findings to stakeholders such as manufacturers, maintenance crews, and regulatory agencies.

In the second stage, a technique will be developed to characterize the detected defects. This technique will provide valuable information about each defect, including its size, depth, and orientation. Our proposed processes operate automatically, but processing time is currently a limitation. To address this, we plan to optimize our code in the future. Our results show promise for use in inspection systems and have broad applicability across various industrial applications, not limited to lithium battery surface inspection.

References

1. Qian, Q., et al.: The role of structural defects in commercial lithium-ion batteries. Cell Reports Phys. Sci. **2**(9) (2021)
2. Aamir, S.Y., Aaqib, M.: Real-Time Object Detection in Occluded Environment with Background Cluttering Effects Using Deep Learning. IWACIII2021
3. Qi, C.R., Chen, X., Litany, O., Guibas, L.J.: Imvotenet: Boosting 3d object detection in point clouds with image votes. In: Proceedings of the IEEE/CVF Conference on Computer Vision and Pattern Recognition, pp. 4404–4413 (2020)
4. Mirani, M.I.K., Chen, T., Khan, M.A.A., Aamir, S.M., Menhaj, W.: Object Recognition in Different Lighting Conditions at Various Angles by Deep Learning Method. arXiv preprint arXiv:2210.09618 (2022)
5. Khameneifar, F., Feng, H.Y.: Extracting sectional contours from scanned point clouds via adaptive surface projection. Int. J. Prod. Res. **55**, 4466–4480 (2017)
6. Dandage, H.K., Lin, K.-M., Lin, H.-H., Chen, Y.J., Tseng, K.-S.: Surface defect detection of cylindrical lithium-ion battery by multiscale image augmentation and classification. Int. J. Mod. Phys. B **35**(14n16), 2140011 (2021)
7. Wu, K., Tan, J., Li, J., Liu, C.: Few-shot learning approach for 3D defect detection in lithium battery. In: Journal of Physics: Conference Series, vol. 1884, no. 1, p. 012024. IOP Publishing (2021)
8. Zong, Y., et al.: An intelligent and automated 3D surface defect detection system for quantitative 3D estimation and feature classification of material surface defects. Opt. Lasers Eng. **144**, 106633 (2021)
9. Changlu, X., Li, L., Li, J., Wen, C.: Surface defects detection and identification of lithium battery pole piece based on multi-feature fusion and PSO-SVM. IEEE Access **9**, 85232–85239 (2021)
10. Lanza, G., Koelmel, A., Peters, S., Sauer, A., Stockey, S.: Automated optical detection of particles and defects on a Li-Ion-cell surface using a single-point analysis. In: 2013 IEEE International Conference on Automation Science and Engineering (CASE), pp. 675–680. Madison, WI, USA (2013). https://doi.org/10.1109/CoASE.2013.6653950
11. Himmelsbach, M., Luettel, T., Wuensche, H.J.: Real-time object classification in 3D point clouds using point feature histograms. In: 2009 IEEE/RSJ International Conference on Intelligent Robots and Systems, pp. 994–1000. St. Louis, MO, USA (2009). https://doi.org/10.1109/IROS.2009.5354493
12. Nasrollahi, M., Bolourian, N., Hammad, A.: Concrete surface defect detection using deep neural network based on lidar scanning. In: Proceedings of the CSCE Annual Conference, pp. 12–15. Laval, Greater Montreal, QC, Canada (2019)
13. Liu, X., Wu, L., Guo, X., Andriukaitis, D., Krolczyk, G., Li, Z.: A novel approach for surface defect detection of lithium battery based on improved K-Nearest Neighbor and Euclidean clustering segmentation (2023)
14. Lancaster, P., Salkauskas, K.: Surfaces generated by moving least squares methods. Math. Comput. **37**(155), 141–158 (1981)

15. Levin, D.: The approximation power of moving least squares. Math. Comput. Am. Math. Soc. **67**(224), 1517–1531 (1998)
16. Levin, D.: Mesh-independent surface interpolation. In: Geometric Modeling for Scientific Visualization, pp. 37– 49. Springer (2004)
17. Dey, T.K., Li, G., Sun, J.: Normal estimation for point clouds: a comparison study for a Voronoi based method. In: Point Based Graphics, 2005. Euro Graphics /IEEE VGTC Symposium Proceedings, pp. 39–46. IEEE (2005)
18. Rusu, R.B., Cousins, S.: 3D is here: point Cloud Library (PCL). In: IEEE International Conference on Robotics and Automation (ICRA). Shanghai, China (2011)

Reducing Communication Consumption in Collaborative Visual SLAM with Map Point Selection and Efficient Data Compression

Weiqiang Zhang(✉), Lan Cheng, Xinying Xu, and Zhimin Hu

Taiyuan University of Technology, Taiyuan 030024, China
iszhangwq@163.com

Abstract. Efficient data communication is a challenging problem for Collaborative Visual Simultaneous Localization and Mapping (CVSLAM), particularly in bandwidth-limited applications. To resolve this problem, we propose a communication load reduction method. We first propose a map point culling strategy by considering maximum pose-visibility and spatial diversity, to eliminate redundant map information in CVSLAM. Then, we employ a Zstandard (Zstd) compression algorithm to compress visual information so as to reduce the required communication bandwidth. To exhibit the efficiency of the suggested approach, we implement this method in a centralized collaborative monocular SLAM (CCM-SLAM) system. Extensive experimental evaluations indicate that our method can reduce communication overhead by approximately 49% while maintaining map accuracy and real-time performance.

Keywords: Collaborative SLAM · Communication bandwidth · Mappoints selection · Zstd compression

1 Introduction

Visual Simultaneous Localization and Mapping (VSLAM) represents a crucial challenge in the field of robotics, as estimating self-motion and constructing maps are essential for facilitating autonomous navigation. As VSLAM has achieved maturity and robustness in single-robot scenarios, there is now a growing interest in multi-robot systems for SLAM. Involving multiple robots with visual sensors in a task can improve the efficiency of a mission and increase the robustness of the system, which is also known as collaborative VSLAM(CVSLAM). CVSLAM promises a wide array of robotic applications [2], examples including conducting inspections of vast industrial complexes and carrying out search-and-rescue operations. However, in scenarios where communication resources are limited, the capability for communication between the intelligent agent and the central server can easily become a limiting factor affecting collaborative system performance.

This research was funded by the National Natural Science Foundation China, grant number 62073232, Foundation for Scientific Cooperation and Exchanges of Shanxi Province, grand number 202104041101030, and the Natural Science Foundation of Shanxi Province, grant number 201901D211079.

B. Xin et al. (Eds.): IWACIII 2023, CCIS 1931, pp. 15–23, 2024.
https://doi.org/10.1007/978-981-99-7590-7_2

Existing work is dedicated to reducing the amount of data, including feature/frame selection, keyframe/3D point culling, etc. [3] aiming to reduce the amount of data while slightly sacrificing accuracy, or compressing the map as much as possible with some compact representation [4]. However, their optimization goal is to reduce data size but pay less attention to the accuracy of the resulting map. Inspired by the work of D. Van [12], which used arithmetic binary encoding for compressing data to be transmitted, we further investigate the application of lossless compression algorithms in the field of CVSLAM. To minimize bandwidth communication demands in a mature collaborative SLAM system while keeping mapping precision, two essential strategies are proposed: compressing the continuous visual data gathered by each agent and utilizing an effective point reduction method that can be seamlessly integrated into any visual SLAM process based on features. Our contributions include (Fig. 1):

- An efficient lossless compression algorithm based on Zstandard(Zstd) is employed in a collaborative SLAM system—CCM-SLAM.
- An algorithm for point sparsification based on the principle of minimum-cost maximum-flow is proposed to control the number of map points while applying a graph representation of the map points and the pairs of camera poses optimized for the highest visibility of points to maintain pose accuracy.
- Bandwidth efficiency is analyzed concerning pose accuracy and running speed on public datasets.

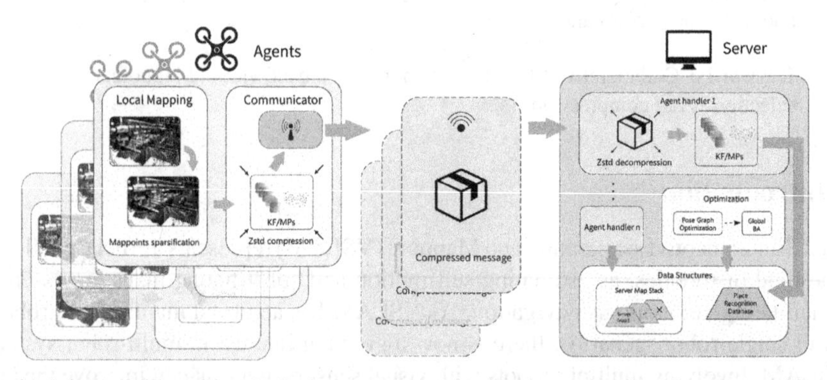

Fig. 1. Schematic illustration. Robotic agents maintain a local mapping thread, and a sparsification step for map points is processed before all local bundle adjustments (BA) in the local mapping thread. The compressed map information is transmitted to a server running agent handlers, where the compressed map information is decompressed and restored into map information that can be recognized by subsequent processing.

2 Proposed Method

One of the crucial elements that affect the bandwidth consumption in the CVSLAM system is the number of map points triangulated in the local map constructed by agents. The map point information occupies most of the map information. As CVSLAM proceeds, the size of data to be transferred escalates dramatically, so we explore two strategies to control data size while maintaining the map accuracy: 1) extracting only relevant points that contribute most to pose optimization, or 2) directly compressing map point information using compression algorithms.

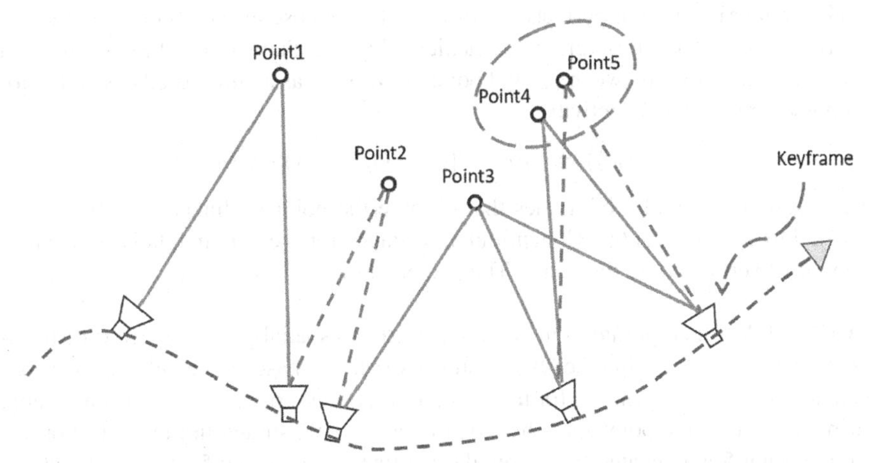

Fig. 2. Sample path (dotted grey) featuring approximated keyframe positions (black camera). Five keyframes share five points in the 3D space. The solid green and dashed red lines indicate the corresponding observations.

2.1 Mappoints Culling Strategy

Our objective is to minimize data volume while preserving pose accuracy. To efficiently select points among frame pairs, we construct a bipartite graph $G = (v, e)$ structure, including source vertice, sink vertice, and two separate sets of vertices, based on their connectivity to either the point or frame. The main challenge addressed by our proposed method is determining a subset of points that minimally impact the local and global bundle adjustment. When all interest points cluster near a corner in the visual field, the pose adjustment issues exhibit steep edges, complicating efficient resolution. If the baseline between two camera frames for a single point approach none, depth estimation and point adjustment become more difficult. An ideal map point should possess the following characteristics: 1) observation in the maximum number of frames, 2) diversified distribution across image space, and 3) maximized distance between the centers of any two camera frames [1].

We suggest a novel approach utilizing a directed graph representation, in which nodes correspond to point and pose pairs, to address the aforementioned issues in a unified algorithm. In the bipartite graph, vertices on the point side correspond to one of the m points on the map and are labeled as v_{pi}, where $p_i \in P = \{p_1, p_2, \cdots, p_m\}$. The collection of point vertices is signified as V_P. On the succeeding layer, vertices embody all feasible frame pairs, interconnected through V_P. A vertex $v_{f_{i,j}} \in V_F$ characterizes the coupling of f_i and f_j, where $f_i, f_j \in F = \{f_1, f_2, \cdots, f_k\}$, and k symbolizes the aggregate number of keyframes within the map. The source vertex, labeled as v_{so}, solely maintains outgoing edges linked with V_P On the other hand, the sink vertex, referred to as v_{si}, exclusively sustains incoming edges connected from V_F. In this setup, we allocate costs and capacities that quantify desirable characteristics for each edge.

Points exhibiting strong connectivity suggest durable local features and elevated visibility, rendering them as perfect choices since they impose stringent conditions across numerous poses on the pose graph. To achieve this, we characterize the cost function C_c for the connections between v_{so} and point vertices v_{pi} as a diminished cost value for v_pi possessing high connectivity:

$$C_c\left(e\left(v_{so}, v_{pi}\right)\right) = c_m(n) = [(\mathrm{n}+1)/(n-1) \cdot c_m(n+1)] \tag{1}$$

Here, n signifies the tally of frames that observe the point p_i, linked with the vertex v_{pi} bound to the source edge. Concurrently, m stands for the maximal tally of n values spanning all vertices v_p in the graph. The computation of $C_m(n)$ is recursive and $c_m(m)$ = 1.

CCM-SLAM incorporates a method of selecting a spatially homogeneous distribution of interest points during the ORB feature extraction phase. Nonetheless, a distribution that starts as homogeneous for features doesn't necessarily guarantee a comparable distribution during the point sparsification process. As illustrated in Fig. 2, The observation of point 5 is indicated by the red dashed line. Points 4 and 5 are spatially close, indicating spatial redundancy, and such points are the targets we want to eliminate. Therefore, we establish the concept of spatial cost C_s to retain or potentially enhance the distribution of features throughout the point sparsification process.

$$C_s\left(e\left(v_{pi}, v_{f_{jk}}\right)\right) = \left[log_{10}\left(N\left(p_i, f_j\right)N\left(p_i, f_k\right) + 1\right)\right], \tag{2}$$

where $N\left(p_i, f_j\right)$ indicates the count of adjacent key points p_i within the frame f_j.

Lastly, we take into account the baseline distance for each pair of frames. We apply the baseline cost C_b to the edges existing between v_F and v_{si}:

$$C_b\left(e\left(v_{f_{jk}}, v_{si}\right)\right) = \lceil\frac{10}{0.1 \cdot d\left(f_j, f_k\right) + 1}\rceil, \tag{3}$$

where $d\left(f_j, f_k\right)$ is the L_2 norm of (O_{f_j}, O_{f_k}) and O_{f_j} is the camera center of the frame f_j. In Fig. 2, the frame pair observing point 2 shows a short baseline, and correspondingly the C_b between them would be large.

We address the graph problem described above by employing a minimum cost maximum flow algorithm [7], which calculates the maximum flow from v_{so} to v_{si} and minimizes the overall cost:

$$C = \sum\nolimits_e f(e)c(e), \tag{4}$$

where $f(e)$ represents the flow on the edge e. Post flow computation, we only consider the point p_i if the flow on the edge $e\left(v_{so}, v_{pi}\right)$ exceeds a predetermined threshold θ_f. As confirmed by Goldberg's algorithm[6], the upper limit of time complexity can be bounded within $O(n^2 m \cdot \log(n \cdot C))$, wherein n represents the aggregate number of vertices, m indicates the comprehensive number of edges, and C is synonymous with the highest input cost.

2.2 Zstd Compression Algorithm

In this study, we introduce a unique method designed for compressing data transmitted between ROS nodes in the CCMSLAM system using the Zstandard(Zstd) compression algorithm. Similar to the majority of Deflate compression algorithms, Zstd functions in two separate phases, each of which can be performed independently. The initial phase involves LZ77 compression, which is then succeeded by the entropy encoding phase. The compression outcomes from the initial phase are organized into twin data flows: one catering to the literal stream and the other to the sequence stream. The literal stream integrates data that doesn't coincide with any sequences previously detected and stored in the history buffer, while the sequence stream comprises of data matches found within the history buffer. Every matching outcome is depicted using a trio of numbers: the match's offset, its length, and the count of literal bytes that come before the match. During the second phase, Huffman coding and FSE coding6 are employed to compress the literal and sequence streams, respectively.

Algorithm 1 The implementation of Zstd compression

Input *msgMap*, an object of type *ccmslam_msgs::Map*, *pMsg*, a constant pointer to an object of type *ccmslam_msgs::Map*
Output: *msgMap*
1: **if** *msgMap* contains relevant data **then**
2: increment *mOutMapCount*, *mOutMapCount* ← *msgMap.mMsgId* and Set *msgMap.header.stamp* to current time;
3: **if** *mbCompFlag* = True **then**
4: configuring compression parameters,serialize and compress *MapPoints* and *MPUpdates* data;
5: publish *msgMap*;
6: **function** MapCbServer(*pMsg*)
7: create a mutable copy of *pMsg*, *pMutableMsg* ← *pMsg*;
8: configuring decompression parameters, decompress and deserialize *MapPoints* and *MPUpdates* data;
9: **end if**
10: **end if**

The core concept behind Zstd is its utilization of a dictionary-based approach to compress data. A dictionary is a pre-built data structure containing common patterns and phrases found in the input data. During compression, Zstd searches for matching patterns and phrases in the input data and replaces them with dictionary indices. This method significantly reduces the quantity of data that requires preservation, resulting in higher compression ratios. In addition to its dictionary-based approach, Zstd employs other techniques to enhance compression and decompression speeds. For instance, it uses a preprocessing technique to minimize the amount of data that needs to be processed, thereby accelerating compression and decompression.

Since the majority of visual information transmitted by ROS nodes comprises data characterizing map points, there is a substantial number of similar segments present in this data. This characteristic aligns well with the requirements of Zstd compression, thereby achieving an optimal compression ratio and reducing bandwidth demands.

We assess the performance of the suggested approach on CCM-SLAM, which can be readily modified for any collaborative visual SLAM application. Since CCM-SLAM [5] was proposed, it has been used as a reference centralized collaborative monocular visual because of its efficient system architecture and robust communication strategy.

In summary, the overall implementation procedure of the Zstd is presented in Algorithm 1.

3 Experimental Results

3.1 Implementation Details

Since CCM-SLAM5 was proposed, it has been used as a reference centralized collaborative monocular visual because of its efficient system architecture and robust communication strategy.

Specifically, our approach consists of two steps: map point sparsification, executed before each local Bundle Adjustment (BA) in CCM-SLAM's local mapping component, and Zstd compression, applied before sending each ROS message queue. We modify the original CCM-SLAM implementation by altering the keyframe insertion strategy. Since keyframe decision-making in CCM-SLAM hinges on the quantity of tracking points and the status of the local mapping thread, the frequency of local BA execution fluctuates when using single-threaded operations or executing point sparsifica-tion. Consequently, we employ a deterministic keyframe inclusion criterion, which is grounded on changes in rotation and translation, to evaluate the impact of our proposed point sparsification. To solve the minimum-cost maximum-flow graph problem, we apply Google Optimization Research Instruments [8], which is a portable, efficient, and open-source software suite designed for combinatorial optimization tasks. Our experiments were conducted on an Ubuntu 20.04 system with an i7-11800H CPU@2.3GHz and 16 GB of RAM.

3.2 Evaluation Metric

To assess the accuracy of the projected trajectory, we utilize the root mean square (RMS) of the absolute trajectory error (ATE) [11] as our metric for pose accuracy determination. This metric mirrors the overall consistency of the forecasted trajectory by measuring the absolute gap between the predicted pose and the actual ground truth pose. We prefer ATE over relative pose error (RPE), indicative of the trajectory's local precision, as ATE is more fitting for evaluating the accuracy of the final visual map. ATE signifies the average divergence from the ground truth pose and provides a more holistic evaluation of the total accuracy of the trajectory. For gauging the precision of rotational pose, we employ the same method as with ATE, which we term as *ATEr*.

For the absolute trajectory error matrix at timestamp, E_i, defined as

$$E_i = {Q_i}^{-1} P_i, \tag{5}$$

where Q_i is the ground truth pose and P_i is the projected pose at timestamp i aligned in the identical coordinate frame, RMS *ATEr* is computed as:

$$ATE_r = \left(\frac{1}{n}\sum\nolimits_{i=1}^{n} (|\angle rot(E_i)|)^2\right)^{\frac{1}{2}}, \tag{6}$$

based on the implementation in [9]. $\angle rot(E_i)$ is rotation error.

3.3 Performance Evaluation

To assess the effectiveness of the proposed approach, we utilize the publicly accessible EuRoC dataset [10], which includes precise ground-truth location information obtained from a Leica Total Station. The EuRoC dataset comprises video sequences filmed by an AscTec FireFly UAV equipped with a front-facing camera, flying repeatedly over an industrial environment. Each sequence is handled by a distinct agent running concurrently while communicating with the online server.

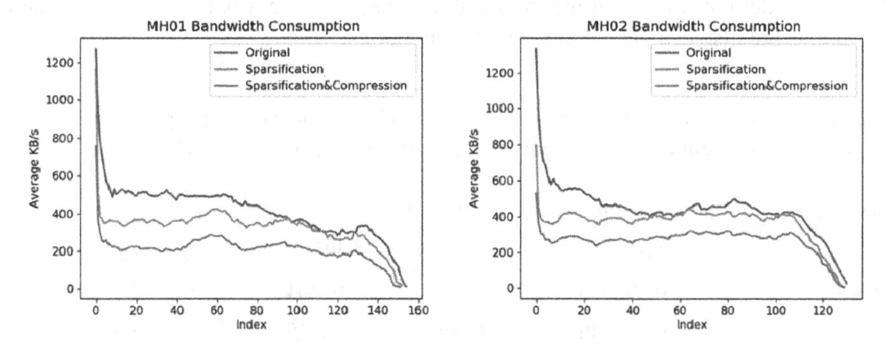

Fig. 3. Bandwidth consumption on MH01 and MH02 sequences.

Experimental results from Machine Hall 01 and 02 in the EuRoC dataset are presented in Fig. 3. On the MH01 sequence, after executing the map point sparsification step, we can observe that the bandwidth of transmitted data decreases by approximately 21%. Subsequently, by performing Zstd compression, the bandwidth of transmitted data is further reduced by approximately 37%. As a result, the utilization of the proposed method results in an approximately 49% reduction in bandwidth consumption.

Fig. 4. Visualization of map building results. **Left**: Original. **Right**: After map point sparsification.

To demonstrate the real-time performance, we compared the running time before and after the adoption of the proposed method in CCM-SLAM. We notice that compressing the data before sending messages to a ROS node takes only about 0.2 ms, and the decompression step on the server side takes even less time. After running the entire MH01 sequence (which is relatively long, lasting 180 s), the proposed method adds only about 0.3 s, indicating the impact on running time can be neglected.

Figure 4 presents the results of Rviz visualization for map building. We can see that after the map point sparsification operation, the sparse map displays fewer map points. However, the map-building accuracy is not compromised, as shown in Table 1. The results indicate that after adding the process of map points culling, the trajectory error even decreased slightly in both two sequences with easy and difficult modes. This observation demonstrates that the map culling strategy can not only reduce redundant map points but also can get rid of some ill map points.

Table 1. Comparison of tracking accuracy with and without adoption of the proposed method. MH_easy, medium, difficult refer to sequences of varying difficulty in EuRoC datasets.

Methods	Run	Cumulative tracking RMSE(m)			
		MH01_easy	MH02_easy	MH03_medium	MH04_difficult
Original	#1	0.220250	0.219224	0.298704	0.312763
	#2	0.190246	0.148006	0.270645	0.268856
	#3	0.203911	0.165844	0.279278	0.304720
	#4	0.295000	0.269902	0.254714	0.234604
	#5	0.193424	0.277909	0.252690	0.317555
	Avg	0.22735175	0.183615	**0.265984**	0.312763
Ours	#1	0.195233	0.172601	0.320508	0.279053
	#2	0.235074	0.194220	0.310153	0.269612
	#3	0.165582	0.123891	0.226154	0.241966
	#4	0.203613	0.182351	0.255476	0.232161
	#5	0.106816	0.095150	0.235284	0.192830
	Avg	**0.181264**	**0.183057**	0.287992	**0.217398**

4 Conclusion

In conclusion, our proposed method of efficient data compression and map point selection offers a promising solution to the communication overhead problem in collaborative SLAM, particularly in large-scale environments with limited communication data rates. By applying the Zstandard compression algorithm and our map point selection strategy, we have demonstrated a significant reduction in communication overhead while maintaining map accuracy and SLAM performance through experiments. Future work may focus on exploring the potential of our method in other collaborative robotics settings and optimizing the selection of map points for different SLAM systems.

References

1. Park, Y., Bae, S.: Keeping less is more: point sparsification for visual SLAM. In: Proceedings of the IEEE/RSJ International Conference on Intelligent Robots and System, pp. 7936–7943 (2022)
2. Zou, D., Tan, P.: CoSLAM: collaborative visual SLAM in dynamic environments. IEEE Trans. Pattern Anal. Mach. Intell. **35**, 354–366 (2013)
3. Van Opdenbosch, D., Aykut, T., Alt, N., Steinbach, E.: Efficient map compression for collaborative visual SLAM. In: 2018 IEEE Winter Conference on Applications of Computer Vision, WACV, pp. 992–1000 (2018)
4. Baroffio, L., Canclini, A., Cesana, M., Redondi, A., Tagliasacchi, M., Tubaro, S.: Coding local and global binary visual features extracted from video sequences. IEEE Trans. Image Process. **24**(11), 3546–3560 (2015)
5. Schmuck, P., Chli, M.: CCM-SLAM: robust and efficient centralized collaborative monocular simultaneous localization and mapping for robotic teams. J. Field Robot. **36**(4), 763–781 (2019)
6. Jarek, D.; Asymmetric numeral systems: entropy coding combining speed of Huffman coding with compression rate of arithmetic coding. arXiv: Information Theory (2013): n. pag
7. Goldberg, A.V.: An efficient implementation of a scaling minimum-cost flow algorithm. J. Algorithms **22**(1), 1–29 (1997)
8. Google operations research tools. https://github.com/google/or-tools
9. Grupp, M.: evo: python package for the evaluation of odometry and slam. https://github.com/MichaelGrupp/evo (2017)
10. Burri, M., et al.: The euroc micro aerial vehicle datasets. Int. J. of Robot. Res. (2016)
11. Sturm, J., Engelhard, N., Endres, F., Burgard, W., Cremers, D.: A benchmark for the evaluation of rgb-d slam systems. In: IEEE/RSJ International Conference on Intelligent Robots and Systems (IROS) (2012)
12. Van Opdenbosch, D., Aykut, T., Alt, N., Steinbach, E.: Efficient map compression for collaborative visual SLAM. In: 2018 IEEE Winter Conference on Applications of Computer Vision (WACV), pp. 992–1000, Lake Tahoe, NV, USA (2018)

Optimal Information Fusion Descriptor Fractional Order Kalman Filter

Xiao Liang, Guangming Yan, Yanfeng Zhu, Tianyi Li, and Xiaojun Sun(✉)

Electrical Engineering Institute, Heilongjiang University, Harbin 150080, China
sxj@hlju.edu.cn

Abstract. The fractional order Kalman filtering theory is an extension and extension of traditional integer order Kalman filters, which can solve the state estimation problem of fractional order systems. At present, the descriptor fractional order systems have been widely applied in many fields, such as circuit and sensor fault diagnosis. However, there is currently little research on the filtering problem of descriptor fractional order systems. This paper will focus on a fractional order descriptor system with canonical form. Firstly, the non singular linear transformation method is applied to transform the descriptor fractional order system into two normal fractional order subsystems. Then, based on projective theory, a fractional order Kalman state filter with correlated noise subsystems is derived. For multisensor descriptor fractional order systems, the globally optimal weighted measurement fusion algorithm is applied to derive the optimal information fusion fractional order Kalman filter. Simulation results verify the effectiveness and feasibility of the proposed algorithm.

Keywords: Information fusion · fractional order systems · fractional order Kalman filtering · weighted measurement fusion · optimal filtering

1 Introduction

Kalman filter algorithm is a common state estimation method. This method has been widely concerned and applied since it was put forward in the 1960s. Its feature is that it can predict the estimated value of the next moment or even the next moment according to the current data through iteration, so as to solve the dynamic target estimation problem with interference [1]. In addition, Kalman filter has also been widely applied in the engineering field, such as integrated navigation [2–4], target tracking [5, 6], fault diagnosis and detection [7, 8]. At present, many improved Kalman filtering algorithms emerge in an endless stream. From the initial solution of linear system problems to the present solution of nonlinear system problems, simple classical Kalman filtering at the very beginning, Later, adaptive Kalman filtering with unknown noise statistics or model parameters and other uncertain information [9], self-correcting Kalman filtering [10,

This work was supported by the National Natural Science Foundation of China, project approval number: 61104209; Supported by the Special Fund for Basic Scientific Research of Provincial Colleges and Universities, Project approval number :2020-KYYWF-0098.

B. Xin et al. (Eds.): IWACIII 2023, CCIS 1931, pp. 24–36, 2024.
https://doi.org/10.1007/978-981-99-7590-7_3

11], robust Kalman filtering theory [12, 13] and so on. In recent years, with the development of fractional-order calculus theory, the filter problem of fractional-order system has gradually attracted people's attention. The state estimation of fractional-order system is a new idea and new direction of the development of the theory of state estimation, and also meets the actual production demand.

Fractional calculus was proposed by Leibniz and L'Hospital in 1965, and then Liouville and Riemann proposed the definition of fractional derivatives. However, it was initially studied only by engineers, and it was not until the late 1960s that fractional calculus was gradually developed, and it was discovered that fractional derivatives would allow more accurate descriptions of systems in simulation modeling or stability analysis. With the passing of time, the control system theory gradually develops, the traditional calculus theory is not enough to meet the needs of production and research. Fractional calculus became more active in the theory of control systems and played an indispensable role. Literature [14] and [15] proposed fractional Kalman filtering algorithm and extended fractional Kalman filtering algorithm, and analyzed specific cases to discuss the possibility of applying these algorithms to fractional system parameters and fractional estimation, and their algorithms have been applied to image processing, signal transmission and other fields [16, 17]. But compared with the traditional Kalman filter, its application is not so wide. At present, generalized fractional systems have been widely used in the fields of circuit [18, 19] and sensor fault estimation [20]. The fusion estimation problem of generalized fractional systems discussed in this paper has certain theoretical significance and potential application value.

In this paper, a typical fractional-order singular system is firstly transformed into two normal fractional-order subsystems by non-singular linear transformation method. Then, a fractional-order Kalman state filter with correlated noise subsystems is derived based on projective theory. For multi-sensor generalized fractional-order systems, the global optimal weighted observation fusion algorithm is applied to derive the optimal information fusion fractional-order Kalman filter, and the simulation results verify the effectiveness of the proposed algorithm.

2 Problem Formulation

Consider the following linear generalized fractional stochastic system

$$M\Delta^{\gamma}x(k+1) = \Phi x(k) + Bw(k) \tag{1}$$

$$x(k) = \Delta^{\gamma}x(k) - \sum_{j=1}^{k}(-1)^{j}\gamma_{j}x(k-j) \tag{2}$$

$$y(k) = Hx(k) + v(k) \tag{3}$$

where $x(k+1) \in R^{n}$ is the state of the system, $\Phi \in R^{n\times n}$ is the target variance coefficient matrix, $B \in R^{n\times r}$ is the target noise matrix, Δ^{γ} is a fractional operator, γ is fractional order, H is the observation matrix of the observation equation, $v(k)$ is the observed noise, M, H is the constant matrix of corresponding dimension.

Assumption 1: $M \in R^{n \times n}$ is a singular square matrix, means rank$M = n_1 < n$, $\det M = 0$.

Assumption 2: The system is regular, means $\exists z \in C$, so $\det(zM - \Phi) \neq 0$.

Assumption 3: $w(k) \in R^r$ and $v(k)$ are zero mean uncorrelated white noise:

$$\mathrm{E}\left\{\begin{bmatrix} w(k) \\ v(k) \end{bmatrix}\begin{bmatrix} w^{\mathrm{T}}(k) \; v^{\mathrm{T}}(k) \end{bmatrix}\right\} = \begin{bmatrix} Q_w & 0 \\ 0 & Q_v \delta_{ij} \end{bmatrix} \tag{4}$$

where E is the expected value, T is the transpose symbol.

Assumption 4: The system is fully observable, So there's the matrix K that makes:

$$rank\begin{bmatrix} zM - (\Phi - KH) \\ H \end{bmatrix} = n_1, \, rank\begin{bmatrix} M \\ H \end{bmatrix} = n \tag{5}$$

Assumption 4 leads to the existence of two non-singular square matrices R, W that make [77]

$$RMW = \begin{bmatrix} I_{n_1} & 0 \\ 0 & 0 \end{bmatrix}, R(\Phi - KH)W = \begin{bmatrix} Y_1 & 0 \\ 0 & I_{n_2} \end{bmatrix} \tag{6}$$

where $n_1 + n_2 = n$.Introducing block matrix representation:

$$RK = \begin{bmatrix} K_1 \\ K_2 \end{bmatrix}, RB = \begin{bmatrix} B_1 \\ B_2 \end{bmatrix}, HW = [H_1 \; H_2] \tag{7}$$

and introduce state:

$$x(k) = W\begin{bmatrix} x_1(k) \\ x_2(k) \end{bmatrix} \tag{8}$$

where $x_1(k) \in R_{n_1}, x_2(k) \in R_{n_2}$.

Equation (3) is multiplied by K and subtracted by Eq. (1)

$$M\Delta^{\gamma}x(k+1) = (W - KH)x(k) + Ky(k) + \overline{w}(k) \tag{9}$$

$$\overline{w}(k) = Bw(k) - Kv(k) \tag{10}$$

P is left multiplied by formula (9), and formula (6)–(8) is used to derive the observable model as follows:

$$\begin{bmatrix} I_{n_1} & 0 \\ 0 & 0 \end{bmatrix}\begin{bmatrix} \Delta^{\gamma}x_1(k+1) \\ \Delta^{\gamma}x_2(k+1) \end{bmatrix} = \begin{bmatrix} Y_1 & 0 \\ 0 & I_{n_2} \end{bmatrix}\begin{bmatrix} x_1(k) \\ x_2(k) \end{bmatrix} + \begin{bmatrix} \overline{K}_1 \\ \overline{K}_2 \end{bmatrix}y(k) + \begin{bmatrix} B_1 \\ B_2 \end{bmatrix}w(k) \tag{11}$$

$$y(k) = [H_1 \; H_2]\begin{bmatrix} x_1(k) \\ x_2(k) \end{bmatrix} + v(k) \tag{12}$$

This leads to two reduced order subsystems:

$$\Delta^{\gamma} x_1(k+1) = Y_1 x_1(k) + \overline{K}_1 y(k) + B_1 w(k) \tag{13}$$

$$x_2(k) = -\overline{K}_2 y(k) - B_2 w(k) \tag{14}$$

$$y(k) = H_1 x_1(k) + H_2 x_2(k) + v(k) \tag{15}$$

by substituting Eq. (14) into Eq. (15), a subsystem with different local dynamic transformation types and the same local state $x_1(k)$ is derived:

$$\Delta^{\gamma} x_1(k+1) = Y_1 x_1(k) + \overline{K}_1 y(k) + B_1 w(k) \tag{16}$$

$$z(k) = H_1 x_1(k) + \tau(k) \tag{17}$$

therein defined:

$$z(k) = (I_m + K_2 H_2) y(k) \tag{18}$$

$$\tau(k) = v(k) - H_2 B_2 w(k) \tag{19}$$

from (17), (18)

$$y(k) = (I_m + \overline{K}_2 H_2)^{-1} [H_1 x_1(k) + \tau(k)] \tag{20}$$

for the transformed conventional subsystem (16), if a non-zero term is added to the right side and substituted into formula (20), it can be obtained:

$$\begin{aligned}\Delta^{\gamma} x_1(k+1) =& Y_1 x_1(k) + \overline{K}_1 (I_m + \overline{K}_2 H_2)^{-1} \times [H_1 x_1(k) + \tau(k)] + B_1 w(k) + \\ & \Lambda[z(k) - H_1 x_1(k) - \tau(k)] \\ =& \left[Y_1 + \overline{K}_1 (I_m + \overline{K}_2 H_2)^{-1} H_1 - \Lambda H_1\right] x_1(k) + \\ & \Lambda z(k) + \left[\overline{K}_1 (I_m + \overline{K}_2 H_2)^{-1} \tau(k) + B_1 w(k) - \Lambda \tau(k)\right]\end{aligned} \tag{21}$$

where Λ is the undetermined matrix and can be set:

$$\Phi_1 = Y_1 + \overline{K}_1 (I_m + \overline{K}_2 H_2)^{-1} H_1 - \Lambda H_1 \tag{22}$$

$$\phi(k) = \overline{K}_1 (I_m + \overline{K}_2 H_2)^{-1} \tau(k) + B_1 w(k) - \Lambda \tau(k) \tag{23}$$

Then equation of state (21) can be reduced to

$$\Delta^{\gamma} x_1(k+1) = \Phi_1 x_1(k) + \Lambda z(k) + \phi(k) \tag{24}$$

However, the observation equation is still Eq. (17). According to assumption 3, we can know:

$$E[\phi(k)] = 0$$

then have

$$
\begin{aligned}
E\left[\phi(k)\tau^T(j)\right] &= E[\overline{K}_1(I_m+\overline{K}_2H_2)^{-1}\tau(k)+B_1w(k)-\Lambda\tau(k)] \\
&= \overline{K}_1(I_m+\overline{K}_2H_2)^{-1}Q_\tau+E\left[B_1w(k)\tau^T(j)\right]-\Lambda Q_\tau \\
&= \overline{K}_1(I_m+\overline{K}_2H_2)^{-1}Q_\tau-B_1Q_w(H_2B_2)^T-\Lambda Q_\tau
\end{aligned} \tag{25}
$$

and Q_τ is the reciprocal covariance matrix of $\tau(k)$:

$$
Q_\tau = E\left[\tau(k)\tau^T(j)\right] = Q_v + H_2B_2Q_w(H_2B_2)^T \tag{26}
$$

so you can take the undetermined matrix

$$
\Lambda = \left[\overline{K}_1(I_m+\overline{K}_2H_2)^{-1}Q_\tau - B_1Q_w(H_2B_2)^T\right]Q_\tau^{-1} \tag{27}
$$

therefore, there is $E\left[\phi(k)\tau^T(j)\right]=0$, that is, $\phi(k)$ is unrelated to $\tau(j)$, and the autocovariance matrix of $\phi(k)$ is easily obtained:

$$
\begin{aligned}
&E\left[\phi(k)\phi^{\mathrm{T}}(j)\right] = \overline{K}_1(I_m+\overline{K}_2H_2)^{-1}Q_\tau \times \left[\overline{K}_1(I_m+\overline{K}_2H_2)^{-1}\right]^T \\
&+B_1Q_wB_1^{\mathrm{T}}+\Lambda Q_\tau\Lambda^{\mathrm{T}}
\end{aligned} \tag{28}
$$

substitute into Eq. (27) to get

$$
\begin{aligned}
E\left[\phi(k)\phi^T(j)\right] &= \overline{K}_1(I_m+\overline{K}_2H_2)^{-1}Q_\tau \times \left[\overline{K}_1(I_m+\overline{K}_2H_2)^{-1}\right]^T + B_1Q_wB_1^{\mathrm{T}}+ \\
&\left[\overline{K}_1(I_m+\overline{K}_2H_2)^{-1}Q_\tau - B_1Q_w(H_2B_2)^T\right] \\
&\times\left\{\left[\overline{K}_1(I_m+\overline{K}_2H_2)^{-1}Q_\tau - B_1Q_w(H_2B_2)^T\right]Q_\tau^{-1}\right\}^T
\end{aligned} \tag{29}
$$

So $\phi(k)$ has zero mean white noise, and the variance is

$$
\begin{aligned}
Q_\phi &= \overline{K}_1(I_m+\overline{K}_2H_2)^{-1}Q_\tau\left[\overline{K}_1(I_m+\overline{K}_2H_2)^{-1}\right]^T + B_1Q_wB_1^{\mathrm{T}} \\
&+[\overline{K}_1(I_m+\overline{K}_2H_2)^{-1}Q_\tau - B_1Q_w(H_2B_2)^T]\times \\
&\left\{\left[\overline{K}_1(I_m+\overline{K}_2H_2)^{-1}Q_\tau - B_1Q_w(H_2B_2)^T\right]Q_\tau^{-1}\right\}^T
\end{aligned} \tag{30}
$$

which is independent of white noise $\tau(k)$.Generalized fractional filtering problem is to calculate the minimum variance estimation $\hat{x}(k|k)$ of state $x(k)$ based on the observed value $y(1),\cdots,y(k)$ of data obtained by multiple sensors.

3 Kalman Filter for Single Sensor Generalized Fractional Order System

Theorem 1 For the observable singular system (1)–(3), under hypothesis 1-(4), the reduced-order subsystem (17) and (24) has a local recursive fractional Kalman filter of $x_1(k)$.

$$
\hat{x}_1(k|k) = \hat{x}_1(k|k-1) + K_1(k)[z(k) - H_1\hat{x}_1(k|k-1)] \tag{31}
$$

$$\Delta^{\gamma}\hat{x}_1(k|k-1) = \Phi_1\hat{x}_1(k-1|k-1) + \Lambda z(k-1) \tag{32}$$

$$\hat{x}_1(k|k-1) = \Delta^{\gamma}\hat{x}_1(k|k-1) - \sum_{j=1}^{k}(-1)^j\gamma_j\hat{x}_1(k-j|k-j) \tag{33}$$

$$P_1(k|k-1) = (\Phi_1+\gamma_1)P_1(k-1|k-1)\times(\Phi_1+\gamma_1)^T + \sum_{j=2}^{k}\gamma_j P_1(k-j|k-j)\gamma_j^T + Q_\phi \tag{34}$$

$$P_1(k|k) = (I_n - K_1(k)H_1)P_1(k|k-1) \tag{35}$$

$$K_1(k) = P_1(k|k-1)H_1^T \times \left[H_1P_1(k|k-1)H_1^T + Q_\eta\right]^{-1} \tag{36}$$

put in the initial value $\hat{x}_1(0|0) = \rho_{01}, P_1(0|0) = P_{01}$.

Proof: It can be obtained from literature [13] that

$$\begin{aligned}&\hat{x}_1(k|k-1) = \text{proj}(x_1(k)|z(1),\cdots,z(k-1)) = \text{proj}[\Phi_1x_1(k-1)+\Lambda z(k-1)+\phi(k-1)-\\&\sum_{j=1}^{k}(-1)^j\gamma_jx_1(k-j)|z(1),\cdots,z(k-1)] = \Phi_1\text{proj}[x_1(k-1)|z(1),\cdots,z(k-1)]+\\&\Lambda\text{proj}[z(k-1)|z(1),\cdots,z(k-1)] - \sum_{j=1}^{k}(-1)^j\gamma_j\text{proj}[x_1(k-j)|z(1),\cdots,z(k-1)]\end{aligned} \tag{37}$$

available at this time

$$\hat{x}_1(k|k-1) = \Phi_1\hat{x}_1(k-1|k-1) + \Lambda z(k-1) - \sum_{j=1}^{k}(-1)^j\gamma_j\hat{x}_1(k-j|k-j) \tag{38}$$

So we get (32) and (33) easily.

According to literature [13], the one-step optimal linear prediction $\hat{z}(k|k-1)$ of $z(k)$ can be obtained, i.e.

$$\begin{aligned}\hat{z}(k|k-1) &= \text{proj}[z(k)|z(1),\cdots,z(k-1)]\\&= \text{proj}[H_1x_1(k)+\tau(k)|z(1),\cdots,z(k-1)]\\&= H_1\hat{x}_1(k|k-1)\end{aligned} \tag{39}$$

thus easy to obtain

$$\hat{x}_1(k|k) = \hat{x}_1(k|k-1) + K_1(k)\varepsilon(k) \tag{40}$$

among them $\varepsilon(k) = z(k)-\hat{z}(k|k-1), K_1(k) = E[x_1(k)\varepsilon^T(k)][E(\varepsilon(k)\varepsilon^T(k)]^{-1}$ is called Kalman

filter gain. Define $\tilde{x}_1(k|k-1) = x_1(k) - \hat{x}_1(k|k-1)$, then $E[x_1(k)\varepsilon^T(k)] = E[(\hat{x}_1(k|k-1) + \tilde{x}_1(k|k-1))$.

$\times(H_1\tilde{x}_1(k|k-1)+\tau(k))^T]$, By projective orthogonality we have $\hat{x}_1(k|k-1)\perp\tilde{x}_1(k|k-1)$, $\tau(k)\perp\hat{x}_1(k|k-1)$, $\tau(k)\perp\tilde{x}_1(k|k-1)$, then

$$E[x_1(k)\varepsilon^T(k)]=P_1(k|k-1)\overline{H}^T \tag{41}$$

In the same way, we can get

$$E[\varepsilon(k)\varepsilon^T(k)]=E[(z(k)-\hat{z}(k|k-1))\times(z(k)-\hat{z}(k|k-1))^T]=\overline{H}P_1(k|k-1)\overline{H}^T+Q_\xi \tag{42}$$

where $P_1(k|k-1)=E[(x_1(k)-\hat{x}_1(k|k-1))(x_1(k)-\hat{x}_1(k|k-1))^T]$ is the prediction error variance matrix.It can be obtained from (40) that

$$\hat{x}_1(k|k)=\hat{x}_1(k|k-1)+P_1(k|k-1)H_1^T\times\left[H_1P_1(k|k-1)H_1^T+Q_\eta\right]^{-1}\varepsilon(k) \tag{43}$$

$K_1(k)=P_1(k|k-1)H_1^T[H_1P_1(k|k-1)\times H_1^T+Q_\eta]^{-1}$ is denoted as the gain matrix of fractional Kalman filter, then formulas (31) and (36) can be obtained.

$$\begin{aligned}x_1(k)-\hat{x}_1(k|k-1)&=\Phi_1x_1(k-1)+\Lambda z(k-1)+\phi(k-1)-\sum_{j=1}^{k}\left[(-1)^j\gamma_jx_1(k-j)\right]\\&-\Phi_1\hat{x}_1(k-1|k-1)-\Lambda z(k-1)+\sum_{j=1}^{k}\left[(-1)^j\gamma_j\hat{x}_1(k-j|k-j)\right]=(\Phi_1+\gamma_1)\times\\&[x_1(k-1)-\hat{x}_1(k-1|k-1)]-\sum_{j=2}^{k}(-1)^j\gamma_j(x_1(k-j)-\hat{x}_1(k-j|k-j))+\phi(k-1)\end{aligned} \tag{44}$$

where $E[(x_{1m}-\hat{x}_{1\,m|m-1})(x_{1n}-\hat{x}_{1\,n|n-1})^T]=0$, $m\neq n$.It follows that:

$$\begin{aligned}P_1(k|k-1)&=E[(x_1(k)-\hat{x}_1(k|k-1))\times(x_1(k)-\hat{x}_1(k|k-1))^T]\\&=(\Phi_1+\gamma_1)P_1(k|k)(\Phi_1+\gamma_1)^T+\sum_{j=2}^{k}\gamma_jP_1(k-j|k-j)\gamma_j^T+Q_\phi\end{aligned} \tag{45}$$

where, Q_ϕ is the autocovariance matrix of $\phi(k)$, which is given by Eq. (28) and can be proved by Eq. (34).

It can be obtained from Eqs. (17) and (31) that

$$\begin{aligned}x_1(k)-\hat{x}_1(k|k)&=x_1(k)-\{\hat{x}_1(k|k-1)+K_1(k)\times[z(k)-H_1\hat{x}_1(k|k-1)]\}\\&=x_1(k)-\hat{x}_1(k|k-1)-K_1(k)\times[H_1x_1(k)+\tau(k)-H_1\hat{x}_1(k|k-1)]\\&=[I_n-K_1(k)H_1](x_1(k)-\hat{x}_1(k|k-1))-K_1(k)\tau(k)\end{aligned} \tag{46}$$

$$\begin{aligned}P_1(k|k)&=E[(x_1(k)-\hat{x}_1(k|k))(x_1(k)-\hat{x}_1(k|k))^T]\\&=[I_n-K_1(k)H_1]P_1(k|k-1)\times[I_n-K_1(k)H_1]^T+K_1(k)Q_\tau K_1^T(k)\\&=(I_n-K_1(k)H_1)P_1(k|k-1)\end{aligned} \tag{47}$$

Theorem 2: Fractional subsystem 2 has a local recursive fractional Kalman filter under Eqs. (14) and (15).

$$\hat{x}_2(k|k) = -(I + \overline{K}_2 H_2)^{-1} \overline{K}_2 H_1 \hat{x}_1(k|k) \tag{48}$$

Prove: Applying theorem 1, it can be proved easily by Eqs. (14) and (15).

4 Observational Fusion Kalman Filter for Generalized Fractional-Order Systems

The observation fusion of generalized fractional order system is carried out for the normalized subsystem, so the normalized subsystem of multi-sensor generalized fractional order system is considered as follows.

$$\Delta^{\gamma} x_1(k+1) = \Phi_1 x_1(k) + \Lambda_i z_i(k) + \phi_i(k) \tag{49}$$

$$z_i(k) = H_{1i} x_i(k) + \tau_i(k) \tag{50}$$

$$z_i(k) = (I_m + \overline{K}_2 H_{2i}) y_i(k) \tag{51}$$

$$\tau_i(k) = v_i(k) - H_{2i} B_2 w(k) \tag{52}$$

$$H_{1i} = G_i \overline{H} i = 1, \cdots, L \tag{53}$$

where $x_i(k) \in R^n$ is the state quantity and $z_i(k) \in R^m$ is the observation of the ith sensor,$\tau_i(k) \in R^{m_i}$ is observed noise, $H_{1i} \in R^r$ is observed white noise,Φ_1 、$\overline{K}_2$ 、$\overline{H}$ is a known constant matrix of appropriate dimension, and the observed matrix $\overline{H}_i$ has the same $m \times n$ dimensional right factor $\overline{H}$,and

$$\Delta^{\gamma} x_1(k+1) = \begin{bmatrix} \Delta^{\gamma} x_{1i}(k+1) \\ \vdots \\ \Delta^{\gamma} x_{1n}(k+1) \end{bmatrix}$$

Assumption 5: $\phi_i(k) \in R^r$ and $\tau_i(k) \in R^n$ are mutually independent white noises with zero mean and variance matrices $Q_{\phi i}$ and Q_{τ_i}, respectively, and

$$\mathrm{E}\left\{\begin{bmatrix} \phi_i(k) \\ \tau_i(k) \end{bmatrix}\left[\phi_i^T(k)\ \tau_j^{\mathrm{T}}(k)\right]\right\} = \begin{bmatrix} Q_{\phi i}\delta_{ij} 0 \\ 0\, Q_{\tau i}\delta_{ij} \end{bmatrix}\delta_{tk} \tag{53}$$

where E is the mean symbol and T is the transpose symbol,$\delta_{tt} = 1$, $\delta_{tk} = 0\,(t \neq k)$。

Assumption 6: $(\Phi_1\, H_{1i})$ is a completely observable pair.

Assumption 7: The matrix $\sum_{i=1}^{L} [G_i^{\mathrm{T}} R_{\xi i}^{-1} G_i]^{-1}$ is invertible.

The centralized fusion observation equation can be obtained from Eqs. (50)–(54) that

$$z_0(k) = H_0 x(k) + \tau_0(k) \quad (55)$$

$$z_0(k) = [z_1^{\mathrm{T}}(k), \cdots, z_L^{\mathrm{T}}(k)]^T \quad (56)$$

$$H_0 = [H_0^{\mathrm{T}}, \cdots, H_L^{\mathrm{T}}]^T \quad (57)$$

$$\tau_0(k) = [\tau_1^{\mathrm{T}}(k), \cdots \tau_L^{\mathrm{T}}(k)]^T \quad (58)$$

The fused observation noise $\tau_0(k)$ has a variance matrix

$$Q_{\tau 0} = \mathrm{diag}(Q_{\tau_1}, \cdots, Q_{\tau_L}) \quad (59)$$

Equation (55) can be regarded as an observation model for $H_1 x_1(k)$, so the weighted least squares (WLS) method can be applied to estimate $H_1 x_1(k)$ as

$$z(k) = \left[G_0^{\mathrm{T}} Q_{\tau 0}^{-1} G_0\right]^{-1} G_0^{\mathrm{T}} Q_{\tau 0}^{-1} z_0(k) \quad (60)$$

The weighted observation fusion equation can be obtained by substituting Eq. (60) into (55)

$$z(k) = H_1 x(k) + \tau(k) \quad (61)$$

And it has fused observation noise

$$\tau(k) = \left[G_0^{-1} Q_{\tau_0}^{-1} G_0\right]^{-1} G_0^{-1} Q_{\tau_0}^{-1} \tau_0(k) \quad (61)$$

It has the minimum error variance matrix

$$Q_\tau = \left[G_0^{-1} Q_{\tau_0}^{-1} G_0\right]^{-1} \quad (63)$$

5 Simulation Study

Generalized fractional systems have important applications in circuits[18, 19] and sensor fault estimation[20]. Here, the canonical form of a generalized fractional order circuit system is considered.

$$\begin{bmatrix} 1 & 0 \\ 0 & 0 \end{bmatrix} \begin{bmatrix} \Delta^\gamma x_1(k+1) \\ \Delta^\gamma x_2(k+1) \end{bmatrix} = \begin{bmatrix} 0.1 & 0 \\ 0 & 1 \end{bmatrix} \begin{bmatrix} x_1(k) \\ x_2(k) \end{bmatrix} + \begin{bmatrix} 0 \\ 1 \end{bmatrix} y(k) + \begin{bmatrix} 1 \\ 0 \end{bmatrix} w(k) \quad (64)$$

$$y(k) = [10] \begin{bmatrix} x_1(k) \\ x_2(k) \end{bmatrix} + v(k) \quad (65)$$

where $w(k)$ and $v(k)$ are uncorrelated white noise with zero mean andvariance $Q = 0.01$ and $Q_v = 0.02$,respectively,$n_1 = 0.01$,the problem is to find a generalized fractional Kalman filter $\hat{x}(k|k) = [\hat{x}_1(k|k), \hat{x}_2(k|k)]^T$ for state $x(k)$.The simulation results are shown in Fig. 1.

According to theorem 1, the model given above is simulated and analyzed. Figs. 1 and 2 show the filtering results of state values and estimated values of subsystem 1 and subsystem 2. As can be seen from the figure, the estimated value can almost keep up with its state truth value, which is comparable to the effect of Model II. The error is basically within 0.1, so it can be seen that the generalized fractional-order filtering algorithm is feasible. Of course, the effect can be improved by adjusting the parameters. Then, weighted observation fusion was explored for the above model, and simulation analysis was carried out. Based on (64) and (65), different local observation noise error variances were considered respectively as: $Q_{v1} = 0.02$, $Q_{v2} = 0.1$, $Q_{v3} = 1$, As shown in Fig. 3, observation fusion is carried out on subsystem 1 after normalization, and curves of state truth value and observation estimation are made. By analogy with Fig. 1 of local valuation, there are obvious changes.

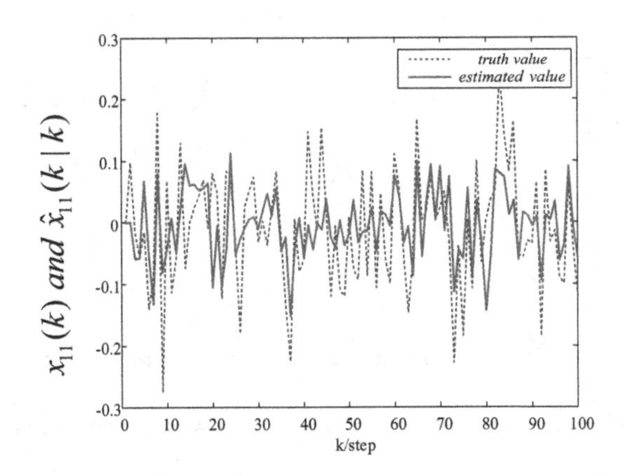

Fig. 1. Comparison of true and estimated values for Subsystem 1

In order to further compare and analyze the estimation accuracy of local fusion, on the basis of (63) and (64), different local observation noise error variances are considered as: $Q_{v1} = 0.02, Q_{v2} = 0.1$, $Q_{v3} = 1$,The comparison graph of mean square error obtained by 100-step Monte-Carlo simulation of the three fusion estimates is shown in Fig. 4. MSEm, MSCI123, MSEguance and MSEjizhong represent the mean square error curves of suboptimal weighted state fusion, SCI fusion, weighted observation fusion and centralized fusion, respectively. At k = 60,then

$$\begin{aligned}
\text{MSEm} &= 0.007304659 \\
\text{MSCI123} &= 0.00650606 \\
\text{MSEguance} &= 0.00647497 \\
\text{MSEjizhong} &= 0.00647497
\end{aligned}$$

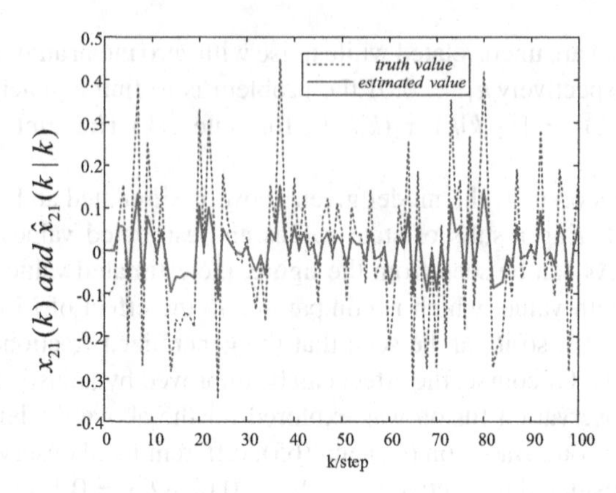

Fig. 2. Comparison between true value and estimated value of subsystem 2

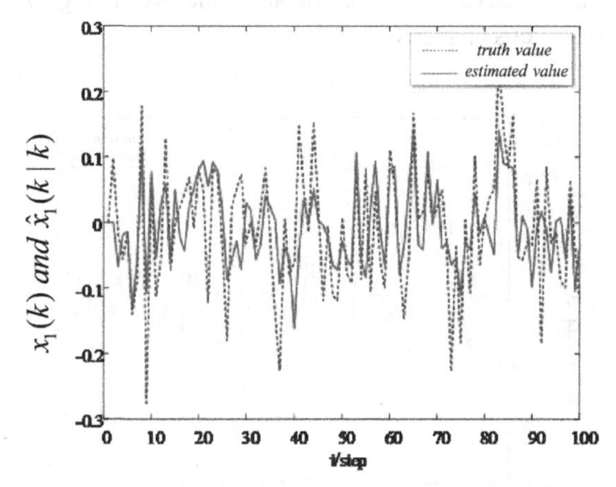

Fig. 3. Comparison of the state truth value of subsystem 1 with the weighted observation fusion estimate

As can be seen from the figure, the actual estimation accuracy of SCI fusion is similar to that of distributed suboptimal state fusion, but lower than that of weighted observation fusion. The estimated accuracy of weighted observation fusion is the same as that of centralized fusion, which is a globally optimal weighted fusion algorithm.

Note 1 Based on subsystem 1 after weighted fusion, the corresponding state estimator of subsystem 2 can also be obtained from Theorem 2. According to the optimality of the state fusion estimation of subsystem 1, it is easy to know that the corresponding state estimator of subsystem 2 also has the same optimality.

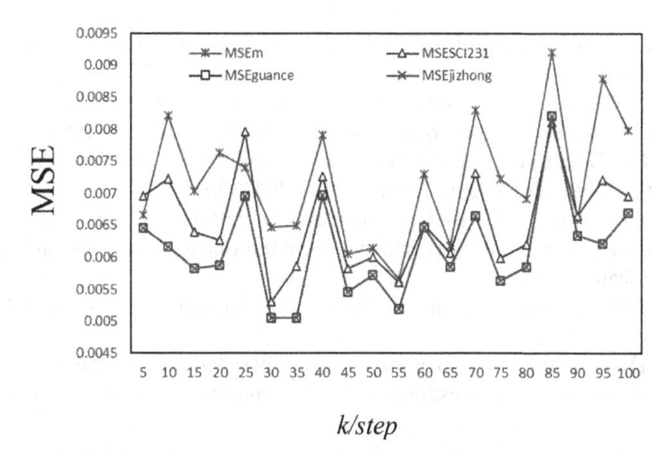

Fig. 4. Comparison curve of mean square error of three fusion estimates

6 Conclusions

A fractional Kalman state filter for fractional descriptor systems is proposed in this paper. For multi-sensor generalized fractional order system, the global optimal weighted observation fusion algorithm is applied to derive the optimal information fusion fractional order Kalman filter. The proposed algorithm has global optimality and equivalent estimation accuracy compared with the centralized fusion algorithm, but the computational complexity is greatly reduced, which is convenient for engineering applications. A simulation example proves the effectiveness and feasibility of the method.

References

1. Kalman, R.E.: A new approach to linear filtering and prediction problems. J. Basic Eng. **82**(1), 35–45 (1960)
2. Cao, J., Yu, C., Xia, Y., Ji, X.: Clock synchronous GNSS/SINS tightly combined adaptive filtering algorithm. In: Proceedings of the 8th China Annual Conference on Satellite Navigation—S10 Multi-source Fusion Navigation Technology, pp. 109–113. Shanghai (2017)
3. Wang, W., Cong, N., Wu, J.: Design of robust GPS/INS integrated navigation filtering algorithm. J. Harbin Eng. Univ. **42**(02), 240–245 (2021)
4. Hong, Z.: Research on improved algorithm of unscented Kalman filter and its application in GPS/INS integrated navigation, pp. 45–58. East China University of Technology (2019)
5. Chi, J.N., Qian, C., Zhang, P., et al.: A novel ELM based Adaptive Kalman Filter Tracking Algorithm. Neurocomputing **128**, 42–49 (2014)
6. Chen, X., Fan, Y., Ma, Z.: Kalman filter tracking algorithm simulation based on angle expansion. In: 2021 4th International Conference on Algorithms, Computing and Artificial Intelligence, pp. 1–5. Sanya (2021)
7. Zhao, H.: Fault Diagnosis of Multi-Sensor Integrated Navigation Based on Federated Kalman Filter, pp. 31–44. Beijing Institute of Technology (2017)
8. Du, Z., Li, X., Zhen, Z., Mao, Q.: Fault prediction based on fusion of strong tracking square-root cubature Kalman filter and autoregressive model. Control Theory Appl. **31**(08), 1047–1052 (2014)

9. Mehra, R.H.: On the identification of variances and adaptive kalman filtering. IEEE Trans. Autom. Control **15**(2), 175–184 (1970)
10. Deng, Z.L., Gao, Y., Li, C.B., et al.: Self-tuning decoupled information fusion wiener state component filters and their convergence. Automatica **44**(3), 685–695 (2008)
11. Ran, C.J., Tao, G.L., Liu, J.F., et al.: Self-tuning decoupled fusion Kalman predict-or and its convergence analysis. IEEE Sens. J. **9**(12), 2024–2032 (2012)
12. Yu, H., Jing, Z.: Jianke Lv.Robust Kalman Filtering algorithm based on Outlier Detection. In: 11th National Conference on Signal and Intelligent Information Processing and Applications, pp. 44–48. Zhanjiang (2017)
13. Deng, Z.: Theory and application of information fusion estimation, pp. 463–479. Science Press, Beijing (2012)
14. Sicrociuk, D., Dzielinski, A.: Fractional Kalman filter algorithm for the states, parametiers and order of fractional system estimation. Int. J. Appl. Math. Comput. Sci. **16**(1), 129–140 (2006)
15. Andrze, D., Dominik, S.: Adaptive feedback control of fractional order discrete state-space system. In: International Conference on Intelligent Agents, Web Technologies and Internet Commerce, vol. 1, pp. 804–809. Vienna, Austria (2005)
16. Chen, D., Bufa, H.: Application of adaptive fractional differential in facial image processing. Mech. Manufac. Autom. **46**(06), 137–141 (2017)
17. Wang, T.: Application of Kalman Filter in image processing. Sci. Technol. Inform. (Acad. Ed.) **032**, 575–576 (2008)
18. Kaczorek, T.: Descriptor fractional linear systems with regular pencils. Asian J. Control **15**(4), 1051–1064 (2013)
19. Kaczorek, T., Ruszewski, A.: Application of the drazin inverse to the analysis of pointwise completeness and pointwise degeneracy of descriptor fractional linear continuous-time systems. Int. J. Appl. Math. Comput. Sci. **30**(2), 219–223 (2020)
20. Jmal, A., Naifar, O., Makhlouf, A.B., Derbel, N., Hammami, M.A.: Sensor fault estimation for fractional-order descriptor one-sided lipschitz systems. Nonlinear Dyn. **91**(3), 1713–1722 (2018). https://doi.org/10.1007/s11071-017-3976-1

Multi-sensor Data Fusion Algorithm for Indoor Fire Detection Based on Ensemble Learning

Lei Wang and Jia Zhang(✉)

Beijing Institute of Technology, Beijing, China
zhangjia@bit.edu.cn

Abstract. Among all the disasters, fire is one of the most catastrophic events that frequently and universally threaten public safety and social development. In recent years, indoor fires have resulted in an increasing number of casualties and extensive damages. Therefore, in order to mitigate the impact of indoor fire disasters, it is crucial to propose an indoor fire detection method that can quickly and accurately detect the fire. In the research, a multi-sensor data fusion algorithm based on LogitBoost ensemble learning was designed for indoor fire detection. To improve detection accuracy and robustness, the proposed model utilizes an S-G filter and Min-Max normalization method, then uses Logitboost to synergistically integrate four classifiers, including Naïve Bayes, backpropagation neural network (BPNN), support vector machine (SVM) and k-nearest neighbor (KNN) classifier. A dataset containing various fire scenarios from the National Institute of Standards and Technology (NIST) was adopted to evaluate the effectiveness of the proposed algorithm. The experimental results demonstrated that the proposed method outperforms single models in terms of accuracy, stability, and efficiency.

Keywords: Fire Detection · Ensemble Learning · Multi-sensor Fusion

1 Introduction

In recent years, the continuous advancement of science and technology has led to increasingly complex modern building constructions, which concurrently has resulted in greater fire safety hazards. According to a report from the Department of Emergency Management Fire and Rescue of China on April 3, 2022, a total of 219,000 fires were reported across the country in the first quarter of the year, with residential areas being the most prominent location for fires, accounting for 38% of the total number of fires and 80.5% of the total number of fatalities [1]. It is evident that indoor fire accidents pose a significant threat to people's lives and property safety. Furthermore, minimizing the response time of fire detection systems is a crucial factor in reducing fire losses, as it increases the chances of extinguishing a fire [2]. Therefore, exploring a rapid response and high accuracy fire detection system is of paramount importance.

Currently, there are mainly three types of fire detection methods: traditional single sensor algorithms, multi-sensor fusion algorithms, and vision-based algorithms.

B. Xin et al. (Eds.): IWACIII 2023, CCIS 1931, pp. 37–49, 2024.
https://doi.org/10.1007/978-981-99-7590-7_4

The traditional single sensor algorithm detects temperature, smoke, sound, light, and other parameters monolithically. This method employs a single source of information from the fire scene for threshold judgment, which is highly susceptible to external environmental interference, resulting in missing or false alarms. Furthermore, the alarm signal is generated late, making fire rescue more difficult and increasing the risk of fire damage.

The multi-sensor fusion algorithm incorporates data fusion theory into the fire detection system. This method comprehensively analyzes and intelligently processes the multi-source information collected by different sensors, which effectively compensates for the uncertainty and one-sidedness caused by the traditional single sensor detection method. As a result, more reliable and faster detection results can be obtained. It has become one of research hotspots in the fire detection field. Multi-sensor fusion algorithm fuses temperature, smoke, combustible gases, and other parameters. The main algorithms include Bayesian network, neural network algorithm, support vector machine (SVM), fuzzy algorithm, etc. Nakıp proposed a trend prediction neural network (TPNN) model [3], and on this basis, proposed a recursive trend prediction neural network (rTPNN) model [4] which captures the trend of time series data from multi-sensors through trend prediction and horizontal prediction. Anand proposed a multi-sensor information fusion algorithm based on a feedforward neural network (FNN) for the delay-free prediction of forest fire storms [5]. Wen proposed a self-organizing T-S fuzzy neural network infrared flame detection algorithm, which overcomes local minimums by adjusting parameters through gradient descent. And consequentially improves the robustness of the model effectively [6]. Zheng combined BP neural network and D-S evidence theory in an underground fire scenario [7]. Ren proposed a fault identification method based on multi-information fusion for the fault arc problem of the electrical fire in low-voltage distribution system[8]. Although the algorithms in the above literatures have respective advantages in fire detection, most of the models are trained and tested using laboratory data and do not fully consider the temporal dimension information of sensor data. In the early stage of the fire, the sensor data shows a steady upward or downward trend in the long term but shows a random opposite trends or even irregular fluctuations in the short term.

The vision-based algorithm mainly detects smoke and flame images and processes them using neural networks, SVM, Markov model, expert systems, etc. Smoke image detection is based on features such as irregular motion and grayscale of smoke, while flame image detection relies on flame color, flame shape, and dynamic characteristics. Premal proposed a forest fire detection algorithm using a rules-based color model to classify fire pixels [9]. Mahmoud used the SVM method to divide the fire area, effectively improving the accuracy of the algorithm [10]. Pritam proposed a fire detection system based on LUV colour space and hybrid transforms [11]. Saima used the Grad-CAM method to visualize and locate flames in images, and built a CNN network based on an attention mechanism to detect fires[12]. Li combined multi-scale feature extraction, implicit depth supervision, and channel attention mechanisms to construct a fire detection system that achieved a better balance between accuracy, model size and speed[13]. Mukhriddin used a deep CNN model and an improved YOLOv4 object detector to identify smoke and flames[14]. However, in actual indoor fire applications, surveillance cameras may have blind areas due to furniture occlusion, resulting in missed fire reports. Moreover, vision-based fire detection algorithms generally do not detect combustible gases and are slow to detect smoldering fires.

In summary, this paper proposed a multi-sensor data fusion algorithm based on a LogitBoost ensemble learning model that uses time-dimensional multi-sensor data from real fire scenes to solve the problem of false alarms caused by irregular fluctuations of fire data in a short period. This method effectively improves the accuracy of fire detection.

2 Data Selection and Analysis

2.1 Data Seletion

The open-source fire data from NIST Test FR 4016 [15] was adopted in this work to test the effectiveness of the proposed algorithm. The dataset includes 27 experiments (SDC01-SDC15, SDC30-SDC41) conducted in a manufactured home to investigate the performance of different sensors in fire detection.

Table 1. Details of training set.

Scenario	Fire Type	Fire Location	Fire Material	Ignition Time(s)	No-Fire Data Count	Fire Data Count
SDC02	Flaming	Living area	Chair	84	366	142
SDC04	Smoldering	Bedroom	Mattress	8	149	797
SDC05	Flaming	Bedroom	Mattress	87	166	56
SDC08	Smoldering	Bedroom	Mattress	131	253	1886
SDC11	Smoldering	Living area	Chair	53	361	2264
SDC12	Flaming	Kitchen	Cooking oil	233	544	710
SDC13	Flaming	Kitchen	Cooking oil	100	390	880
SDC31	Smoldering	Living area	Chair	4931	3338	1660
SDC34	Smoldering	Living area	Chair	218	534	1852
SDC35	Flaming	Living area	Chair	94	546	75
SDC36	Flaming	Bedroom	Mattress	32	1794	974
SDC37	Smoldering	Bedroom	Mattress	130	485	926
SDC38	Flaming	Bedroom	Mattress	35	787	568
SDC39	Flaming	Bedroom	Mattress	22	357	55
SDC40	Smoldering	Bedroom	Mattress	522	665	1425
SDC41	Flaming	Kitchen	Cooking oil	110	310	972

Since SDC03, SDC30, and SDC32 were aborted due to ignition failure, the remaining 24 experiments were selected as dataset to investigate the performance of different

Table 2. Details of test set.

Scenario	Fire Type	Fire Location	Fire Material	Ignition Time(s)	No-Fire Data Count	Fire Data Count
SDC01	Smoldering	Living area	Chair	377	100	1355
SDC06	Smoldering	Bedroom	Mattress	83	151	1179
SDC07	Flaming	Bedroom	Mattress	59	392	169
SDC09	Flaming	Bedroom	Mattress	31	424	536
SDC10	Flaming	Living area	Chair	100	566	167
SDC14	Flaming	Bedroom	Mattress	3398	2407	407
SDC15	Flaming	Living area	Chair	271	441	106
SDC33	Flaming	Living area	Chair	88	1352	72

sensors in fire detection. The experiment recorded in detail the entire process data of various kinds of sensor data including temperature, smoke, CO, CO2, and O2 concentration. In this work, temperature, CO concentration, and smoke concentration is used as input feature parameters.

The training set contains 16 groups of experimental data and test set contains 8 groups of experimental data. Details of training set and test set are shown in Tables 1 and 2. As is shown in the table, the training set includes 11405 no-fire data count and 15242 fire data count, and the test set includes 5833 no-fire data count and 3391 fire data count. Among all the tests, SDC09, SDC14, and SDC36 were conducted with the bedroom door closed, while the rest were open.

2.2 Data Processing and Analysis

In various combustion states, different fire detection parameters exhibit distinct characteristics. Therefore, selecting appropriate fire characteristic parameters is essential for accurately and efficiently identifying the fire state. Moreover, since real experiments are subject to external interference, the obtained data exhibit random and irregular fluctuations, so it is necessary to preprocess the data before the subsequent model training and testing.

Select Input Feature Parameters. In this work, the following three fire characteristic parameters are used as the identification basis.

Temperature. One of the easiest indicators to measure and can be used as an important characteristic parameter of fire. Expressed as $x = (x_1, x_2, \ldots, x_{24})^T$, where $x_i = (x_{i1}, x_{i2}, \ldots, x_{ip})^T, i = 1, 2, \ldots, 24$, the unit is Celsius and p is the data length, as shown in the sixth column of Tables 1 and 2.

CO Concentration. Increases rapidly when a fire occurs, especially when smoldering, easy to float to the roof due to low density, and easy to be detected. Expressed as $y = (y_1, y_2, \ldots, y_{24})^T$, where $y_i = (y_{i1}, y_{i2}, \ldots, y_{ip})^T, i = 1, 2, \ldots, 24$, the unit is volume fraction $\times 10^6$.

Smoke Concentration. One of the most obvious phenomena in early fire, is of great significance to describe the state of fire. Expressed as $z = (z_1, z_2, \ldots, z_{24})^T$, where $z_i = (z_{i1}, z_{i2}, \ldots, z_{ip})^T$, $i = 1, 2, \ldots, 24$, the unit is %/ft.

Data Processing

Filtering. The S-G (Savitzky-Golay) filtering method is adopted to filter the temperature, CO concentration and smoke concentration. It is based on polynomial least square fitting in the time domain which can preserve the shape and width of the signal while filtering noise. The smoothing formula of S-G filtering method is shown in $\bar{y}_i = \sum_{n=-M}^{+M} y_{i+n} h_n$, $i = 1, 2, \ldots, 24$ (1). Taking CO concentration as an example, for a set of data within the window $y_i[n]$, $n = -M, \ldots, 0, \ldots, M$, The value of n is 2M + 1 consecutive integer values, M is the half width of the approximate interval, M = 5, frame length is 11, and a third-order fitting polynomial $f(n) = \sum_{k=0}^{3} b_{nk} n^k = b_{n0} + b_{n1} n + b_{n2} n^2 + b_{n3} n^3 = h_n y_i[n]$ is constructed to fit the data, the smoothing coefficient is $h_n = \underset{h_n}{\operatorname{argmin}} \sum_{n=-M}^{M} (f(n) - y_i[n])^2$ and the filtered data is $\bar{y}_i$. Similarly for temperature and smoke concentration filtering, the S-G-filtered dataset is expressed as $\left[\, \bar{x}_i \; \bar{y}_i \; \bar{z}_i \,\right]$.

$$\bar{y}_i = \sum_{n=-M}^{+M} y_{i+n} h_n, \; i = 1, 2, \ldots, 24 \tag{1}$$

The comparison of CO concentration data before and after filtering is shown in Fig. 1. It can be seen from the figure that the zigzag fluctuations of the filtered curve is reduced and smoothed obviously, indicating the S-G filter can effectively reduce data noise.

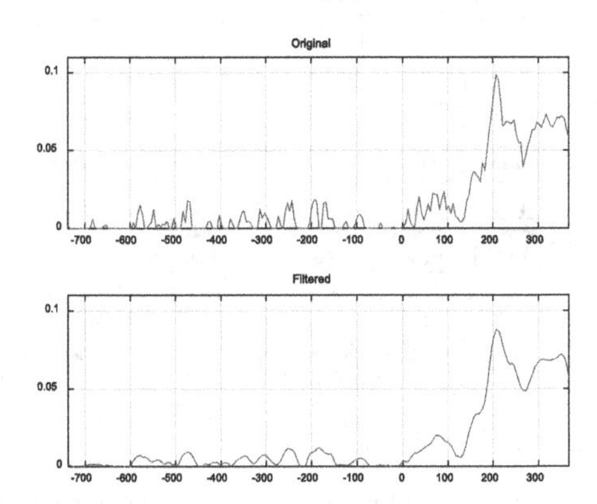

Fig. 1. Comparison of CO concentration data before and after filtering

Normalization. To mitigate the influence of larger attribute values (such as temperature) on smaller values (such as CO concentration) and eliminate the adverse effects caused by singular sample data, all characteristic parameter values are normalized using $\left[\, \tilde{x}_i \; \tilde{y}_i \; \tilde{z}_i \,\right] = \left[\, \frac{\bar{x}_i - \bar{x}_{i\,min}}{\bar{x}_{i\,max} - \bar{x}_{i\,min}} \; \frac{\bar{y}_i - \bar{y}_{i\,min}}{\bar{y}_{i\,max} - \bar{y}_{i\,min}} \; \frac{\bar{z}_i - \bar{z}_{i\,min}}{\bar{z}_{i\,max} - \bar{z}_{i\,min}} \,\right]$, $i = 1, 2, \ldots, 24$ (2) and mapped to the interval (0,1). $\left[\tilde{x}_i\ \tilde{y}_i\ \tilde{z}_i\right]$ is the feature value after normalization.

$$\left[\tilde{x}_i\ \tilde{y}_i\ \tilde{z}_i\right] = \left[\frac{\overline{x}_i-\overline{x}_{imin}}{\overline{x}_{imax}-\overline{x}_{imin}}\ \frac{\overline{y}_i-\overline{y}_{imin}}{\overline{y}_{imax}-\overline{y}_{imin}}\ \frac{\overline{z}_i-\overline{z}_{imin}}{\overline{z}_{imax}-\overline{z}_{imin}}\right], i = 1, 2, \ldots, 24 \quad (2)$$

The data obtained through S-G filtering and Min-Max normalization composed the fire dataset in this study. Temperature, smoke concentration, CO concentration, and label are considered the sample of data for training and testing.

3 Algorithm Analysis and Evaluation

3.1 Architechture of Algorithm

Ensemble learning is a machine learning method that involves training a series of basic models and processing the output result of each model through the ensemble principle [16, 17]. There are three common types of ensemble learning: bagging, boosting, and stacking. Bagging and boosting are homogeneous ensemble models, which often involve standalone weak learners of the same type. Stacking is a heterogeneous ensemble model, which typically employs standalone learners of different types. The stacking ensemble principle is to combine multiple models in different layers, and iteratively learns the classification deviation of the previous model [18]. Stacking ensemble algorithm can integrate different types of models and classification features, leading to better performances.

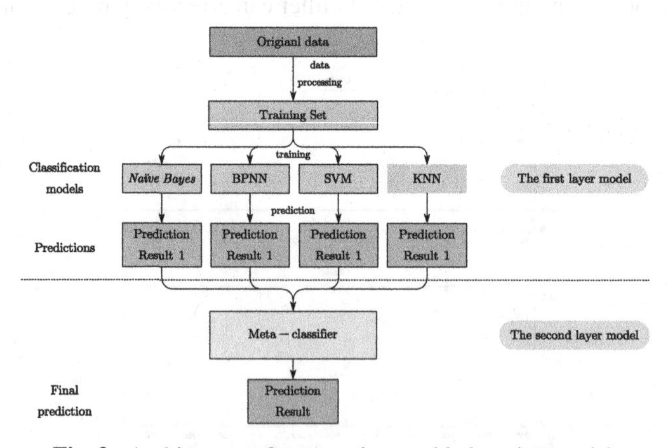

Fig. 2. Architecture of proposed ensemble learning model.

The architecture of the proposed multi-sensor fusion algorithm based on ensemble learning is illustrated in Fig. 2. The stacking ensemble learning model in this work is designed as a two-layer structure. The first layer has four types of basic classifiers including Naïve Bayes, BPNN, SVM, and KNN models. All of the models were trained on the training set to generate the classification results of each model. The second layer is the LogitBoost ensemble learning model. It combines the four classification results of the first layer as new feature parameters, and uses the newly constructed feature parameters and labels to train the ensemble model with the same initial weight value. Eventually outputs the final classification result.

3.2 Research Methodology

The First Layer. The first layer contains 4 base-classifiers, including Naive Bayes, BPNN, SVM and KNN classifiers. Temperature, smoke concentration and CO concentration $\left[\tilde{x}_i\ \tilde{y}_i\ \tilde{z}_i\right], i = 1, \ldots, 16$ were used as input features, and the fire state $k_i \in \{1, 2\}$ was used as the label value, where 1 represented the state without fire and 2 represented the state with fire. Four basic classifier models were obtained by training the classifier using the training set respectively. The model was tested using 3227 sets of data from the test set $\left[\tilde{x}_i\ \tilde{y}_i\ \tilde{z}_i\right], i = 17, \ldots, 24$, and outputed 4 sets of predictions $o_{i,1}, o_{i,2}, o_{i,3}, o_{i,4}, i = 1, \ldots, 3391$.

A brief description of the principle of each classifier model is as follows.

Naïve Bayes Model. Naïve Bayes constructs a classifier using the attribute condition independent assumption, and for a certain feature, the classification output probability can be expressed as (3), where k is the value of the fire state label, $P(\tilde{x}|k_i), P(\tilde{y}|k_i), P(\tilde{z}|k_i)$ is conditional probabilities and $P(k_i) = \frac{k_i}{k_1+k_2}$ is prior probability, both of which can be calculated from the sample data. The Naïve Bayes model outputs the prediction results$o_{i,1} = [o_{1,1}, \ldots, o_{i,1}, \ldots, o_{3391,1}]^T, i = 1, \ldots, 3391$.

$$P(k_i|\tilde{x}, \tilde{y}, \tilde{z}) = \frac{P(\tilde{x}|k_i)P(\tilde{y}|k_i)P(\tilde{z}|k_i)P(k_i)}{P(\tilde{x})P(\tilde{y})P(\tilde{z})}, i = 1, 2 \tag{3}$$

BPNN Model. The output of BPNN uses forward propagation and the errors use back propagation. In this work, the BPNN model has 2 input neurons, 2 hidden layers, and 1 output neuron. The first hidden layer has 10 neurons and the second hidden layer has 4 neurons. The BPNN model outputs the prediction results $o_{i,2} = [o_{1,2}, \ldots, o_{i,2}, \ldots, o_{3391,2}]^T, i = 1, \ldots, 3391$.

SVM Model. The SVM algorithm is one of the most robust classification and regression algorithms. The main objective of SVM algorithm in binary classification is to get the minimum hyperplanes that have maximum distance from the training data set. The SVM model outputs the prediction results $o_{i,3} = [o_{1,3}, \ldots, o_{i,3}, \ldots, o_{3391,3}]^T, i = 1, \ldots, 3391$.

KNN Model. The KNN is based on the training instance categories of k nearest neighbors and predictions are made through majority voting, etc. In this work, the value of k is 1. The KNN model outputs the prediction results $o_{i,4} = [o_{1,4}, \ldots, o_{i,4}, \ldots, o_{3391,3}]^T, i = 1, \ldots, 3391$.

The Second Layer. The second layer of the proposed algorithm is an ensemble learning layer, which combines the four basic classifiers of the first layer to construct an integrated fire detection model. The stacking ensemble method is utilized to combine the predictions from multiple well-performing machine learning models to make better predictions than any single model.

The ensemble learning algorithm used in this work is LogitBoost. It addresses the two shortcomings of AdaBoost, which respectively are sensitive to anomalies due to excessive punishment for misclassification points, and cannot predict the probability of a category. LogitBoost was employed as meta-learning model to detect fire, and a general framework of ensemble classifiers was constructed by combining five decision tree based classifiers. LogitBoost uses a logarithmic loss function that iterates Newton-Raphson at each step. The method is applied to the error function to update the model, and the logistic regression model is employed to combine the predictions of all the weak classifiers to get the final prediction result.

Let the prediction results of the four basic classifiers in the first layer be $o_{3391\times 4} = \left[o_{i,1}, o_{i,2}, o_{i,3}, o_{i,4}\right] = \begin{bmatrix} o_{1,1} & \cdots & o_{1,4} \\ \vdots & \ddots & \vdots \\ o_{3391,1} & \cdots & o_{3391,4} \end{bmatrix}, i = 1, \ldots, 3391$. Each column of the prediction result matrix is taken as a meta-features vector, and the fire state is taken as the label value. LogitBoost model is used to train the meta-features, and a model mapping from the meta-features to the ground-truth is obtained. The fire detection model is then trained using LogitBoost algorithm.

For training set$D = \left\{\left(o_{1,j}, k_1\right), \ldots, \left(o_{i,j}, k_i\right), \ldots, \left(o_{3391,j}, k_{3391}\right)\right\}, i = 1, \ldots, 3391, j = 1, 2, 3, 4$, where $k_i \in \{1, 2\}$ indicates the fire status label. 1 is no fire state, and 2 is fire state. Let the ensemble function be$F\left(o_{i,j}\right)$, the probability of $k_i = 1$ is$p\left(k_i = 1|o_{i,j}\right) = \frac{e^{F(o_{i,j})}}{e^{F(o_{i,j})}+e^{-F(o_{i,j})}}$, the logarithmic loss function is$L\left(k_i, F\left(o_{i,j}\right)\right) = -k_i\log\left(p\left(k_i = 1|o_{i,j}\right)\right) - (1-k_i)\log\left(p\left(k_i = 1|o_{i,j}\right)\right) = \log(1+e^{-2F(o_{i,j})})$. The goal of the LogitBoost algorithm is to use the decision tree algorithm$f\left(o_{i,j}\right)$ as the basic learner to calculate $F\left(o_{i,j}\right) = arg \min_{F(o_{i,j})} L\left(k_i, F\left(o_{i,j}\right)\right)$, the specific steps are as follows:

- Step 1. For each set of data, set the initial weight $w_i = \frac{1}{3391}$, the initial value of the ensemble function $F_0\left(o_{i,j}\right) = 0$, and the initial probability of $k_i = 1$ is $p\left(k_i = 1|o_{i,j}\right) = \frac{1}{2}$.
- Step 2. For iteration step $m = 1, \ldots, M$,repeat:
  1) For $i = 1, \ldots, 3391$, calculate working response value $z_{m,i} = \frac{k_i - p_{m-1}(o_{i,j})}{p_{m-1}(o_{i,j})(1-p_{m-1}(o_{i,j}))}$ and weight $w_{m,i} = p_{m-1}(o_{i,j})(1-p_{m-1}(o_{i,j}))$;
  2) Fit the function $f_m\left(o_{i,j}\right)$ by a weighted least-squares regression of $z_{m,i}$ to $o_{i,j}$ with weights $w_{m,i}$ using the decision tree approach. The derivative of the loss function is $\frac{\partial L(k_i, F_{m-1}(o_{i,j}))}{\partial F_{m-1}(o_{i,j})} = z_{m,i}$, $\frac{\partial^2 L(k_i, F_{m-1}(o_{i,j}))}{\partial F^2_{m-1}(o_{i,j})} = w_{m,i}$, thus $f_m\left(o_{i,j}\right) = arg \min_{f_m(o_{i,j})} L\left(k_i, F_{m-1}\left(o_{i,j}\right)\right)$ can be obtained;
  3) For $i = 1, \ldots, 3391$, update $F_m\left(o_{i,j}\right) \leftarrow F_{m-1}\left(o_{i,j}\right) + \frac{1}{2}f_m\left(o_{i,j}\right), p_m\left(k_i = 1|o_{i,j}\right) \leftarrow \frac{e^{F_m(o_{i,j})}}{e^{F_m(o_{i,j})}+e^{-F_m(o_{i,j})}}$.
- Step 3. Output ensemble classifier $sign\left[F\left(o_{i,j}\right)\right] = sign\left[\sum_{m=1}^{M} f_m\left(o_{i,j}\right)\right]$.

Test the ensemble model using the meta-features of testing data and thus the final prediction results areobtained as $[\tilde{o}_1, \tilde{o}_2, \ldots, \tilde{o}_{3391}]^T$.

3.3 Evaluation Index

By comparing the model predicted label with the true label value, the performances of the algorithms were evaluated on the basis of accuracy, precision, recall, F1 score, receiver operating characteristic (ROC) curve, and area under the ROC curve (AUC).

The confusion matrix is shown in Table 3, where TP and TN respectively denotes the number of correctly predicted positive values and negative values, by analogy FP and FN denote the prediction is false.

Table 3. Confusion matrix.

True value	Predicted value	
	Positive	Negative
Positive	TP	FP
Negative	FN	TN

The indexes are as (4)–(7).

$$Accuracy = \frac{TN + TP}{TN + FP + TP + FN} \tag{4}$$

$$Precision = \frac{TP}{TP + FP} \tag{5}$$

$$Recall = \frac{TP}{TP + FN} \tag{6}$$

$$F1score = \frac{2 \times Precision \times Recall}{Precision + Recall} \tag{7}$$

Accuracy is the percentage of the correctly predicted samples in the total samples in the whole test set, which intuitively reflects the overall classification ability of algorithms. Precision is the percentage of the correctly predicted positive samples to all actual positive samples, which can reflect the ability to distinguish negative samples. Recall is the percentage of correctly predicted positive samples to all predicted positive samples, which can represent the ability to identify positive samples. F1 score is the harmonic average of precision and recall, which can characterize the comprehensive performance measurement of algorithms in precision and recall. The value of the preceding indicators ranges from 0 to 1. The closer the indicator is to 1, the more robust the model is.

$$TPR = \frac{TP}{TP + FN} \tag{8}$$

$$FPR = \frac{FP}{FP + TN} \tag{9}$$

The vertical axis of the ROC curve is true positive rate (TPR), which represents the proportion of the correctly predicted positive samples to all predicted positive samples, as shown in (8). The horizontal axis of the ROC curve is false positive rate (FPR), which stands for the proportion of incorrectly predicted positive samples to all predicted negative samples, as shown in (9). The ROC curve,which is not affected by the imbalance of samples, can embody the classification performance of algorithms at each sample point. The closer the curve is to the top left, the better the performance is. AUC is the area under the ROC curve, which can more accurately reflect the overall classification ability of the algorithms in the form of detailed figures.

3.4 Experimental Results

The performance and ROC curve of each algorithm in this work is respectively shown in Table 4 and Fig. 3.

Table 4. The performance of each algorithm.

	Accuracy	Precision	Recall	F1 score	AUC
Naïve Bayes	0.9252	0.8932	0.9928	0.9403	0.9103
BPNN	0.8755	0.8943	0.8961	0.8952	0.8759
SVM	0.8955	0.8635	0.9789	0.9176	0.8864
KNN	0.9222	0.9082	0.9666	0.9365	0.9112
Ensemble model	0.9371	0.9139	0.9871	0.9491	0.9592

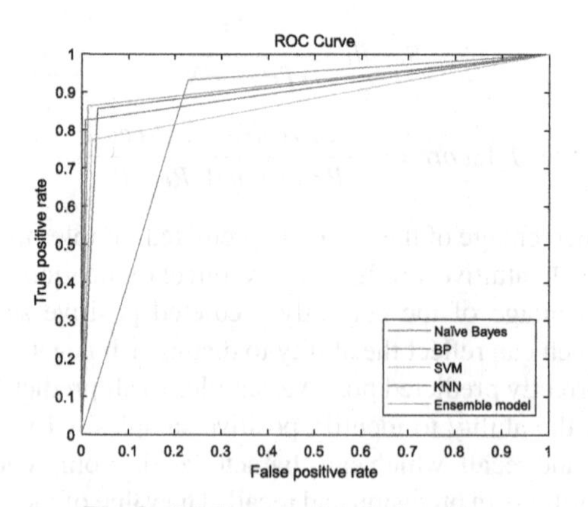

Fig. 3. ROC curve of each algorithm.

The prediction performance of the single models is shown in Fig. 4 and the prediction performance of ensemble model is as Fig. 5. The upper left is the confusion matrix, with rows corresponding to real classes and columns corresponding to prediction classes. The upper right graph is the column-normalized column summary, showing the percentage of the number of observations that are correctly and incorrectly classified for each prediction category. The lower-left graph is the row-normalized row summary, showing the percentage of the number of observations that are correctly and incorrectly classified for each real category.

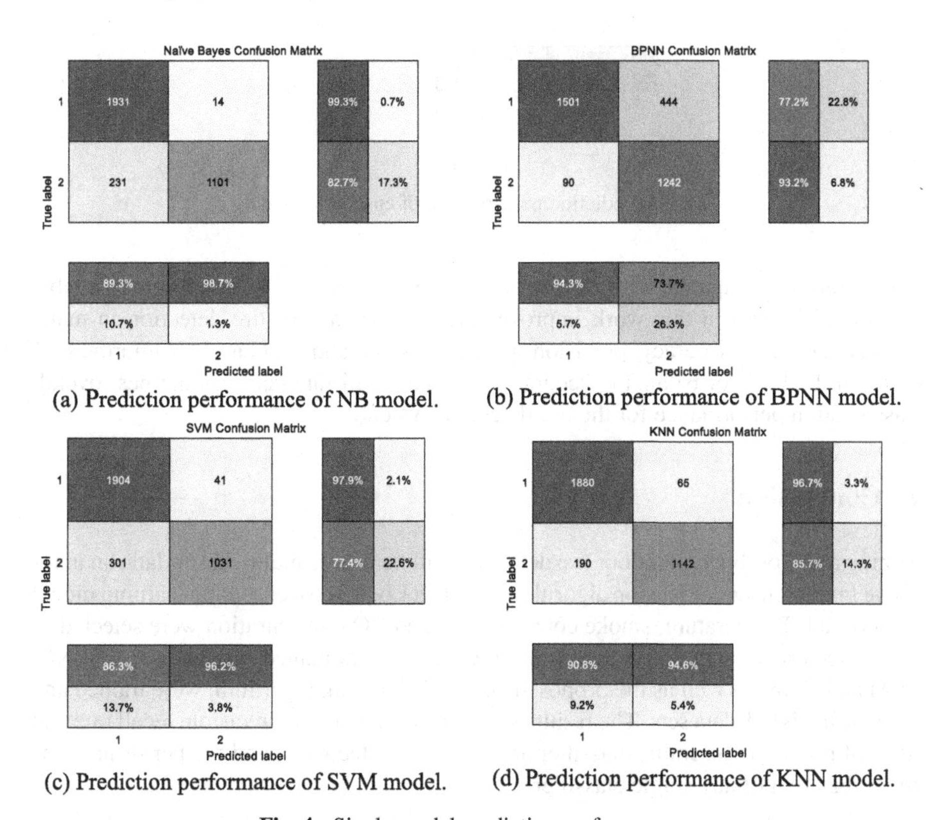

(a) Prediction performance of NB model. (b) Prediction performance of BPNN model.

(c) Prediction performance of SVM model. (d) Prediction performance of KNN model.

Fig. 4. Single model prediction performance.

It can be observed from Table 4 that the performance of Naïve Bayes is the best of the four basic classifiers. It has a high recall reaching 99.28%, however, the precision is only 89.32%, which means the false positive rate is high, leading to potential missinng alarm rate.The BPNN model has the best recognition for on-fire state while the recognition accuracy for no-fire state is low, leading to higher false alarm rate. The proposed ensemble model well balanced the precision and recall while improving the multi-faceted performance including accuracy, precision, recall, F1 score, and AUC area. The ROC curve of ensemble model is closest to the upper left corner as Fig. 3 shows, indicating the overall classification effect is remarkable. Meanwhile, the performances of Naïve Bayes and KNN are highly similar, which are more excellent than the BPNN and SVM. The value of the AUC accurately quantifies the above analysis. AUC of ensemble model is 0.9592, higher than the rest of the algorithms.

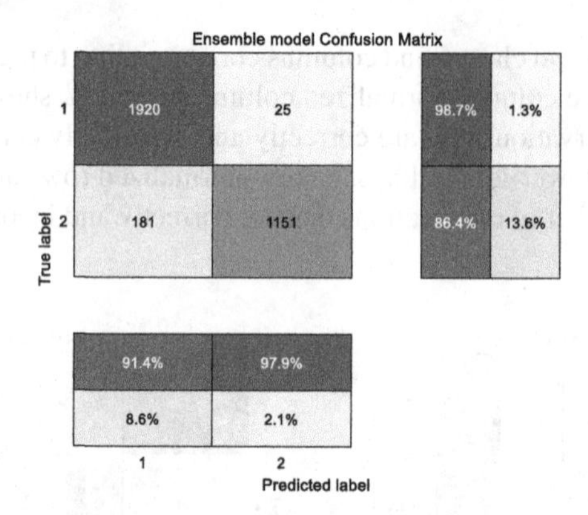

Fig. 5. Prediction performance of ensemble model.

Therefore, compared with other comparison algorithms, the proposed ensemble learning algorithm in this work improves the performance of fire detection in many aspects, including accuracy, precision, recall, F1 score and AUC area. It improves F1 score on the basis of balancing accuracy rate and recall rate, and has the best overall classification performance for the fire detection system.

4 Conclusion

Aiming at the problem of indoor fire detection with real-time multi-sensor data, an innovative multi-sensor data fusion algorithm based on LogitBoost ensemble learning model is proposed. Temperature, smoke concentration and CO concentration were selected as fire characteristic parameters, and four single classifiers named Naive Bayes, BPNN, SVM and KNN, as well as the proposed ensemble learning algorithm, were trained and tested using NIST data sets. The results show that the accuracy, precision, recall rate and AUC of proposed ensemble classifier are improved effectively, and the performance is better than that of any single classifier.

References

1. Fang, X.: A total of 219 000 fire deaths and 625 deaths in the first quarter of China. China Fire Control **557**(04), 21 (2022)
2. Qureshi, S., Ekpanyapong, M., Dailey, N., et al.: QuickBlaze: Early fire detection using a combined video processing approach. Fire Technol. **52**(5), 1293–1317 (2016)
3. Nakıp, M., Güzeliş, C.: Multi-Sensor fire detector based on trend predictive neural network. In: 11th International Conference on Electrical and Electronics Engineering, pp. 600–604. Bursa, Turkey (2019)
4. Nakip, M., Güzelíş, C., Yildiz, O.: Recurrent trend predictive neural network for multi-sensor fire detection. IEEE Access **9**, 84204–84216 (2021)
5. Anand, S, Keetha, K.: FPGA implementation of artificial neural network for forest fire detection in wireless Sensor Network. In: 2nd International Conference on Computing and Communications Technologies, pp. 265–270. Chennai, India (2017)

6. Ziteng, W., Linbo, X., Hongwei, F., Yong, T.: Infrared flame detection based on a self-organizing TS-type fuzzy neural network. Neurocomputing **337**, 67–79 (2019)
7. Zianyu, Z., Yi, L., et al.: Fire detection scheme in tunnels based on multi-source information fusion. In: 18th International Conference on Mobility, Sensing and Networking, pp. 1025–1030. Guangzhou, China (2022)
8. Ren, X., Li, C., et al.: Design of multi-information fusion based intelligent electrical fire detection system for green buildings. Sustainability **13**(6), 3405 (2021)
9. Premal, E., Vinsley, S.: Image processing based forest fire detection using YCbCr colour model. In: International Conference on Circuits. Power and Computing Technologies, pp. 1229–12372014. Nagercoil, India (2014)
10. Mahmoud, M., Ren, H.: Forest fire detection and identification using image processing and SVM. J. Inform. Process. Syst. **15**(1), 159–168 (2019)
11. Pritam, D., Dewan, H.: Detection of fire using image processing techniques with LUV color space. In: 2nd International Conference for Convergence in Technology, pp. 1158–1162. Mumbai, India (2017)
12. Saima, M., Fayadh, A.: Attention based CNN model for fire detection and localization in real-world images. Expert Syst. Appl. **189**, 116114 (2022)
13. Songbin, L., Qiangdong, Y., Peng, L.: An efficient fire detection method based on multiscale feature extraction, implicit deep supervision and channel attention mechanism. IEEE Trans. Image Process. **29**, 8467–8475 (2020)
14. Mukhiddinov, M., Abdusalomov, B., Cho, J.: Automatic fire detection and notification system based on improved YOLOv4 for the blind and visually impaired. Sensors **22**(9), 3307 (2022)
15. Bukowski, W., Peacock, D., et al.: Performance of Home Smoke Alarms, Analysis of the Response of Several Available Technologies in Residential Fire Settings. Technical Note https://tsapps.nist.gov/publication/get_pdf.cfm?pub_id=100900. Accessed 12 June 2023
16. Arya, N., Saha, S.: Multi-modal classification for human breast cancer prognosis prediction: proposal of deep-learning based stacked ensemble model. IEEE/ACM Trans. Comput. Biol. Bioinf. **19**(2), 1032–1041 (2022)
17. Jiang, J., Zhu, W.: Electrical load forecasting based on multi-model combination by stacking ensemble learning algorithm. In: IEEE International Conference on Artificial Intelligence and Computer Applications. pp. 739–743. Dalian, China (2021)
18. Smith, G.: Fire-detection and alarm systems. Wiring Install. Suppl. **3**, 9–11 (1977)

Research on Water Surface Environment Perception Method Based on Visual and Positional Information Fusion

Qin Na, Zhe Zuo, Ning Xu, ZhenYu Zhang(✉), and Yi Lu

Beijing Institute of Technology, Beijing 100081, People's Republic of China
zuzeus@bit.edu.cn

Abstract. The water surface environment characterised by complexity and variability, is heavily influenced by weather. To address this problem, this paper proposes a water surface environment perception network based on the fusion of visual and positional information, and proposes an encoder-decoder based semantic segmentation neural network for classifying the pixel points of the input image into three categories: water, sky and environment (obstacles).

Keywords: Semantic segmentation · Positional information · Attentional mechanisms

1 Introduction

A good surface environment perception method is an important guarantee to help surface ships realize autonomous unmanned navigation in waters. It is difficult to understand the complex and variable features of different water surface objects on the basis of traditional methods, while research concerning deep learning methods lacks some practical and a priori knowledge in traditional methods.

To address this problem, this paper proposes a water surface environment sensing technique in accordance with the fusion of visual and positional information, which combines the advantages of both traditional methods and deep learning methods. In the model structure, residual network acts as an encoder to extract the information and features of different scale images. An attention mechanism and a feature fusion module are used in the decoder enabling the network to focus on locally focused information and feature fusion at different scales. The bit-position information is encoded into feature vectors of the neural network and fused with it, and the features of the encoder and decoder merge at different stages of the decoder. After that, the model designed in this paper is compared with the latest SOTA model in the field of semantic segmentation, in order to qualitatively and quantitatively analyze the advantages and disadvantages of different models, and to confirm the effectiveness and advantages of Swan-Net in the field of water surface environment sensing.

Compared with the existing studies, the platform position information obtained from the inertial measurement unit (IMU) is applied as the priori

B. Xin et al. (Eds.): IWACIII 2023, CCIS 1931, pp. 50–62, 2024.
https://doi.org/10.1007/978-981-99-7590-7_5

knowledge of the water boundary and encoded into feature channels to be fused with different scale image features. After fusing the position information, the network can effectively improve the accuracy of water boundary estimation in low-contrast environments.

2 Swan-Net

In this paper, a semantic segmentation model named Swan-Net is designed. The concept and local substructure of the design are mainly refer SOTA networks in the field of semantic segmentation in recent years. The network is also contains an a priori water boundary knowledge encoding of the positional information, which is obtained through an inertial measurement unit (IMU). The overall structure of the model is shown in Fig. 1.

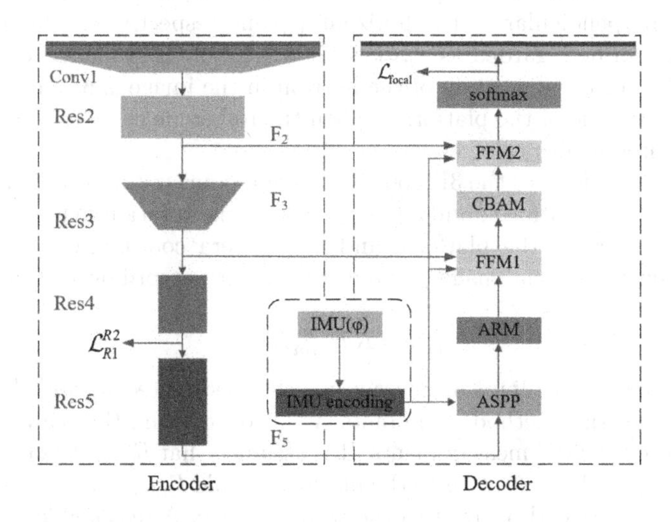

Fig. 1. General structure of the model

The general structure of the model is shown in the figure above, which consists mainly of an encoder and a decoder. The model accepts an input image of size $480 \times 640 \times 3$ (height × width × number of channels) and assigns each of these pixels a category label, water, sky or environment. The final output is of the same resolution as the input, avoiding the loss of detailed information as much as possible.

2.1 Feature Extraction Module

The main function of the encoder is to accept the input image and extract its features at different scales. The main composition is a ResNet101 [1] neural network. That is mainly composed of a pre-convolutional layer (Conv1), four residual convolutional blocks (Res2,Res3,Res4,Res5), unlike traditional ResNet101, the dilated convolutions is used in Res4 and Res5. Through controlling the sampling step of the convolution kernel, dilated convolutions is achieved so as to increase the convolution kernel field of perception and reduce the number of parameters. It can also be realized by inserting zeros in the middle of the convolution kernel or by leaving the convolution kernel unchanged and sampling the input at equal intervals.

2.2 Position Information Feature Encoding

The IMU usually consists of a three-axis accelerometer and a three-axis gyroscope. By fusing and solving the accelerometer and gyroscope data, it can measure the current attitude information of the platform. The attitude information includes roll angle, pitch angle and yaw angle, referring to the angel of tilt relative to the horizontal plane, the angel of front-to-back tilt, and the angel of rotation of the axis perpendicular to the horizontal plane respectively. This positional information can be regarded as a priori knowledge of the semantic components in the image, since the position of the horizon in the image is usually related to the current attitude of the platform. When the roll angle changes, the tilt angle of the sea antenna also changes.

Suppose X^{usv}denotes the 3D coordinates of a point in the coordinate system of the amphibious platform, and let R_{cam}^{usv} denote the rotation matrix describing the rotation between the platform and the camera coordinate system. Point X_i^{usv} is projected to the image plane of the camera according to the following Equation:

$$\lambda_c x_i = K R_{cam}^{usv} X_i^{usv} \tag{1}$$

K is the camera calibration matrix, which is estimated during the calibration process. In this method, the points X_i^{usv} constituting the sea antenna are obtained from the IMU measurements. It is assumed that R_{cam}^{usv} denotes the rotation matrix of the IMU relating to the platform, while R_{imu} denotes the rotation matrix of the IMU with respect to the water surface. A reasonable assumption can be made that the Z-axis of the IMU and the camera are approximately aligned. In principle, these geometric relationships are sufficient to compute the vanishing point and can be used directly to estimate the sea antenna.

However, it has been shown that projecting vanishing points into the input image leads to inaccuracies, due to the fact that the vanishing points are likely to be projected outside the image boundaries and the calibrated radial aberration model can reliably estimate the aberrations of points located only inside the image. The sea antenna can therefore be obtained by projecting two points, $\{X_1^{imu}, X_2^{imu}\}$being two points in the XZ plane of the IMU coordinate system that are located at a horizontal angle $\pm\alpha_h$ and at a finite distance Z=l_{dist}. These points are rotated into a plane parallel to the water surface by following Equation:

$$X_i^{usv} = R_{imu} \left(R_{usv}^{imu}\right)^{-1} X_i^{imu} \tag{2}$$

X_i^{imu} and X_i^{usv} denote the points before and after the rotation, respectively. The rotated points $\{X_1^{imu}, X_2^{imu}\}$ are projected into the image using Eq. 1 while considering radial distortion. The sea antenna is estimated by fitting the projected radial distortion points to a line. Through the above method, the attitude angle information of the platform can be converted into a projection of the sea antenna in the image, in order to fuse the projection information into the neural network.

2.3 Feature Fusion Module

The task of the decoder is to fuse the image features extracted by the encoder module with the information from the IMU, and after feature refinement and upsampling, to produce the final semantic segmentation output. The decoder accepts features from the three modules in the encoder (Res2, Res3, Res5) as well as the encoded IMU feature channels, utilising both the more detailed information from the high resolution features and the global semantic information captured at lower resolutions. First, the output features from the last layer of the encoder are fed into a spatial pyramid pooling module (ASPP) with 4 dilated convolutions, and the output of the ASPP module passes through the attention refinement module, which reweights the features.

The feature fusion module in the decoder aims to integrate the F3 and F2 features from the encoder with the IMU information features to achieve effective fusion of the different path features and to take full advantage of the different features.

The CBAM module calculates both attention on the channel and attention on the space. The inclusion of this module will effectively help the decoder network to learn spatially focused features.

ARM Module. The attention refinement module ARM is derived from BiSeNet [3], a convolutional neural network module for image classification and semantic segmentation, and is primarily used to enhance the representational power of feature maps. It automatically learns relevant features in the input feature map and applies these features to enhance the representational power of the model. The ARM module enables the enhancement or weakening of the feature representation at different locations by performing a channel-by-channel weighted summation of the input feature map. By introducing the ARM module, the accuracy and robustness of the model can be significantly improved, especially when dealing with areas such as detail and edges in an image. The overall structure of the ARM module is shown in Fig. 2.

The tensor output from the ASPP module goes one way through the raw feature input channel without any processing. The other way goes through the attention vector computation channel, which first goes through a global averaging pooling layer. In global averaging pooling, for each channel feature map, the features of all pixels are averaged into a single value that represents the statistical features of the entire feature map. The feature maps are then subjected to batch normalisation [4] and Sigmoid units [5] to compute the attention vector.

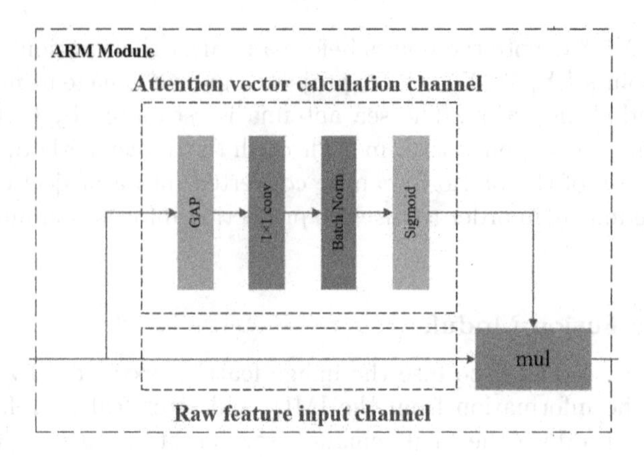

Fig. 2. ARM Module diagram

After computing the attention vector, it is multiplied with the original feature map on a channel-by-channel point-by-point basis. The original features will be re-weighted by the attention vector, enhancing the important features and weakening the less important ones, thus making the extracted features more directional.

FFM Module. The main role of the feature fusion module FFM is to fuse feature maps from multiple scales and levels to obtain richer and more complete image feature information. Typically, the FFM module consists of several branches, each of which uses different convolution kernels and pooling strategies to extract features at different scales and then integrate these features together. Feature maps at different scales and levels complment each other to obtain more comprehensive and accurate image information.

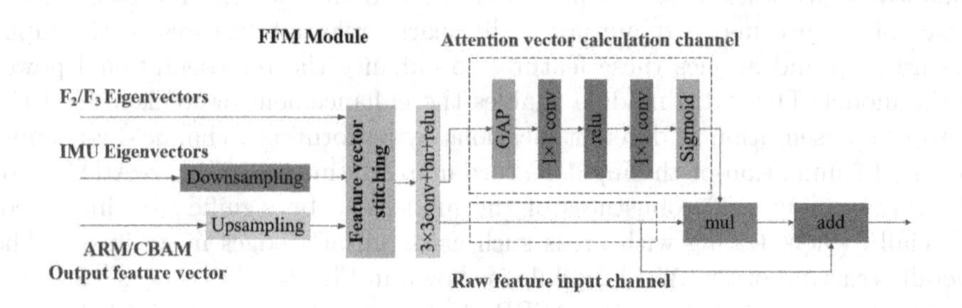

Fig. 3. Feature Fusion Module diagram

The FFM module used in this paper is shown in Fig. 3. First the FFM module stitches the output of the ARM module with the ASPP module, followed by further feature extraction using a 3×3 convolution, batch regularisation and the

ReLU activation function [6]. Thereafter a similar attention vector re-weighting method as in the ARM1 module was utilised to further extract and fuse features from different paths.

CBAM Module. The design of the CBAM module is derived from the literature [7]. The purpose of this module is mainly used to enhance the perceptual field of each location in the feature graph, thus advancing the performance of the network. The overall structure of CBAM is shown in Fig. 4. The CBAM module consists of two main parts: the channel attention module and the spatial attention module. The former is mainly used to perform attention computation in the channel dimension in order to learn the importance of different channels in the feature map adaptively, so as to better utilize the information of different channels. The latter is used to perform attention computation in the spatial dimension in order to adaptively learn the importance of different positions in the feature map for making use of the information of different positions in the feature map.

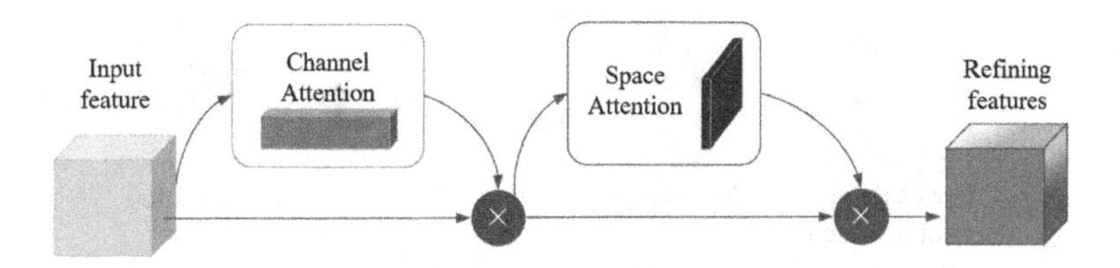

Fig. 4. CBAM Attention Module diagram

The results of the channel attention module are shown in Fig. 5. In the channel attention module, a one-dimensional vector is obtained by compressing the feature map in the spatial dimension. In the compression, both global average pooling and global maximum pooling are considered. The average pooling and maximum pooling can be used to aggregate the spatial information of the feature map, send to a shared network, compress the spatial dimension of the input feature map, sum and merge element by element then normalize the weights using the Sigmoid activation function, to obtain the channel attention map.

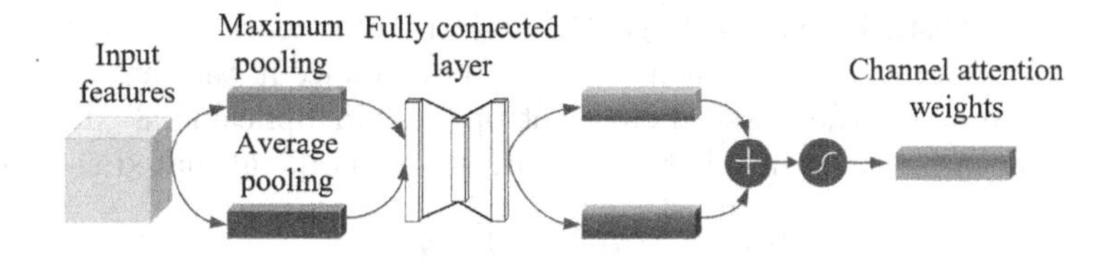

Fig. 5. CBAM Channel Attention

The channel attention computation process can be expressed by the following equation:

$$\begin{aligned} M_c(F) =&\sigma(MLP(AvgPool(F)) + MLP(MaxPool(F))) \\ =&\sigma(W_1(W_0(F^c_{avg})) + W_1(W_0(F^c_{max}))) \end{aligned} \tag{3}$$

where F denotes the input feature map, w_0 and w_1 represent the parameters of the two layers in the multilayer perceptron model, and $\sigma(\cdot)$denotes the Sigmoid activation function.

The results of the spatial attention module are shown in Fig. 6, differing from the channel attention module in the dimensionality of the processed features. The spatial attention module uses average pooling and maximum pooling in the channel dimension to put the two obtained H×W×1 feature descriptions stitched together according to the channels. Then, after a 7×7 convolutional layer is reduced to a single channel and a Sigmoid activation function is used to obtain the weight vector.

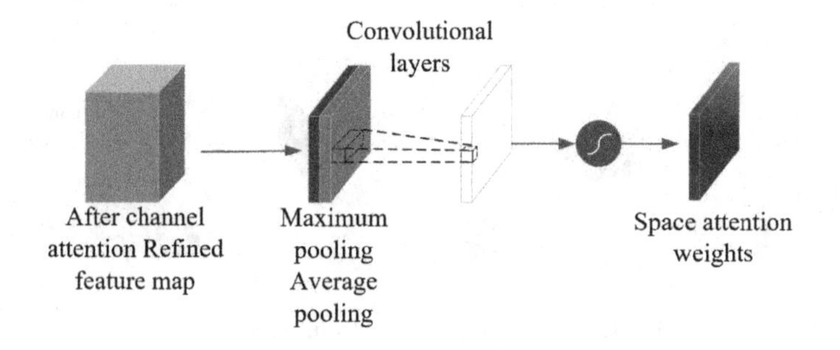

Fig. 6. CBAM Spatial Attention

The channel attention calculation process can be expressed by the following equation:

$$\begin{aligned} M_s(F) =&\sigma(f^{7\times7}([AvgPool(F); MaxPool(F)])) \\ =&\sigma(f^{7\times7}([F^s_{avg}; F^s_{max}])) \end{aligned} \tag{4}$$

2.4 Loss Function

The aquatic environment differs from the usual environmental dataset in that although some obstacles may be large, the majority of pixels in a typical aquatic scene belong to water or sky, which leads to an imbalance in the categories, and this imbalance makes the classical cross-entropy loss inapplicable [8]. Therefore, a weighted Focal Loss applicable to segmentation is used, calculated as follows:

$$L_{foc} = -\alpha_t(1 - p_t)^{\gamma} log(p_t) \tag{5}$$

where p_i denotes the prediction probability of the model for the sample, α_t denotes the category weights, and γ is a moderator. When $\gamma = 0$, Focal Loss degenerates to the ordinary cross-entropy loss function. When $\gamma > 0$, Focal Loss decreases the loss contribution of easy-to-categorize samples and increases the weights of hard to categorize samples, thus making these samples receive more attention.

3 Experimental Methods and Analysis of Results

The experimental framework in this paper uses Pytorch 1.8, CUDA 11.4, TITAN XP graphics card, cosine annealing strategy for learning rate setting, adam optimizer [9] to update the model parameters during training, and 50 training rounds. The training data are input using a data normalization strategy that normalizes the pixel values of each channel of the input image to a mean of 0 and a variance of 1. The benefit of normalization is to ensure that the pixel values of all channels are in the same range of values, preventing a particular channel from having too much influence on the model training.

3.1 Training Dataset

The publicly available datasets used in this paper for training and testing are: MaSTr1325 [10], MODD2, SMD [11] and USV Inland [12], and a dataset MyDataset collected during the experiments in this paper. Images in the dataset are shown in Fig. 7.

The network designed in this paper, as well as all other networks used for comparison, is trained on the MaSTr1325 dataset, which contains 1325 unique images taken over a 24-month period. Three semantic components are manually annotated on a pixel-by-pixel basis: water, sky and environment (obstacle).

In this paper, data augmentation is used in the training. Two types of data enhancement are chosen to suit the water environment: horizontal mirroring and luminance transformations. An elastic distortion is also applied to the water component of the training image to artificially simulate waves and curls, increasing the diversity of local textures in the training set. The effect of data enhancement is shown in figure. The final result after applying data enhancement is a total of 48724 training images.

The performance of the model is evaluated in MyDataset, as well as two publicly available ocean datasets, MODD2 and SMD, and an inland unmanned vessel dataset, USV Inland. MODD2 is recorded in the Adriatic coastal area, consisting of 28 different time series, all collected by the camera and synchronized with IMU measurement times. It is recorded using an unmanned surface vessel. SMD dataset is recorded at different locations in the port of Singapore, which consists of 66 sequences containing the following. The USV Inland is more different from the previous dataset and is the first inland unmanned boat dataset with multiple sensors and weather conditions in real scenarios.

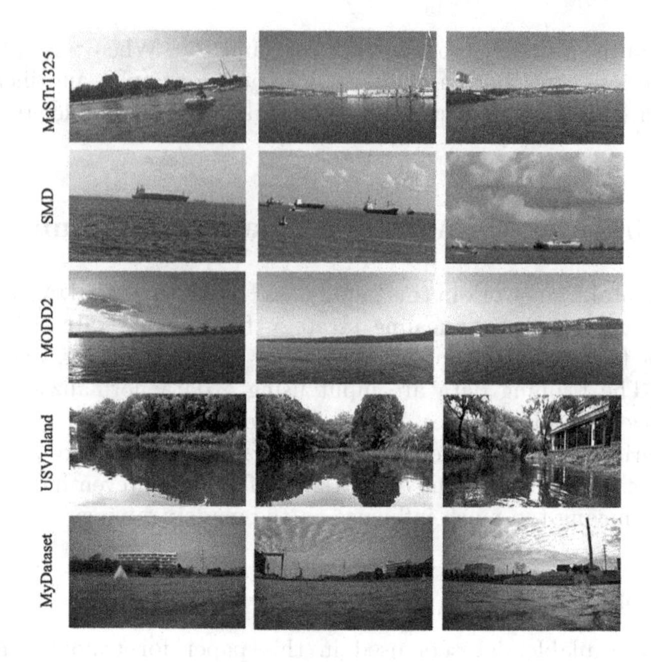

Fig. 7. Training dataset

3.2 Model Structure Ablation Experiment

In deep neural networks, ablation experiments are performed by removing certain component parts of the network and observing the change in the performance of the neural network to determine the effect of each substructure on the model performance. In the Swan network designed in this paper, the model that does not contain ASPP module, ARM module and FFM module is noted as Baseline. In Baseline the decoder only upsamples and splices the final output with the features from the encoder at different stages. The following table shows the impact of different structures on the model.

The Table 1 shows that the ARM, FFM, and ASPP structures of the model all positively influence the final F1 score, and that the first ASPP module has the greatest improvement in model performance, followed by the FFM and ARM modules.

3.3 Performance Comparison of Different Models

To confirm the effectiveness of the model designed in this paper, this section compares some previous research works and the accuracy metrics of the latest semantic segmentation models on the same dataset. A total of three current state-of-the-art segmentation networks RefineNet [12], BiSeNet [13] and U-Net [14] are compared, which achieve the best results in segmentation tasks in either the self-driving car domain or the medical domain, with different encoder and

Table 1. Model substructure ablation experiment

Method	F1-score
Baseline	73.3
Baseline+ASPP	78.7
Baseline+ASPP+ARM	83.0
Baseline+ASPP+ARM+FFM1	86.8
Baseline+ASPP+ARM+FFM1+CBAM	89.7
Baseline+ASPP+ARM+FFM1+CBAM +FFM2	93.3

decoder architectures. The Table 2 summarizes the number of different model parameters and inference times.

Table 2. Comparison of the number of parameters and inference time of different models

Model	N_{param}	δ_t (ms)	FPS
RefineNet	85.7M	130	7
U-Net	28.0M	45	22
BiSeNet	47.5M	68	15
Swan	66.5M	100	10

The Table 3 summarizes the results of all models tested on MODD2 and SMD. Swan greatly outperforms all competing networks in terms of water edge estimation. The second best is RefineNet, with an accuracy about two pixels lower, followed closely by BiSeNet.

Table 3. Test results of different models

Model	TP_{100} (times)	FP_{100} (times)	F1(%)
U-Net	39.9	17.5	69.2
BiSeNet	48.4	12.1	83.8
RefineNet	49.0	2.2	91.6
Swan	51.1	3.8	93.3

The traditional target detection task generally compares the prediction results of the model with the real annotation when evaluating the metrics, and if the intersection ratio of the output rectangular box to the real annotated box reaches a certain threshold and the classification is correct, it is considered as

TP. This approach is fine for detecting small obstacles (e.g., buoys, small boats) on the water surface. However, large vessels are usually integrated with the water boundary, and the model can segment the vessels but the output cannot identify this part of the vessels.

In this paper, we use the method proposed in the literature [15] in calculating the obstacle detection index. We determine whether the detection is correct by calculating the proportion of correctly classified pixel points in the labeled area, and when the proportion exceeds the set threshold of 20%, it is considered as a TP detection, otherwise it is considered as a FP detection.

In terms of TP and FP metrics, Swan receives the best recall score because of the highest true positive (TP) detection rate, followed by RefineNet and BiSeNet.

The Fig. 8 shows the comparison of prediction results of different networks. From the Fig. 8, it can be concluded that in the presence of mist on the sea antenna (first row of the figure), U-Net and BiSeNet are more inaccurate in estimating the water edge, which usually results in over- and underestimation. RefineNet is better, but the expected water is still underestimated, while Swan neither overestimates nor underestimates its position, and Swan has a significant improvement.

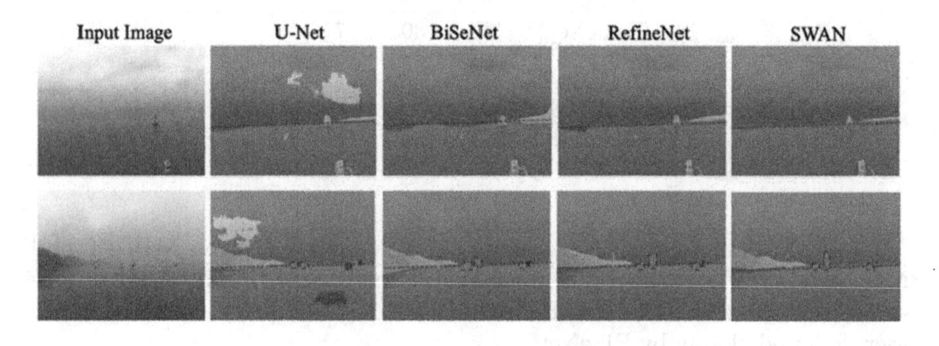

Fig. 8. Comparison of different model forecasts

In the presence of small objects in the distance of the image and in the absence of contrast (second row of the Fig. 8), Swan detects smaller obstacles more accurately than the other networks, and the other three networks showe missed detections, while Swan accurately detectes all the objects in the figure, And it has the most TP detections and achieves the best recall, followed by RefineNet, then BiSeNet and U-Net.

When evaluating the value of a model for engineering applications, the metric for detecting obstacles is usually an important part. A high Precision means that the model can detect the obstacles more accurately, while a high Recall means that the model is better able to detect more obstacles. In practice, in order to achieve uninterrupted autonomous navigation, it is necessary to maintain both a certain recall to ensure that all obstacles can be found as fully as possible and that the platform does not collide.

It is also necessary to maintain a certain level of accuracy to prevent the model from misreporting too many obstacles and affecting the normal operation and obstacle avoidance of the platform, so there is also a trade-off between the number of FP and TP detections, which is measured by the F1 score. The best performing methods based on F1 are Swan, RefineNet, BiSeNet and U-Net. The quality of segmentation masks can be further understood based on the accuracy of obstacle detection overlap thresholds, recall and F1 score plots. The plots of the four best performing networks are shown in Fig. 9.

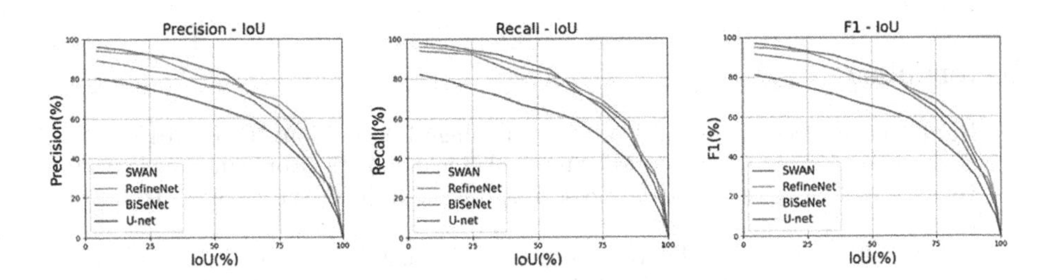

Fig. 9. Plot of model metrics and obstacle overlap threshold

The Swan network proposed in this paper has the highest accuracy and recall scores at medium overlap threshold, followed by RefineNet, which means that Swan detects more obstacles with fewer FPs. Further analysis shows that most of the TP detected by Swan but not detected by other networks are small objects with area less than 900 pixels, proving that Swan is more advantageous for detecting small targets.

However, RefineNet performs better when the overlap threshold is relatively high (above 70%), predicating that the localization of RefineNet is more accurate than that of Swan. However, RefineNet can show poor detection of isolated obstacles and miss some small obstacles, leading to relatively low TP rate, that is dangerous in real autonomous navigation. As the threshold rises above 65%, the curves of all models drop faster, indicating that these models still have shortcomings in localization. Thus accurate obstacle segmentation is very challenging for all models, and the models still have room for improvement upwards.

4 Conclusion

In this paper, we propose a semantic segmentation neural network, Swan-Net, which fuses the pose information to extract the environmental semantic information from the image, and classifies the pixels in the image into three categories: water, sky, and environment, that can be used to provide passable area information for surface ships and vessels. In this paper, we first integrate the positional information into a semantic segmentation neural network as an a priori water boundary approach, encode the platform positional information obtained from the inertial measurement unit (IMU) into a feature channel, and apply it to the decoder to fuse with the image features. The decoder design makes full use of multi-scale feature fusion, channel attention, and spatial attention mechanisms, and the effectiveness of the model design is demonstrated by ablation experiments and comparison experiments. Its advantages in the field of water surface environment perception are illustrated through the comparison with the existing excellent networks in the field of semantic segmentation. The next step needs to improve the segmentation performance of the network on the inland river scenario, so more data collection and application of new model optimization methods needs to be considered to enhance the model performance in this scenario.

References

1. He, K., et al.: Deep Residual Learning for Image Recognition. IEEE (2016)
2. Bovcon, B., Perš, J., Kristan, M.: Stereo obstacle detection for unmanned surface vehicles by IMU-assisted semantic segmentation. Robot. Auton. Syst. **104**, 1–13 (2018)
3. Yu, C., et al.: Bisenet: bilateral segmentation network for real-time semantic segmentation. In: Proceedings of the European Conference on Computer Vision (ECCV), pp. 325–341 (2018)
4. Ioffe, S., Szegedy, C.: Batch Normalization: Accelerating Deep Network Training by Reducing Internal Covariate Shift. JMLR.org (2015)
5. Rumelhart, D.E., Hinton, G.E., Williams, R.J.: Learning representations by back propagating errors. Nature **323**(6088), 533–536 (1986)
6. Krizhevsky, A., Sutskever, I., Hinton, G.: ImageNet classification with deep convolutional neural networks. Adv. Neural Inf. Process. Syst. **25**(2) (2012)
7. Woo, S., et al.: CBAM: convolutional block attention module. In: Proceedings of the European Conference on Computer Vision (ECCV), pp. 3–19 (2018)
8. Krizhevsky, A., Sutskever, I., Hinton, G.: ImageNet classification with deep convolutional neural networks. Adv. Neural Inf. Process. Syst. **25**(2) (2012)
9. Kingma, D., Ba, J.: Adam: a method for stochastic optimization. Comput. Sci. (2014)
10. Bovcon, B., et al.: The mastr1325 dataset for training deep USV obstacle detection models. In: 2019 IEEE/RSJ International Conference on Intelligent Robots and Systems (IROS), pp. 3431–3438. IEEE (2019)
11. Prasad, D.K., et al.: Video processing from electro-optical sensors for object detection and tracking in a maritime environment: a survey. IEEE Trans. Intell. Transp. Syst. 1993–2016 (2017)
12. Cheng, Y., et al.: Are we ready for unmanned surface vehicles in inland waterways the USV inland multi-sensor dataset and benchmark. IEEE Robot. Automat. Lett. **99**, 1 (2021)
13. Lin, G., et al.: RefineNet: multi-path refinement networks for high-resolution semantic segmentation. In: 2017 IEEE Conference on Computer Vision and Pattern Recognition (CVPR). IEEE (2017)
14. Ronneberger, O., Fischer, P., Brox, T.: U-net: convolutional networks for biomedical image segmentation. In: Navab, N., Hornegger, J., Wells, W.M., Frangi, A.F. (eds.) MICCAI 2015. LNCS, vol. 9351, pp. 234–241. Springer, Cham (2015). https://doi.org/10.1007/978-3-319-24574-4_28
15. Bovcon, B., Kristan, M.: WaSR-a water segmentation and refinement maritime obstacle detection network. IEEE Trans. Cybernet. **99** (2021)

Novel Fault Diagnosis Method Integrating D-L2-FDA and AdaBoost

Yang Zhao[1,2], Wei Ke[3], Wei Zhang[3], Yi Luo[4], Qun-Xiong Zhu[1,2], Yan-Lin He[1,2], Yang Zhang[1,2], Ming-Qing Zhang[1,2](✉), and Yuan Xu[1,2](✉)

[1] College of Information Science and Technology, Beijing University of Chemical Technology, Beijing 100029, China

[2] Engineering Research Center of Intelligent PSE, Ministry of Education of China, Beijing 100029, China

mqzhang@buct.edu.cn , xuyuan@mail.buct.edu.cn

[3] School of Applied Sciences, Macao Polytechnic Institute, Macao, SAR 999078, People's Republic of China

[4] Research Institute of Mine Big Data, Chinese Institute of Coal Science, Beijing 100013, China

Abstract. Industrial process safety has always been a concern for engineers and researchers. Fault diagnosis frameworks based on data-driven methods are prevalent and play a vital role in guaranteeing industrial process safety. However, the data collected in actual industrial production regularly exhibits high-dimensional and complex timing characteristics. In this research, a new framework for fault diagnosis is constructed on the strength of dynamic L2-norm normalized fisher discriminant analysis (FDA) integrating with AdaBoost. Firstly, timing characteristic in industrial process is taken into account so that a dynamic fault dataset is constructed. Then, the FDA vectors are normalized by L2-norm and utilized to reduce data dimension which can learn fault patterns in feature extraction. In addition, an ensemble learning method named AdaBoost is adopted for pattern classification. To verify the effectiveness of the proposed method, simulation experiments based on Tennessee Eastman process are carried out and satisfactory results are obtained.

Keywords: D-L2-FDA · AdaBoost · Tennessee Eastman Process · Feature Extraction

1 Introduction

As a significant aspect of industrial production, process industry occupies a greatly important position. Fault diagnosis is a key technology to ensure the

This work was supported in part by the National Natural Science Foundation of China under Grant 61973022 and in part by the Fundamental Research Funds for the Central Universities under BH2321.

B. Xin et al. (Eds.): IWACIII 2023, CCIS 1931, pp. 63–74, 2024.
https://doi.org/10.1007/978-981-99-7590-7_6

safety of industrial production, which is quite vital for the safe operation of industrial control system. Recently, data-driven fault diagnosis based on multivariate statistical technology has been widely studied and applied in recent years.

Principal component analysis (PCA) is a classic data-driven dimensionality reduction method that has been extensively used in industrial monitoring system. The PCA-based feature extraction frameworks can eliminate the correlation between variables of dataset while maximizing the variance [1]. Due to the effectiveness of PCA, it has been widely studied and extended by researchers. For instance, literature [2] proposes a CA-PCA based hybrid method and applies it to multivariate process monitoring. However, PCA and its variants divide the data into the principal subspace and the residual subspace to learn data features from a global perspective, which may lead to missing part of the data information. Therefore, some scholars have proposed new methods to learn local structural features of data such as Local Preserving Projection (LPP), which can achieve dimensionality reduction while preserving data spatial position information [3,4]. Nevertheless, although these methods are effective in extracting features, they are highly dependent on the performance of classifiers when applied to fault diagnosis. Similar to the above methods, Fisher discriminant analysis (FDA) is also a data-driven multivariate statistical method, which has been widely studied because of its unique advantages [5–8]. Also, as a method of data feature selection, FDA has the function of pattern classification and has more advantages than PCA in fault diagnosis [9,10].

In order to obtain the structural characteristics and class information of data, we propose a comprehensive and effective framework for fault diagnosis, namely dynamic L2-normed fisher discriminant analysis integrating AdaBoost method. In actual industrial production, collected data is not uniform and noise is ubiquitous. So, we normalize the collected data before feature learning. At the same time, considering timing and dynamic natures existing in industrial processes, the sample space is processed in a dynamical manner. In this paper, FDA is used for feature selection, and then the feature vectors are normalized by L2 norm to make the feature space uniform. Ultimately, with regard to multiple faults mode, an ensemble learning method that is called AdaBoost is selected for pattern classification. The proposed framework is finally applied to the Tennessee Eastman Process (TEP), and satisfactory results are obtained.

The remainder of this article is organized as follows: in Sect. 2, we introduce the relevant methods briefly. In Sect. 3, there is a systematic introduction to the proposed method. In Sect. 4, we apply the proposed method to TEP and obtain experimental results. In Sect. 5, we summarize the significance of dynamic data processing in industry fault diagnosis and the effectiveness of AdaBoost algorithm.

2 Related Methods

2.1 Fisher Discriminant Analysis

FDA is a supervised learning dimensionality reduction technique that tries to find a linear combination of features of events to be able to characterize or distinguish them. The main idea of the FDA is to select the direction of the projection by extracting the principal characteristics of the data, so that different types of samples are separated as far as possible after projection, and similar samples are converged as possible. The details of the algorithm are described in the literature [11,12].

To summarize, the goal of FDA is to maximize the following Rayleigh entropy,

$$max\ J = \arg\max_{v \neq 0} \frac{\boldsymbol{v}^T \boldsymbol{S}_b \boldsymbol{v}}{\boldsymbol{v}^T \boldsymbol{S}_w \boldsymbol{v}}. \tag{1}$$

where $\boldsymbol{S}_b$ is the dispersion matrix between classes while $\boldsymbol{S}_w$ is the dispersion matrix within classes.

2.2 Ensemble Learning Method AdaBoost

In pattern recognition, a single strong classifier is generally used, such as support vector machine (SVM) [13], Bayesian algorithm [14,15], or deep belief network (DBN) [16]. These methods mentioned above can be applied to TEP to operate fault diagnosis through a single classifier.

In contrast, in the field of data mining, an integrated learning method AdaBoost can be promoted to a strong classification algorithm by integrating several basic classifiers, such as decision trees [17,18]. Even if the performance of the basic classifier is weak, the recognition performance can be enhanced through a certain combination strategy.

3 The Proposed Method

We propose a comprehensive approach of FDA of dynamic data normalization with L2 norm (D-L2-FDA) based on AdaBoost for fault diagnosis and apply it to TEP. AdaBoost is applied for pattern recognition in fault diagnosis. The pleasing results have been obtained in the experiment.

3.1 D-L2-FDA for Feature Extraction

Data Dynamic Processing. The data of the process industry has a strong correlation in time series, and the data collected before often has a certain impact on the data collected later. In addition, the industrial environment is complex and changeable, and manual real-time monitoring is almost impossible. Therefore, the monitoring of process variables is pretty dependent on various sensors, but the feedback of hardware sensors may be delayed, which has caused the lag of data collection to some extent. As the front-to-back correlation of industrial data and the lag of sensors collection is fully considered, a method to dynamically process the existing data is proposed to maximize the extraction of various information contained in the data.

Suppose the original data set contains m variables and each variable has n observations. The data is arranged in matrix $\boldsymbol{X}_0$ as follows,

$$\boldsymbol{X}_0 = \begin{bmatrix} x_{11} & x_{12} & \dots & x_{1m} \\ x_{21} & x_{22} & \dots & x_{2m} \\ \vdots & \vdots & \ddots & \vdots \\ x_{n1} & x_{n2} & \dots & x_{nm} \end{bmatrix}. \tag{2}$$

In matrix $\boldsymbol{X}_0$, each column represents a variable, and each row vector represents the observation values. Therefore, the column values in the matrix are related in time series. In this work, the time lag coefficient $L = 2$, and three column values are selected each time to construct a new set of observations. The augmentation matrix $\boldsymbol{X}$ is formed as follows,

$$\boldsymbol{X} = \begin{bmatrix} (x_{11} \quad x_{21} \quad x_{31}) & (x_{12} \quad x_{22} \quad x_{32}) & \dots & (x_{1m} \quad x_{2m} \quad x_{3m}) \\ (x_{21} \quad x_{31} \quad x_{41}) & (x_{22} \quad x_{32} \quad x_{42}) & \dots & (x_{2m} \quad x_{3m} \quad x_{4m}) \\ \vdots & \vdots & \ddots & \vdots \\ (x_{n-2,1} \; x_{n-1,1} \; x_{n,1}) & (x_{n-2,2} \; x_{n-1,2} \; x_{n,2}) & \dots & (x_{n-2,m} \; x_{n-1,m} \; x_{n,m}) \\ (x_{n-1,1} \quad x_{n,1} \quad x_{n,1}) & (x_{n-2,2} \quad x_{n,2} \quad x_{n,2}) & \dots & (x_{n-1,m} \quad x_{n,m} \quad x_{n,m}) \\ (x_{n,1} \quad x_{n,1} \quad x_{n,1}) & (x_{n,2} \quad x_{n,2} \quad x_{n,2}) & \dots & (x_{n,m} \quad x_{n,m} \quad x_{n,m}) \end{bmatrix} \tag{3}$$

Specially note that in the last two rows of the augmentation matrix, because three completely different observations cannot be formed, appropriate adjustments are made, and two time-collected values are retained in $(n-1)th$ row, while only one in the nth row.

Augmentation Matrix Dimensionality Reduction Using FDA. It can be seen that the dimension of the augmented matrix is high. In order to eliminate redundant variables, FDA is used to reduce data dimension. The dimensionality reduction process is described as follows.

Assume that the fault category space is $\boldsymbol{C}, c \in \boldsymbol{C}$ is the number of categories. Represent the ith statistical variable as a vector $\boldsymbol{x}_i$, n_j is the number of observations in the jth category.

Define $\overline{\boldsymbol{x}}$ as the total mean vector and can be computed as follows,

$$\overline{\boldsymbol{x}} = \frac{1}{n} \sum_{i=1}^{n} \boldsymbol{x}_i \tag{4}$$

Define χ_j as the set of vector $\boldsymbol{x}_i$ that belongs to the jth class, then the internal dispersion matrix of jth class is

$$\boldsymbol{S}_j = \sum_{\boldsymbol{x}_i \in \chi_j} (\boldsymbol{x}_i - \overline{\boldsymbol{x}}_j)(\boldsymbol{x}_i - \overline{\boldsymbol{x}}_j)^T \tag{5}$$

where $\overline{\boldsymbol{x}}_j$ is the mean vector of jth class and can be calculated as,

$$\overline{\boldsymbol{x}}_j = \frac{1}{n_j} \sum_{\boldsymbol{x}_i \in \chi_j} \boldsymbol{x}_i \tag{6}$$

The intra-class dispersion matrix is defined as follows,

$$\boldsymbol{S}_w = \sum_{j=1}^{c} \boldsymbol{S}_j \tag{7}$$

And the dispersion matrix between classes is formulated as

$$\boldsymbol{S}_b = \sum_{i=1}^{n} n_j (\overline{\boldsymbol{x}}_j - \overline{\boldsymbol{x}})(\overline{\boldsymbol{x}}_j - \overline{\boldsymbol{x}})^T \tag{8}$$

The linear transformation vector of FDA can be calculated by solving the stability point of Rayleigh entropy in Eq. (1).

Using the Lagrange multiplier method, the eigenvector is equivalent to solving the generalized eigenvalue of the following equation:

$$\boldsymbol{S}_b \boldsymbol{w}_k = \boldsymbol{\lambda}_p \boldsymbol{S}_w \boldsymbol{w}_k \tag{9}$$

After the generalized eigenvectors are figured out, r FDA vectors are then obtained which can form a dimension reduction matrix $\boldsymbol{W}_r$ by columns.

Normalization of FDA Vectors Using L2 Norm. Usually, the value of each dimension of the vector is mapped between (0, 1) or (-1, 1), and a certain norm of the vector can also be mapped to one.

When the generalized eigenvectors are solved in Eq. (9), compared to the magnitude of the feature vector, the direction of the feature vector is more important. Therefore, in order to simplify the calculation process and speed up the establishment of data-driven models, the eigenvectors obtained can be normalized by the *L*2 *norm*. The operation is to divide the value of each dimension of the eigenvectors by the *L*2 *norm* of the vector. That is

$$\boldsymbol{w}_t = \frac{\tau_1}{\|\boldsymbol{w}_t\|_2} + \frac{\tau_2}{\|\boldsymbol{w}_t\|_2} + \ldots + \frac{\tau_n}{\|\boldsymbol{w}_t\|_2}, \ (t = 1, 2, \ldots, r) \tag{10}$$

where τ_i is the coordinates of $\boldsymbol{w}_t$. Then, these normalized vectors are settled into a set and then a normalized dimensionality reduction matrix is obtained,

$$\boldsymbol{W}_{new} = (\boldsymbol{w}_1, \boldsymbol{w}_2, \ldots, \boldsymbol{w}_r) \tag{11}$$

The matrix $\boldsymbol{W}_{new}$ contains r eigenvectors corresponding to the r larger eigenvalues of $\boldsymbol{X}$, which preserves the effective information of the augmented matrix furthest. Data feature extraction can be carried out according to the following equation,

$$\boldsymbol{Z} = \boldsymbol{W}_{new}^T \boldsymbol{X} \tag{12}$$

3.2 AdaBoost for Fault Diagnosis

In this work, AdaBoost is selected for multi-faults classification. The dimensionality-reduced data set $\boldsymbol{Z}$ is utilized as a training set to initialize AdaBoost, which is performed as Algorithm 1.

The algorithm finally produces a strong classifier $\boldsymbol{\Psi}(\boldsymbol{z})$ that boosted by a group of weak or basic classifier, which is utilized in fault diagnosis.

Algorithm 1 Framework of AdaBoost.

Input:

Training set $\boldsymbol{S} = \{(\boldsymbol{z}_1, y_1), (\boldsymbol{z}_2, y_2), \ldots, (\boldsymbol{z}_n, y_n)\}$ and basic classifier space $\psi \in \boldsymbol{\Psi}$;

Output:

Boosted strong classifier $\boldsymbol{\Psi}(\boldsymbol{z}) = \arg\max_f \sum_{f=1}^{F} D_f(\boldsymbol{z}, p)$;

Initialization:

Training set weights:$\boldsymbol{D}_1 = (w_{11}, w_{12}, \ldots, w_{1n}) = (1/n, 1/n, \ldots, 1/n)$;

1: **for** $f = 1, 2, \ldots, F$(F is the number of selected features), each basic classifier ψ **do**

2: i. Divide the sample space $\boldsymbol{Z}$ to $\boldsymbol{Z}_1, \boldsymbol{Z}_2, \ldots, \boldsymbol{Z}_m$;

3: ii. Calculate under probability distribution $\boldsymbol{G}_u$:
$\boldsymbol{M}_l^i = \boldsymbol{P}(\boldsymbol{z}_i \in \boldsymbol{Z}_j, y_i = l) = \sum_{\substack{\boldsymbol{z}_i \in \boldsymbol{Z}_j \\ y_i = l}} G_u(i),\ l = 1, 2, \ldots, c$;

4: iii. Set the output of the basic classifier on the division of $\boldsymbol{Z}$ mentioned in i.:
$\psi_f(\boldsymbol{z}, p) = ln\boldsymbol{M}_f^i, j = 1, 2, \ldots, m, p = 1, 2, \ldots, c$;

5: iv. Calculate the normalization factor
$Q_f = K \sum_{j=1}^{m} (\prod_{l=1}^{c} \boldsymbol{M}_f^i)^{1/l}$;

6: **end for**

7: **for** ψ_f minimizes Q_f **do**

8: $\xi^{(f)} = \frac{\sum_{i=1}^{n} G_f(i)[y_i \neq T_m(\boldsymbol{z})]}{\sum_{i=1}^{n} G_f(i)}$;

9: **end for**

10: Calculate the weight of each basic classifier:
$\beta(f) = lg\frac{1-\xi^{(f)}}{\xi^{(f)}}$;

11: Update sample weights:
$D_{f+1}(i) = \frac{D_f(i)e^{\beta(f)}[y_i \neq T_m(\boldsymbol{z})]}{Q_f}$;

12: **return** $\boldsymbol{\Psi}(\boldsymbol{z})$.

---

4 Cases Study

In order to verify the effectiveness of the proposed method, we carefully perform the experiment and apply the framework to the Tennessee Eastman process (TEP).

4.1 Tennessee Eastman Process

Tennessee Eastman Process, created by Eastman Chemical Company, is a system simulation based on real industrial processes [3,8,18]. The TEP contains 52 process variables and 21 process faults. The original data set is divided into training sets and testing sets. Each type of fault in the training set contains 480 samples, while each testing set contains 960 samples.

4.2 Faults Selection

In this paper, five unknown faults and one known fault are selected, which are Fault 16, Fault 17, Fault 18, Fault 19, Fault 20, and Fault 11. Table 1 is the detailed information of these faults.

Table 1. The selected faults.

Faults	Description	Type
Fault11	Inlet temperature of reactor cooling water	Random
Fault16	Unknown	—
Fault17	Unknown	—
Fault18	Unknown	—
Fault19	Unknown	—
Fault20	Unknown	—

4.3 Confusion Matrix

In data visualization, the confusion matrix is a very intuitive way of expression. The confusion matrix contains a lot of information and can be used to measure a model comprehensively. Not only can the classification accuracy be calculated from the confusion matrix, but also the accuracy, recall and specificity can be obtained.

Table 2 is a confusion matrix obtained in this experiment that indicates how many samples that diagnosed correctly or incorrectly. From the data contained in the diagonal line of the matrix in Table 2, it can be seen that the proposed model has the highest accuracy rate of 90.25% in the diagnosis of fault 18, the average comprehensive diagnosis accuracy rate is 65.44%.

Table 2. The confusion matrix.

Number(Accuracy%)		Predict label						Sum1
		Fault11	Fault16	Fault17	Fault18	Fault19	Fault20	
True label	Fault11	**420(52.50)**	130	18	0	209	23	800
	Fault16	103	**558(69.78)**	0	0	123	16	800
	Fault17	72	70	**614(76.75)**	0	24	20	800
	Fault18	25	24	0	**722(90.25)**	23	6	800
	Fault19	240	263	0	0	**247(30.88)**	50	800
	Fault20	48	67	0	0	89	**596(74.50)**	800
Sum2		908	1112	632	722	715	711	**65.44%(average)**

4.4 Comparison with Other Methods

In this experiment, our work is compared with L2-FDA-AdaBoost model without data dynamic processing. The comparison result is shown in Fig. 1 and Fig. 2. It can be seen from the curve graph that the proposed method has a higher accuracy rate during model training and the classification performance is stable.

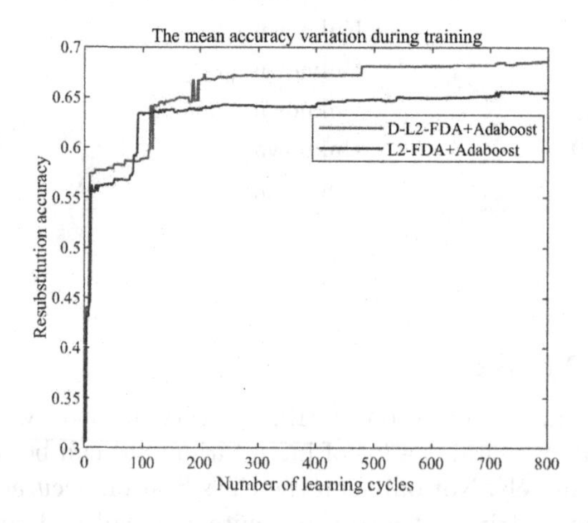

Fig. 1. Data with or without dynamic performance comparison.

As can be seen from the histogram in Fig. 2 that the proposed method has better fault diagnosis effect, and outperforms the model without dynamic processing.

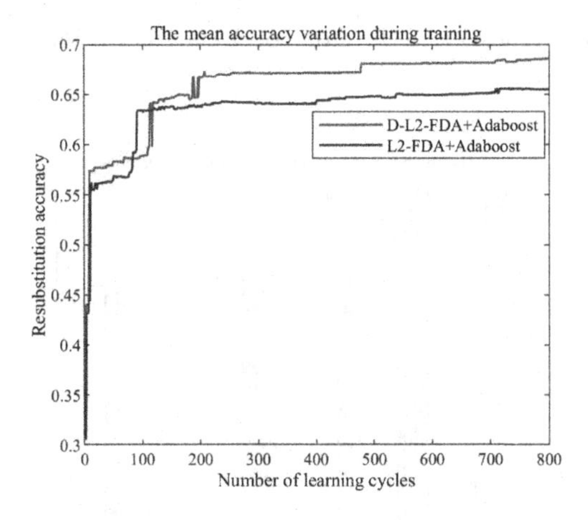

Fig. 2. Data with or without dynamic performance comparison.

Then, the classifier AdaBoost is compared with Bayesian. The comparison results are shown in Fig. 3. Also, as a machine learning method, the Bayesian classifier is inferior to the AdaBoost method in diagnosis accuracy. It can be seen from Fig. 3 that AdaBoost has apparent advantages as an integrated classifier, and the number of correctly classified samples is much higher than the Bayesian classifier.

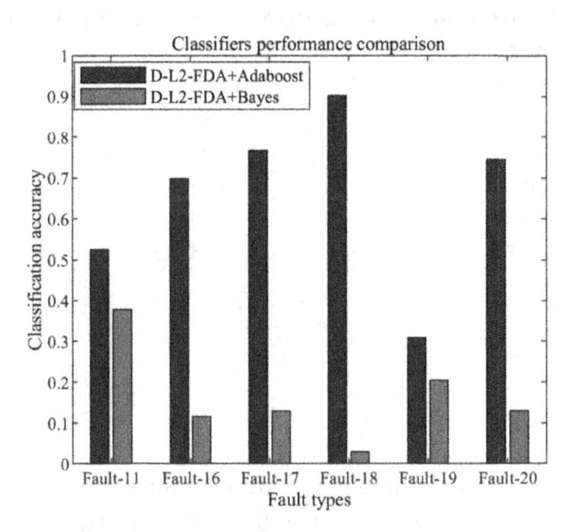

Fig. 3. The classifier performance comparison.

To better illustrate the advantages of the proposed method, the three given methods are compared together and the results are shown in Fig. 4 and Table 3. The experimental results fully indicate that dynamic data processing is crucial for industrial fault diagnosis and the AdaBoost classifier is indeed perform better in TEP fault diagnosis.

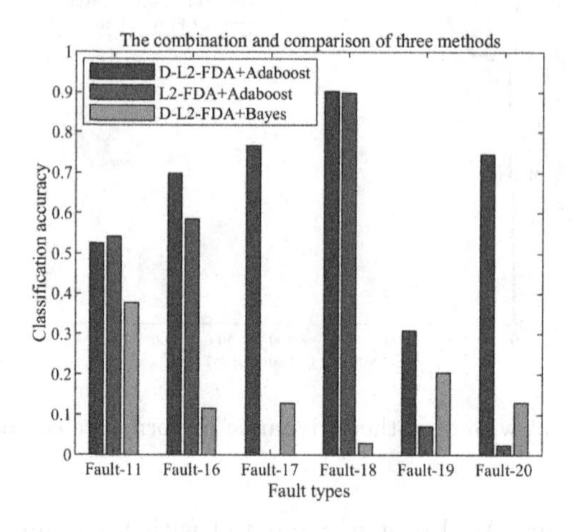

Fig. 4. Comprehensive comparison of fault diagnosis.

Table 3. Comprehensive comparison of fault diagnosis accuracy

Methods	D-L2-FDA+AdaBoost(%)	L2-FDA+AdaBoost(%)	D-L2-FDA+Bayes(%)
Fault11	52.50	37.75	54.25
Fault16	69.75	11.63	58.50
Fault17	76.75	12.88	0.00
Fault18	90.25	2.88	89.75
Fault19	30.88	20.50	7.12
Fault20	74.50	13.00	2.38
Average	**78.93**	19.73	42.40

5 Conclusions

This paper proposes a comprehensive fault diagnosis method based on D-L2-FDA integrating with AdaBoost, which fully considers the dynamic characteristics of the process data. In our work, considering the timing nature of the process data, we process fault dataset dynamically. To improve feature extraction capacity, FDA vectors are normalized by L2 norm. The AdaBoost algorithm that ensembles many weak classifiers is performed for pattern classification, which improves the accuracy of fault diagnosis. Case studies performed in six types faults of TEP, and the simulation results verifies the effectiveness of the proposed framework.

References

1. Wang, T.Z., Xu, H., Han, J.G., Elbouchikhi, E., Benbouzid, M.E.H.: Cascaded H-bridge multilevel inverter system fault diagnosis using a PCA and multiclass relevance vector machine approach. IEEE Trans. Power Electron. **30**(12), 7006–7018 (2015)
2. Xu, Y., Shen, S.Q., He, Y.L., Zhu, Q.X.: A novel hybrid method integrating CA-PCA with relevant vector machine for multivariate process monitoring. IEEE Trans. Control Syst. Technol. **27**(4), 1780–1787 (2019)
3. Xu, Y., et al.: A novel pattern classification integrated global-local preserving projections with improved adaptive rank-order morphological filter for fault diagnosis. Process Safety Environ. Protect. **171**(1), 299–311 (2023)
4. Cai, H., Hao, L., Su, Y.Z.: ISAR target recognition based on two-dimensional locality preserving projection. J. Phys. Conf. **1060**(1), 012006 (2018)
5. Zhu, Z.B., Song, Z.H.: A novel fault diagnosis system using pattern classification on kernel FDA subspace. Exp. Syst. Appl. **38**(6), 6895–6905 (2011)
6. Tang, J., Yan, X.: Neural network modeling relationship between inputs and state mapping plane obtained by FDA-t-SNE for visual industrial process monitoring. Appl. Soft Comput. **60**(C), 577–590 (2017)
7. Li, J.H., Cui, P.L.: Improved kernel fisher discriminant analysis for fault diagnosis. Exp. Syst. Appl. **36**(2), 1423–1432 (2009)
8. Chiang, L.H., Russell, E.L., Braatz, R.D.: Fault diagnosis in chemical processes using Fisher discriminant analysis, discriminant partial least squares, and principal component analysis. Chemometr. Intell. Lab. Syst. **50**(2), 243–252 (2000)
9. Gharavian, M.H., Ganj, F.A., Ohadi, A.R., Bafroui, H.H.: Comparison of FDA-based and PCA-based features in fault diagnosis of automobile gearboxes. Neurocomputing **121**(9), 150–159 (2013)
10. Clarke, M., Duda, R., Hart, P.: Data-driven fault diagnosis using deep canonical variate analysis and fisher discriminant analysis. IEEE Trans. Indust. Inf. **17**(5), 3324–3334 (2021)
11. Chang, C.C.: Fisher's linear discriminant analysis with space-folding operations. IEEE Trans. Pattern Anal. Mach. Intell. **45**(7), 9233–9240 (2023)
12. Liu, J., Jiang, P., Song, C.Y., Xu, H., Hmelnov, A.E.: Manifold-preserving sparse graph and deviation information based fisher discriminant analysis for industrial fault classification considering label-noise and unobserved faults. IEEE Sens. J. **22**(5), 4257–4267 (2022)
13. Kumari, A., Tanveer, M.: Universum twin support vector machine with truncated pinball loss. Eng. Appl. Artif. Intell. **123**(1), 106427 (2023)
14. Zou, L., Zhuang, K.J., Zhou, A., Hu, J.: Bayesian optimization and channel-fusion-based convolutional autoencoder network for fault diagnosis of rotating machinery. Eng. Struct. **280**(1), 115708 (2023)
15. Meng, Q.Q., Zhu, Q.X., Gao, H.H., He, Y.L., Xu, Y.: A novel scoring function based on family transfer entropy for Bayesian networks learning and its application to industrial alarm systems. J. Process Control **76**(1), 122–132 (2019)
16. Yang, J., et al.: Joint pairwise graph embedded sparse deep belief network for fault diagnosis. Eng. Appl. Artif. Intell. **99**(1), 104149 (2021)
17. Cui, Y.Q., Shi, J.Y., Wang, Z.L.: Analog circuit fault diagnosis based on quantum clustering based multi-valued quantum fuzzification decision tree (QC-MQFDT). Measurement **93**(1), 421–434 (2016)

18. Xu, Y., Cong, K.D., Zhang, Y., Zhu, Q.X., He, Y.L.: A novel AdaBoost ensemble model based on the reconstruction of local tangent space alignment and its application to multiple faults recognition. J. Process Control **104**(1), 158–167 (2021)

Structural Health Monitoring of Similar Gantry Crane Based on Federated Learning Algorithm

Zexuan Peng[1], Zhaohui Zhang[1](✉), Xiaoyan Zhao[2], Tianyao Zhang[1], and Qi Wu[1]

[1] School of Automation and Electrical Engineering, University of Science and Technology Beijing, 30# Xueyuan Road, Haidian District, Beijing 100083, China
zhangzhaohui@ustb.edu.cn

[2] Beijing Engineering Research Center of Industrial Spectrum Imaging, University of Science and Technology Beijing, 30# Xueyuan Road, Haidian District, Beijing 100083, China

Abstract. When using Internet of Things (IoT) technology for health monitoring of similar batches of identical gantry cranes, uploading all their structural data to a cloud server would result in significant bandwidth waste and reduced real-time performance. Additionally, due to different usage scenarios and habits among similar batches of identical gantry cranes, existing algorithms can only monitor specific gantry cranes after training is completed. For the health monitoring of identical gantry cranes, this study employed the federated averaging algorithm (FedAvg) and XGBoost algorithm in an edge gateway to detect damage types in gantry crane structures. A small-scale gantry crane was used to simulate various connection damages. A IoT gateway was designed to monitor the structural condition of facilities. Upon receiving structural data, the gateway locally performed anomaly detection. When the edge gateway detected anomalies in the time series, it utilized the XGBoost algorithm for anomaly type classification. Additionally, the edge gateway employed the FedAvg to mitigate the impact of non-independent and imbalanced data caused by different environments and usage patterns. Ultimately, a unified anomaly detection model was obtained to facilitate subsequent usage of identical gantry cranes.

Keywords: Structural health monitoring of Gantry crane · Federated learning · Anomaly Detection

1 Introduction

The gantry crane is widely used in ports, warehouses, or construction sites. As a metal structure, its components may suffer damage due to frequent use or aging over time. Considering the complexity of the structural mechanism model of the gantry crane and the decentralization of similar gantry cranes, data-driven approaches are considered appropriate methods.

In recent years, there have been many studies on health monitoring of cranes using data-driven methods. Xu Yunfei's team from Southeast University [1] used acoustic emission technology and improved kernel entropy component analysis with a moving

B. Xin et al. (Eds.): IWACIII 2023, CCIS 1931, pp. 75–88, 2024.
https://doi.org/10.1007/978-981-99-7590-7_7

window to detect both laboratory-scale and real cranes. Qin Xianrong's team from Tongji University [2] proposed a prediction model based on Dual-Tree Complex Wavelet Transform (DTCWT) combined with Autoregressive Moving Average (ARMA) and Support Vector Regression (SVR), and applied it to structural health monitoring of container cranes. Kim Hoon's team from Incheon National University in Korea [3] utilized Convolutional Neural Networks (CNN) to detect crane defects. Samiksha Yede's team from India [4] employed a Gaussian regression model for monitoring material overload.

The aforementioned studies were all conducted on remote cloud devices, which had poor real-time performance and consumed a significant amount of bandwidth. Furthermore, the algorithms mentioned above lacked model generalization and could only be used on specific cranes. When deploying them on similar cranes of the same type, data needs to be collected again and retraining is required. This paper proposed migrating the computational tasks from the cloud service center to edge gateways closer to the data source. This ensured real-time performance of the model and prevented wastage of edge-side computing resources [5]. On the other hand, the main objective of federated learning is to address machine learning problems in scenarios with multiple devices or data centers, especially on mobile or edge devices [6]. Federated learning can aggregate the datasets from these devices to train a unified model. This approach allows for the utilization of similar batches of gantry cranes of the same type. However, for a specific manufacturer's similar batches of gantry cranes, due to different environments and usage habits, their data often exhibit non-independent and non-identically distributed characteristics. Federated learning can amplify this discrepancy. To address this issue, this paper employed FedAvg.

In this paper, an unsupervised anomaly detection algorithm for gantry cranes of the same type was implemented on edge gateways. The FedAvg was employed to address the issue of non-independent and non-identically distributed data in the gantry cranes. Subsequently, the XGBoost was utilized to classify the detected errors.

2 Monitoring Model and Fault Simulation

2.1 Monitoring Model and Damage Detection

The main monitoring subject of this paper was the gantry crane. In order to better simulate the anomalies in facility structures, a small-scale gantry crane was used as the monitoring object in this paper. The gantry crane is depicted in Fig. 1.

In order to simulate the structural damage of the gantry crane, this paper introduces various types of damages by loosening the connections of each pillar. The method of damage involved loosening the screws that connected two beams or rods, causing slight displacement or oscillation in the beams. Figure 2 shows the damage markers in the facility model, with numbered locations indicating potential damage. Multiple types of damages are set in this paper, and their types and descriptions are listed in Table 1.

2.2 Wireless Edge Gateway Data Acquisition System

This paper was based on data-driven damage detection and classification, so it required the collection of relevant data from the gantry crane. Due to the large size and height

Fig. 1. Physical diagram of a gantry crane.

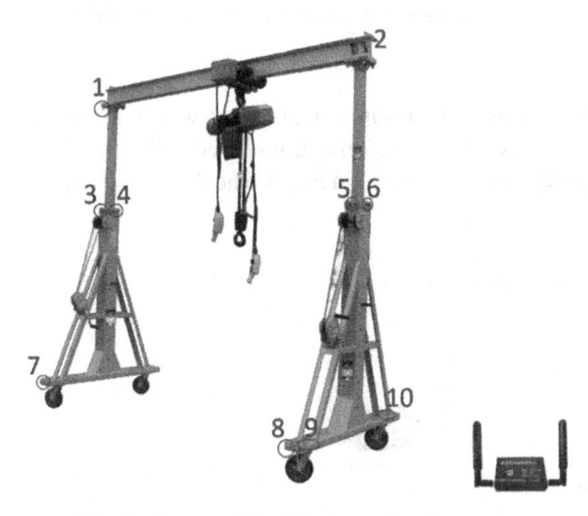

Fig. 2. Damage identification of gantry crane.

of the gantry crane, wired data acquisition methods may be inconvenient during crane operations, limiting the mobility, flexibility, and remote operation capabilities of the crane. In this paper, a wireless transmission mode was adopted. Since the measurement locations were numerous and widely distributed, a wireless edge gateway was designed to collect and process the data.

The gateway system designed in this paper consisted of measurement nodes and an edge gateway. The measurement nodes consisted of 5 acceleration sensors and 1 tension sensor. Compared to other IoT monitoring gateways, the gateway designed in this paper had a high integration level. It only required measuring vibration and tension to monitor the gantry crane, resulting in fewer measurement data dimensions and lower bandwidth requirements for transmission.

Table 1. Damage type and description.

Damage type	Damage description
01H	health
01D	Damage to gantry support points 1
02D	Damage to gantry support points 1 and 2
03D	Damage to gantry support points 1, 2, and 6
04D	Damage to gantry support points 3, 4, 5, and 6
05D	Damage to gantry support points 1, 2, 3, 4, and 5
06D	Damage to gantry support points 1, 2, 3, 4, 5, and 6
07D	Damage to leg support point 7
08D	Damage to leg support points 7 and 8
09D	Damage to leg support points 7, 8, and 9
10D	Damage to leg support points 7, 8, 9, and 10

Once the sensors measured relevant data, the data was transmitted through the edge gateway and further processed before being transferred to the cloud server. The relationship between the gateways is illustrated in Fig. 3. The flow direction of data transmission is depicted in Fig. 4.

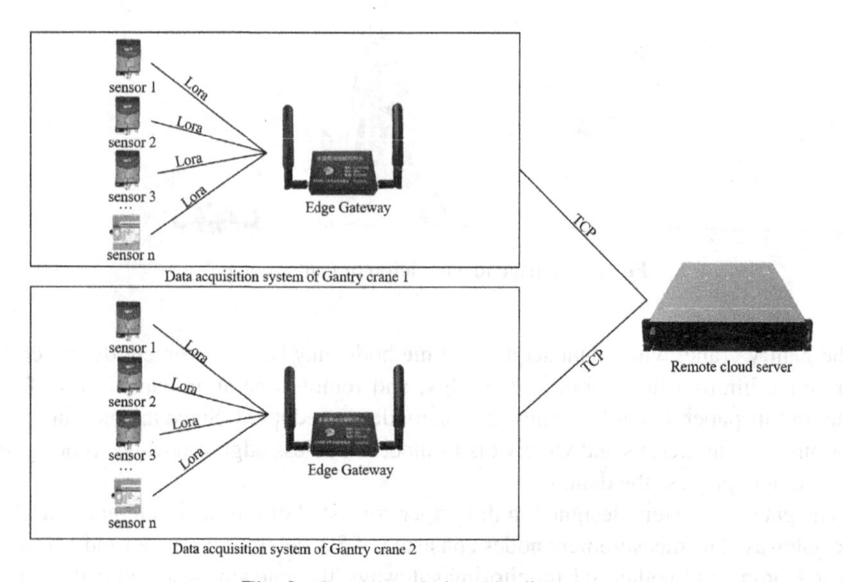

Fig. 3. Gateway relationship diagram.

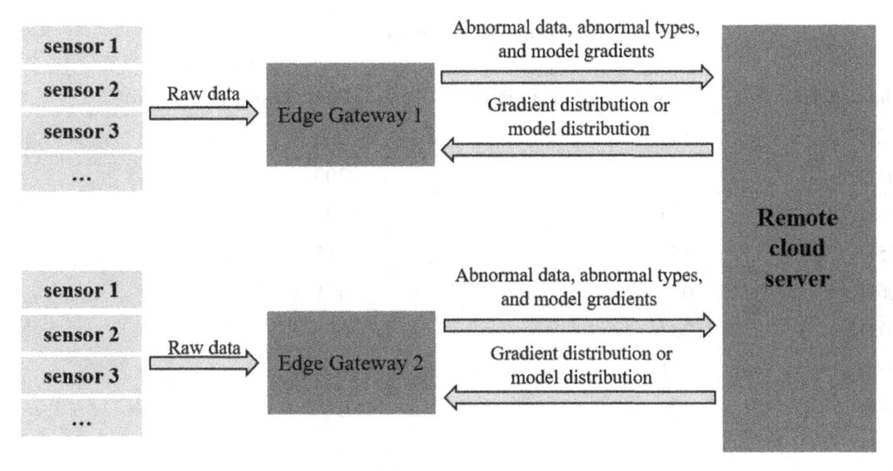

Fig. 4. Data flow diagram.

2.3 Design and Measurement of Load Excitation

In this paper, the load of the gantry crane was used as a self-excited source. To be more representative of real-world scenarios, the loads used had different magnitudes. The numerical values of the load magnitude were measured by the tension sensor on the traction rope. When the object under test was subjected to traction and underwent slight up-and-down stretching at random speeds, the values of the tension sensor changed, thereby initiating the measurement and monitoring process.

Given that the small gantry crane used in this paper was a scalable crane, in order to simulate data with non-independent and unbalanced characteristics, this paper adjusted the height of the gantry crane to simulate differences in usage scenarios and habits of similar cranes. Moreover, there were significant differences between the selected heights, and the height values were determined randomly.

The amount of data for various scenarios in this paper is presented in Table 2. In order to ensure greater diversity in the dataset, the types of damage for the two different heights were not the same, meaning that a certain height lacked a certain type of damage. The actual measurement curves are shown in Fig. 5.

Table 2. Quantity of each status data.

Injury types	Number of data at height 1	Number of data at height 2
Training set	4213	3293
01H	200	601
01D	600	997

(continued)

Table 2. (*continued*)

Injury types	Number of data at height 1	Number of data at height 2
02D	543	996
03D	659	700
04D	774	394
05D	553	515
06D	0	471
07D	233	0
08D	311	0
09D	200	0
10D	336	0

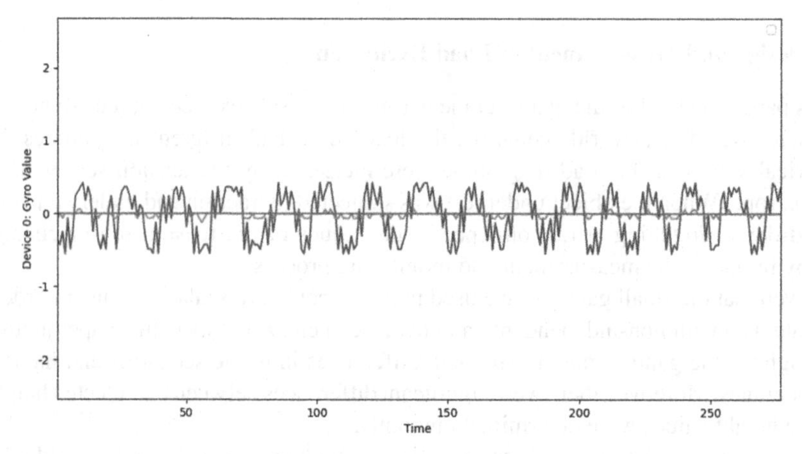

Fig. 5. Actual angular velocity curve.

3 Federated Learning Algorithm for Fault Identification of Gantry Crane

3.1 Algorithm Overview

In real-life situations, the structure of gantry cranes is mostly in a normal state, with minimal abnormal data [7]. If a supervised classification algorithm is to be used to distinguish different types of anomalies, a balanced dataset is required for training [8]. This means that when collecting or supplementing data, a large amount of normal data needs to be filtered. However, since the data is unlabeled, unsupervised anomaly detection algorithms need to be used for data filtering. The unsupervised anomaly detection algorithm we adopted is USAD. This algorithm can perform well even when the data is unlabelled or imbalanced.

In addition, this paper utilized XGBoost [12] for classifying abnormal data. Through testing, it was found that running XGBoost directly on the edge gateway consumed a considerable amount of memory. Therefore, using this algorithm on a large amount of normal data would have resulted in significant performance waste. As a result, XGBoost was only employed on the edge gateway when anomalies occurred.

As mentioned earlier, existing algorithms could only monitor specific structures after training. In order to obtain a unified and general model that could be used for other similar gantry cranes, this paper utilized the federated learning algorithm. The relevant algorithms were deployed on the edge gateway to ensure real-time performance. However, federated learning can amplify the problem of non-independent and identically distributed data, which can have a significant impact on federated modeling [9]. To mitigate this issue, this paper employed the FedAvg [10] in federated learning to alleviate the related effects.

The overall flowchart of the algorithm is shown in Fig. 6. First, the data is input into the unsupervised anomaly detection algorithm USAD [11] to detect abnormal data. Normal data is saved as a self-updating dataset within the edge gateway. Since the USAD algorithm is an unsupervised anomaly detection algorithm, these normal datasets can be used for self-updating. After several rounds of self-updating, the Federated Averaging algorithm (FedAvg) is used to aggregate the models from multiple edge gateways within the server, and at the appropriate time, they are deployed back to each respective gateway to obtain a unified anomaly detection model.

Next, the XGBoost algorithm is used to classify the abnormal data, and XGBoost is only invoked within the edge gateway when an anomaly occurs to save resources. This study assumes that there is already a significant amount of abnormal data within the server. When the edge gateway detects abnormal data, it sends the abnormal data to the server to augment the abnormal dataset. The server periodically updates the XGBoost

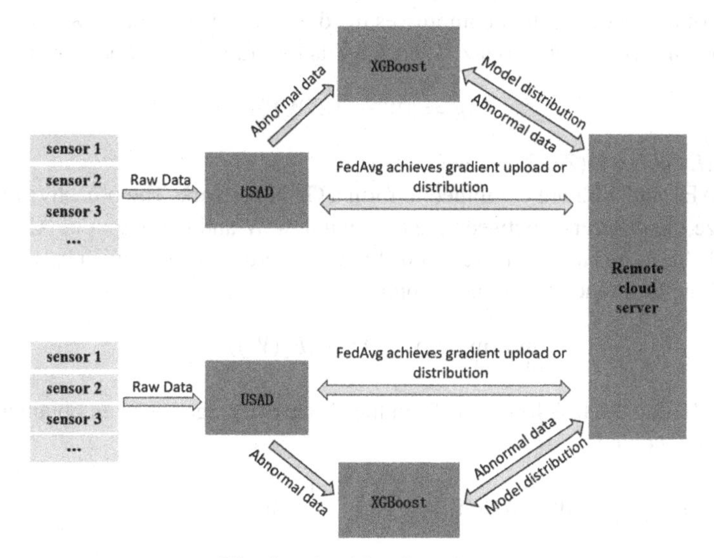

Fig. 6. Algorithm flow chart.

algorithm using the newly augmented dataset and deploys the updated model back to the edge gateways.

3.2 Unsupervised Neural Network USAD Algorithm

When the data from the gantry crane is transmitted from the wireless gateway to the edge device, the edge device will perform detection on the uploaded data. In this paper, unsupervised anomaly detection is achieved using a neural network called USAD (Unsupervised Anomaly Detection on Multivariate TimeSeries).

USAD is a combination of Generative Adversarial Network (GAN) [13] and Autoencoder (AE) [14]. This algorithm utilizes the power of GANs for generating realistic data and the reconstruction capability of autoencoders to detect anomalies in multivariate time series data. By training the USAD model on normal data, it can learn to capture the normal patterns and identify any deviations or anomalies in the incoming data stream.

The collected data mentioned above is multidimensional and can be represented as follows: $\tau = \{x_1, ..., x_T\}$. x_T represents the data generated by all sensors at a specific node at time T. In the USAD algorithm, to simulate the relationship between the current time point and previous time points, a time window W_t is defined. It represents a time series of length K at time t as $W = \{W_1, ..., W_T\}$. If a time window is inputted into the USAD algorithm, an anomaly score can be obtained. For known normal data, this paper establishes a threshold based on the algorithm scores. When unknown time window sequences are inputted, an anomaly score for that time point is obtained. If the score is greater than the threshold, it is considered as an anomaly at that time point.

The USAD network consists of three components: an encoder network E and two decoder networks D1 and D2. The encoder network and the two decoders form AE1 and AE2 respectively. Firstly, the two AE networks are trained to achieve the capability of reconstructing the original window data. Based on the structure of the AE network, it is composed of an encoder E that compresses the data into a latent space z, and a decoder D that reconstructs the data from z. Therefore, its loss function can be represented as:

$$L_{AE} = \|W - AE(W)\|_2 \tag{1}$$

wherein: $AE(W) = D(Z), Z = E(W)$。

Next, AE1 and AE2 are combined to form a GAN network. The objective of AE1 is to minimize the difference between the time window W and the output of AE2. On the other hand, AE2 aims to maximize this difference in order to distinguish between them. Therefore, the loss function for the second stage is:

$$\min_{AE1} \max_{AE2} \|W - AE_2(AE_1(W))\|_2 \tag{2}$$

By combining the loss functions from the above two stages, we can obtain a unified loss function, which is:

$$L_{AE_1} = \frac{1}{n}\|W - AE_1(W)\|_2 + (1 - \frac{1}{n})\|W - AE_2(AE_1(W))\|_2 \tag{3}$$

$$L_{AE_2} = \frac{1}{n}\|W - AE_2(W)\|_2 - (1 - \frac{1}{n})\|W - AE_2(AE_1(W))\|_2 \tag{4}$$

where n represents the training cycle.

In the detection stage, the formula for obtaining the anomaly score is:

$$\Psi(W) = \alpha\|W - AE_1(W)\|_2 - \beta\|W - AE_2(AE_1(W))\|_2 \tag{5}$$

where in: $\alpha + \beta = 1$.

3.3 FedAvg

For different regions, the environments of similar gantry cranes often exhibit differences, resulting in datasets that are non-iid (non-independent and identically distributed) and unbalanced. Common deep learning algorithms assume that the training and testing datasets are both iid and balanced, meaning that the differences between data should not be too significant. If these algorithms are directly applied to such datasets, it will greatly affect the generalization capability of the models.

The paper adopts the FedAvg. In the FedAvg, the server first initializes the global model parameters W_0. Then, all local devices will randomly select a portion of the dataset and compute the local model parameters W locally. And upload the parameters to the server. The server will perform multiple rounds of federated aggregation, denoted by the symbol t. The aggregation process is as follows: firstly, randomly select a set of devices, which can be represented as a set denoted by S_t. The model parameters on a certain device k are denoted as W_{t+1}^k. Then, the server will perform weighted averaging on these parameters:

$$\omega_{t+1} = \sum\nolimits_{k=1}^{K} \frac{\mathrm{n}_k}{n} \omega_{t+1}^k \tag{6}$$

After aggregating the parameters for the current round, the server will multicast the updated parameters to the local devices. Upon receiving the new parameters, the local devices will update their respective local models accordingly.

3.4 XGBoost

This paper utilizes the XGBoost for anomaly classification of data. It is an algorithm adapted from the Gradient Boosting Decision Tree (GBDT) algorithm. The overall ideas of both algorithms are similar, but XGBoost optimizes certain aspects of the GBDT algorithm.

The objective function of XGBoost consists of two components: the loss function and the regularization term. If the training set $\mathrm{T} = \{(x_1, y_1), (x_2, y_2), \cdots, (x_n, y_n)\}$ is given and the loss function is represented as $l(y_i, \hat{y}_i)$, with the regularization term denoted as $\Omega(f_k)$, then the overall objective function can be expressed as:

$$\mathcal{L}(\phi) = \sum_i l(y_i, \hat{y}_i) + \sum_k \Omega(f_k) \tag{7}$$

In this context,$\mathcal{L}(\phi)$ represents the linear space expression, i represents the i-th sample, $\hat{y}_i$ represents the prediction value of the i-th sample by the t-th tree.

Due to $\hat{y}_i = \sum_{i=1}^{n} f_k(x_i) = \hat{y}_i^{(t-1)} + f_t(x_i)$, $\mathcal{L}(\phi)$ can be transformed as

$$\mathcal{L}(\phi) = \sum_i l(y_i, \hat{y}_i^{(t-1)} + f_t(x_i)) + \sum_k \Omega(f_k) \tag{8}$$

The objective function can be optimized using the following steps. First, expand the equation using the second-order Taylor series approximation and remove the constant term to optimize the loss function. Next, expand the regularization term and remove the constant term to optimize the regularization term. Finally, combine the coefficients of the linear and quadratic terms to obtain the final objective function:

$$\mathcal{L}^{(t)} = \sum_{j=1}^{T} [g_i f_t(x_i) + \frac{h_i}{2} f_t^2(x_i)] + \sum_k \Omega(f_k) \tag{9}$$

4 Experiment

4.1 Federated Learning Based Anomaly Detection

This paper selected two different types of gantry cranes with varying heights and conducted the experimental design according to the following steps:

Anomaly Detection Was Performed Separately on Gantry Cranes of Different Heights. In this paper, we first conducted anomaly detection training for height 1 by inputting normal height 1 data into the model for training. After completion, corresponding anomaly scores could be obtained on the test set, and the ROC curve is shown in (a) of Fig. 7 The F1 score for height 1 was 0.978. We performed the same experiment for height 2, and the results shown in (b) of Fig. 7 were obtained. The F1 score for height 2 was 0.952.

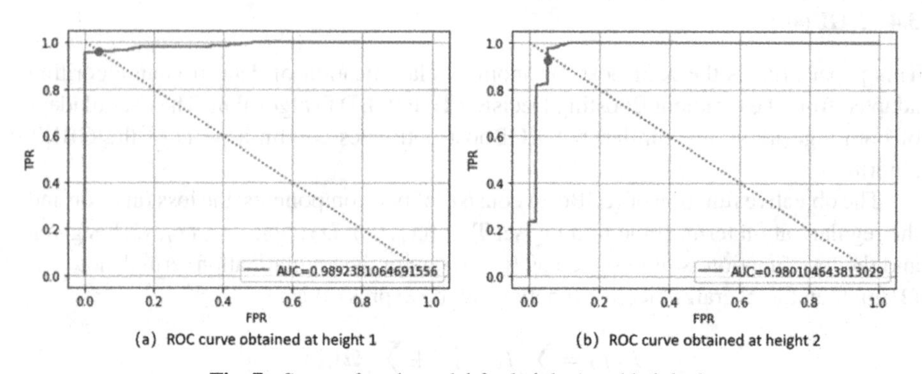

Fig. 7. Score of each model for height 1 and height 2.

The Datasets of Gantry Cranes With Different Heights Were Concatenated, and Anomaly Detection was Performed. The two datasets mentioned above were concatenated to simulate the process of uploading data from various gantry cranes to a centralized location and training a universal model. The results for height 1 and height 2 are shown in Fig. 8.

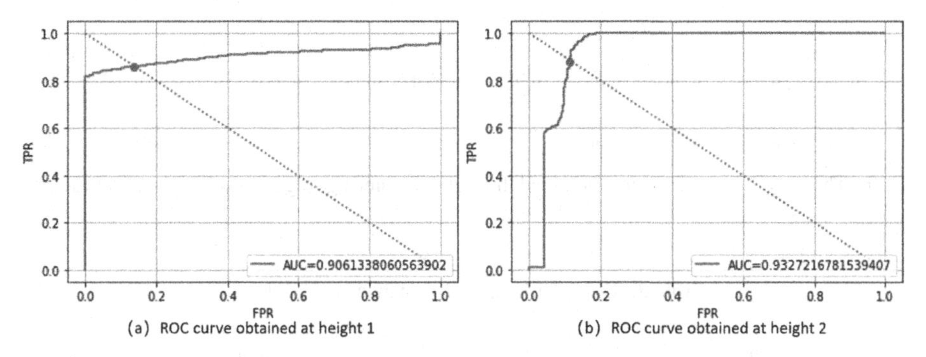

Fig. 8. Overall model scores for height 1 and height 2.

The F1 score for height 1 was 0.922. Due to the presence of too many non-independent and identically distributed data points in the training set, as well as the imbalance in the data quantity, the performance of the model was somewhat reduced. The same experiment was conducted for height 2, resulting in an F1 score of 0.928, indicating a slight decrease in performance as well.

From the above results, it can be observed that training the model after concatenating the datasets leads to a decrease in performance for both heights due to the non-independent and identically distributed nature as well as the data imbalance.

Anomaly Detection was Performed Using the FedAvg. To address the aforementioned issues, this paper employed the FedAvg. As mentioned earlier, the FedAvg can mitigate problems such as non-independent and identically distributed data and data imbalance [15]. Moreover, for similar gantry cranes, the non-independent and identically distributed phenomenon was relatively mild. Therefore, following the steps of the FedAvg, the two datasets were separately trained in this paper, with a total of 5 rounds of training. After each training round, a federated aggregation was performed where the parameters of the two models were sent to the server, aggregated, and then distributed back to their respective models [16]. After 40 rounds of federated aggregation, the corresponding results were obtained. Figure 14 displays the results obtained. The F1 score for height 1 is 0.98, while the F1 score for height 2 is 0.97.

From Fig. 9, it can be observed that using the FedAvg restores the performance to its original state compared to dataset concatenation. It can be inferred that the FedAvg is effective. Tables 3 and 4 show the performance comparison of the three different methods at height 1 and height 2.

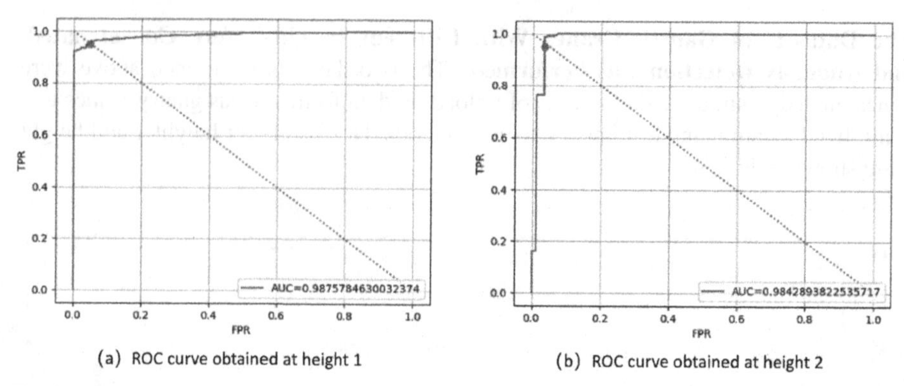

Fig. 9. The overall model's scores obtained through federated aggregation at Height 1 and Height 2.

Table 3. Comparing the performance at Height 1.

Method	AUC	F1 score
Only the dataset for Height 1	0.988	0.978
Dataset concatenation	0.906	0.922
Using FedAvg	0.987	0.970

Table 4. Comparing the performance at Height 2.

Method	AUC	F1 score
Only the dataset for Height 2	0.981	0.952
Dataset concatenation	0.932	0.928
Using FedAvg	0.984	0.973

4.2 Abnormal Classification of Gantry Cranes

After abnormal data was detected at the edge-side gateway, the abnormal data could be inputted into the abnormal classification model located on the same edge gateway. In this paper, the XGBoost was used for classifying abnormal states.

In this paper, a performance comparison was conducted under the same conditions. On an edge gateway without a graphics card, USAD had a peak memory usage of 1175.7783 KiB, while XGBoost had a peak memory usage of 15767.5049 KiB. The memory difference was about 13 times. If a significant amount of memory was occupied for a long time, it would result in considerable waste.

During the training process of the XGBoost algorithm, the abnormal data at different heights mentioned earlier in the text were divided into a training set and a testing set in a ratio of 3:1. A portion of the training set was then fed into the XGBoost model.

Table 5. Different performance metrics of XGBoost for different classes.

Types of damages	Accuracy	Recall	F1 score
02D	0.99	0.96	0.97
01D	0.97	0.98	0.98
03D	0.99	0.96	0.97
04D	0.98	0.98	0.98
05D	0.95	0.98	0.97
06D	0.96	0.96	0.96
07D	0.93	0.99	0.96
08D	0.97	0.97	0.97
09D	0.96	0.95	0.95
10D	0.94	0.97	0.97
Total	0.964	0.969	0.966

After training the XGBoost model, the testing set was inputted to evaluate the model's various performance metrics, as shown in Table 5.

5 Conclusion

For the problem of structural health monitoring of similar gantry crane structures, we validated the effectiveness of an unsupervised anomaly detection algorithm based on federated learning and the XGBoost algorithm. Firstly, a small-scale gantry crane was utilized to simulate structural damages, and an IoT gateway was designed to measure the crane's state during vibrations. We used Unsupervised Anomaly Detection (USAD) algorithm was employed to check if the data was abnormal. Normal data was retained for local self-updating. After a certain number of self-updates, in order to alleviate the non-independent and identically distributed (non-IID) and imbalanced data issues among similar gantry cranes at different heights, federated aggregation was performed several times. This helped mitigate the non-IID phenomenon and imbalanced data, resulting in a robust and generalizable model. For abnormal data, the XGBoost was used for classification by the edge gateway. Assuming that the server had sufficient labeled abnormal data, the pre-trained XGBoost model was utilized to detect anomalies in the data and identify the type of damage present in the gantry crane.

Acknowledgments. This work was supported by the National Key Research and Development Project (2019YFB2101902).

References

1. Li, Y., Xu, F.: Acoustic emission and moving window-improved kernel entropy component analysis for structural condition monitoring of hoisting machinery under various working conditions. Struct. Health Monitor. **21**(4), 1407–1431 (2022)

2. Liu, J., Qin, X., Sun, Y.: Response prediction model for structures of quayside container crane based on monitoring data. J. Perform. Construct. Facil. **35**(4), 1943–1955 (2021)
3. Ranjan, N., Bhandari, S., Hong, Y.-S.: Convolutional neural network based power crane defect analysis. In: IEEE Region 10 Symposium, pp. 1–4. IEEE, Jeju, Korea (2021)
4. Yede, S., Kumar, S., Nimbalkar, M.: Automatic lifting system for crane safety using machine learning. In: 2021 IEEE India Council International Subsections Conference, 2021, pp. 1–4. IEEE, NAGPUR, India (2021)
5. Wu, D.: Empowering smart cities with edge computing: opportunities and challenges. People's Forum·Academic Frontier **193**(09), 18–25 (2020)
6. Yao, X., Zheng, J., Zheng, X.: Joint Optimization of UAV Trajectory and Resources for Federated Learning. Computer Engineering and Applications. http://kns.cnki.net/kcms/detail/11.2127.TP.20230509.1840.007.html. Accessed 29 June 2023
7. Sun, L., Shang, Z., Xia, Y.: Research status and prospects of bridge structural health monitoring under the background of big data. China J. Highway Transp. **32**(11), 1–20 (2019)
8. Wu, H., Chen, X., Fan, G.: An adaptive kernel SMOTE-SVM algorithm for imbalanced data classification. J. Beijing Univ. Chem. Technol. (Natl. Sci. Edn.) **50**(02), 97–104 (2023)
9. Tan, R., Hong, Z., Yu, W.: Decentralized federated learning strategy for non-independent and identically distributed data. J. Comput. Eng. Appl. **59**(01), 269–277 (2023)
10. McMahan, B., Moore, E., Ramage, D.: Communication-efficient learning of deep networks from decentralized data. In: Proceedings of the 20th International Conference on Artificial Intelligence and Statistics, pp. 1273–1282. JMLR, Ft. Lauderdale, FL, USA (2017)
11. Audibert, J., Michiardi, P., Guyard, F.: USAD: unsupervised anomaly detection on multivariate time series. In: Proceedings of the 26th ACM SIGKDD International Conference on Knowledge Discovery & Data Mining, pp. 3395–3404. Association for Computing Machinery, New York, NY, USA (2020)
12. Chen, T., Guestrin, C.: Xgboost: a scalable tree boosting system. In: Proceedings of the 22nd ACM SIGKDD International Conference on Knowledge Discovery and Data Mining, pp. 785–794. Association for Computing Machinery, San Francisco, USA (2016)
13. Goodfellow, I.J., Pouget-Abadie, J., Mirza, M.: Generative adversarial nets. In: Proceedings of the 27th International Conference on Neural Information Processing Systems, pp. 2672–2680. Association for Computing Machinery, Cambridge, MA, USA (2014)
14. Rumelhart, D.E., Hinton, G.E., Williams, R.J.: Learning internal representations by error propagation. Read. Cogn. Sci. **323**(6088), 399–421 (1988)
15. Li, X. Huang, K., Yang, W.: On the convergence of FedAvg on Non-IID data. In: 8th International Conference on Learning Representations, pp. 26–30. OpenReview.net, Addis Ababa, Ethiopia (2020)
16. Liu, B., Zhang, F., Wang, W.: Byzantine robust Federated learning algorithm based on matrix mapping. Comput. Res. Develop. **58**(11), 2416–2429 (2021)

Accelerated Lifetime Experiment of Maximum Current Ratio Based on Charge and Discharge Capacity Confinement

Baoji Wang[1], Boyan Li[2], Qixuan Wang[4], and Lei Dong[2,3](✉)

[1] Gree Altairnano New Energy INC, Zhuhai 519000, China
[2] School of Automation, Beijing Institute of Technology, Beijing, China
leidong@bit.edu.cn
[3] Tangshan Research Institute of Technology, Tangshan, China
[4] School of Software Engineering, Beijing University of Technology, Beijing 100124, China

Abstract. Lithium-ion batteries will undergo continuous aging during the process of charging and discharging. Charging and discharging cycle conditions for lithium-ion batteries are usually an important method to detect the degree of aging of lithium-ion batteries. However, the general aging cycle lasts for thousands of cycles, which is more time consuming. The charge/discharge capacity of the original condition and the accelerated condition are guaranteed to be the same, and the aging path is essentially the same way. The accelerated life test of lithium-ion battery is realized by the constant current rate accelerated operating condition design and the variable current rate accelerated operating condition design with two different constraints, and the accelerated operating condition with the minimum difference between the battery aging path of the original operating condition before acceleration is obtained by the optimization method.

Keywords: Lithium battery · accelerated life test · charge confinement

1 Introduction

Lithium-ion batteries are significantly better than other commonly used batteries in terms of specific energy, specific power and cycle life. Therefore, it has a wide range of applications in the field of intelligent devices and electric vehicles. The aging of lithium-ion batteries will make the consistency of the battery worse, which will cause serious problems for the performance and even safety of the battery and energy storage system.

Aging is an inherent characteristic of lithium-ion power batteries, and its mechanism is complex and depends on the actual operating conditions. In order to verify the energy attenuation caused by aging in long-term use of lithium-ion power batteries, it is necessary to conduct life tests. At present, the lifespan of lithium-ion power batteries is generally several thousand cycles, making conventional lifespan testing time-consuming. It is necessary to develop test methods to accelerate the life of lithium-ion power battery according to its dominant aging mechanism and influencing factors. The main factors

B. Xin et al. (Eds.): IWACIII 2023, CCIS 1931, pp. 89–100, 2024.
https://doi.org/10.1007/978-981-99-7590-7_8

[1–3] affecting the lifespan of lithium-ion power battery include: time (cycle times), temperature, charge and discharge current ratio, state of charge (SOC), etc. These factors are mutually coupled and influence each other.

The aging of lithium-ion batteries is path-dependent [4, 5]. The aging path refers to a certain path that the internal dominant aging mechanism of the lithium-ion power battery passes through in the battery attenuation. Due to the mutual coupling of various influencing factors during the aging process, the superposition of different influencing factors will lead to different attenuation laws of each mechanism parameter, so it is necessary to ensure that the aging paths of each mechanism parameter are the same. The design of the accelerated lifespan conditions should be based on the actual operating conditions of the lithium-ion power battery and its actual aging path of the lithium-ion power battery under real operating conditions. Within the allowable range of error in the actual aging path, the actual operating conditions are simplified or optimized according to certain design principles, so as to achieve the purpose of accelerated lifespan test.

Table 1. Summary of the battery aging model

Type of model	Author	Key formula of the model
Empirical model	Bloom I [6]	$Q_{loss} = A\exp(-\frac{E_a}{RT})t^z$
	Wang J [7]	$Q_{loss} = A\exp(-\frac{E_a+BI}{RT})Ah^z$
	Suri G [8]	$Q_{loss} = (\alpha SOC + \beta)\exp(-\frac{E_a+\eta I}{RT})Ah^z$
Mechanism model	Ramadass P [9]	$J_{SEI} = k_{SEI}\exp\left(\frac{\alpha F}{RT}(\phi_s - \phi_e - U_{SEI} - jR_{\text{film}})\right)$
	Lawder M T [10]	$J_{SEI} = -k_{SEI}c_s\exp\left(\frac{\alpha F}{RT}\left(\phi_s - \phi_e - \frac{\delta}{\kappa_{SEI}}j\right)\right)$
	Atalay S [11]	$J_{SEI} = -Fk_0c_s\exp\left(\frac{\alpha_{e,SEI}F}{RT}\eta_{SEI}\right)$ $J_{Li} = -i_0\exp\left(\frac{\alpha_{e,Li}F}{RT}\eta_{Li}\right)$ $\eta_{SEI} = \phi_s - \phi_e - jR_{\text{film}} - U_{SEI}$

Currently, the aging models of battery can be divided into empirical models and mechanism models. The empirical models generally aims to fit the battery capacity attenuation or the battery internal resistance increase. It can consider multiple main reasons that affect the aging of lithium-ion batteries, such as time t, temperature T, amp-hour throughput Ah and SOC. The advantage of empirical models lies in its small amount of calculation, but they do not consider the internal mechanisms of the battery, so its estimation of the battery aging state only includes the estimation of the external characteristics of the battery, resulting in its incomplete battery life. The mechanism model is generally based on P2D (Pseudo Two Dimensional) model. On the basis of the physical-chemical process of the reaction battery in a short time scale, the battery side reaction equations leading to battery aging are coupled to characterize, so as to establish the aging mechanism model considering the battery side reaction. The empirical and mechanistic models commonly used in existing studies are shown in Table 1.

2 Principle of Maximum Current Rate Acceleration Life Experiment

The principle of impulse equivalence, the principle of area equivalence, is used in the maximum rate acceleration life experiment. In the operating condition curve, the area enclosed by the current curve and the coordinate axis is the charge throughput Q. According to the principle of charge conservation, the area enclosed by the original operating condition and the operating condition accelerated by the maximum current rate always represents the charge throughput in this period, as shown in Fig. 1. Thus the principle of charge conservation is followed.

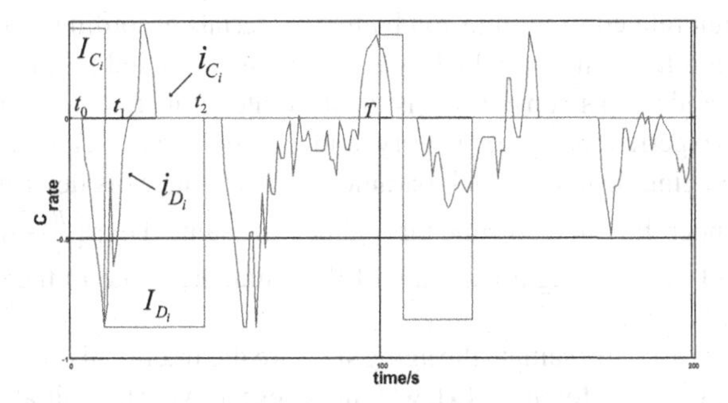

Fig. 1. The maximum current rate acceleration in the original operating condition

As shown in Fig. 1, the original operating condition is divided into sections according to particle size D. In each section, the original operating condition and the accelerating operating condition follow the charge conservation principle of Eq. (1). Since the maximum current magnification of the accelerating operating condition is larger than that of the original operating condition, the accelerating operating condition can have a larger magnification acceleration compared with the original operating condition in terms of time.

$$
\begin{aligned}
Q_{C_i} &= \int_{t_0}^{t_1} I_{C_i} dt = \int_{t_0}^{T} i_{C_i} dt \\
Q_{D_i} &= \int_{t_1}^{t_2} I_{D_i} dt = \int_{t_0}^{T} i_{D_i} dt
\end{aligned} \tag{1}
$$

where Q_{C_i} and Q_{D_i} respectively represent the charging capacity and discharge capacity in a certain segment, i_{C_i} and i_{D_i} are the charging and discharge current in the original operating condition, I_{C_i} and I_{D_i} are the maximum charging and discharge current ratio in the accelerated operating condition, T is the time length of the original operating condition, and t is the time length of the accelerated operating condition.

This acceleration method is suitable for scenarios where simple calculable variables are used as evaluation criteria. Taking the acceleration condition studied at present as an example, the function used to evaluate the degree of aging in the acceleration condition is an empirical formula derived from Arrhenius formula. By fitting the influence of current ratio C_{Rate} on activation energy E, the experimental staff [12] summarized the following formula. From a physical point of view, the high current increases the activation molecules, raising the average energy of all reactant molecules and thereby increasing the reaction rate..

$$Q_{i,loss} = A_i \exp\left(\frac{B_i * |C_{Rate}|^{m_i} - E_i}{RT}\right) Ah^z \quad (2)$$

where, i is the mechanism parameter in the electrochemical model, and $Q_{i,loss}$ is the attenuation of the mechanism parameter i; A_i is the Arrhenius formula referring to the prefactor [13], is also called Arrhenius constant; E_i is the activation energy; $|C_{Rate}|$ is the absolute value of the current multiplier. B_i is the magnification factor coefficient, which is a constant related to lithium-ion battery materials according to Paris's law; m_i is the magnification factor index, which is also a constant related to lithium-ion battery materials; R is the ideal gas constant; T is the absolute temperature; Ah is the ampere-hour throughput experienced by the battery, and z is its index. It can be seen from the equation that the aging degree of each parameter of the battery is determined by C_{Rate} and Ah, that is, under the same on-time throughput, the same $A \exp\left(\frac{B*|C_{Rate}|^m - E}{RT}\right)$ value is guaranteed, and the same aging amount of the parameter state in the current period can be guaranteed.

The parameters used to evaluate the aging state are the fractional order model parameters proposed by Guo in reference [14], which are respectively the initial charge state of positive and negative electrode SOC_p^0, SOC_n^0, the positive and negative electrode capacity Q_p, Q_n, and the positive and negative diffusion constant τ_p, τ_n. The empirical formula parameters corresponding to each parameter [15] are shown in Table 2.

Table 2. Parameter values of empirical formulas for each physical quantity

Physical parameters	A_i	E_i	B_i	m_i
Q_p	671.1	4.113e04	4.187	4
Q_n	545.7	4.225e04	17.94	3
SOC_p^0	438.7	4.180e04	502.3	1
SOC_n^0	635.3	4.133e04	4.187	1
τ_p	896.6	3.980e04	17.94	4
τ_n	650.4	3.507e04	502.5	3

3 Constant Current Rate Acceleration

In order to simulate the aging state of the battery in actual use, the experiment adopted 1800s FUDS [16](Federal Urban Driving Schedule) to verify the effectiveness of the accelerated condition design.

Firstly, accelerated verification is conducted using the constant maximum current rate. As shown in Fig. 2, the original operating condition is accelerated by using a constant rate of 1.2C. It can be observed that the time required for the acceleration operating condition is greatly reduced compared with the original operating condition. However, as shown in Fig. 3, there are errors in each mechanism parameter along the aging path.

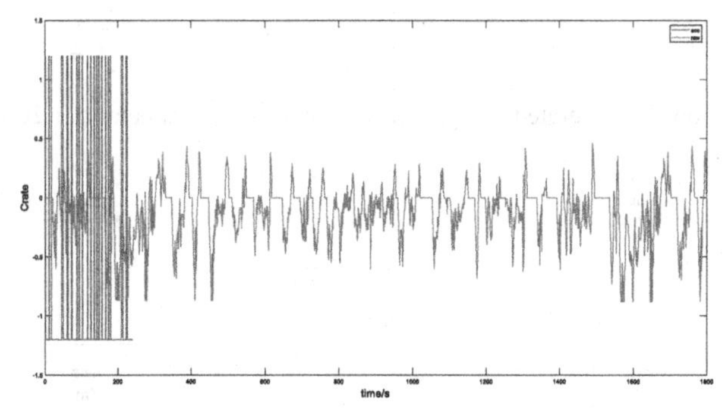

Fig. 2. Comparison between the original and accelerated conditions of 1.2C constant rate acceleration

In addition to comparing the aging path of each parameter, the difference between the original operating condition and the accelerated operating condition can also be seen by comparing the aging rate of each parameter. The aging rate corresponding to each parameter can be obtained from Eq. (2)

$$k_{i,loss} = A_i \exp\left(\frac{B_i * |C_{Rate}|^{m_i} - E_i}{RT}\right) \tag{3}$$

where $k_{i,loss}$ is the aging rate of each parameter, the meaning of other variables is the same as that of Eq. (2).

It can be seen from Fig. 4 that the speed of the accelerated condition is always larger than that of the original condition, so it will lead to differences in the final aging path. Moreover, due to the different sensitivity of each parameter to the current, the aging rate can be made closer to the original aging rate by changing the current rate.

By changing the current rate to 0.8C and comparing the accelerated aging path with the original aging path, the aging path of the accelerated operating condition after the current rate is changed to 0.8C is less different from the aging path of the original operating condition. Therefore, the aging error can be reduced by changing the current rate. However, similarly, if the aging rate is too small, the aging path error will also increase if the difference between the rate and the original operating condition is large (Fig. 5).

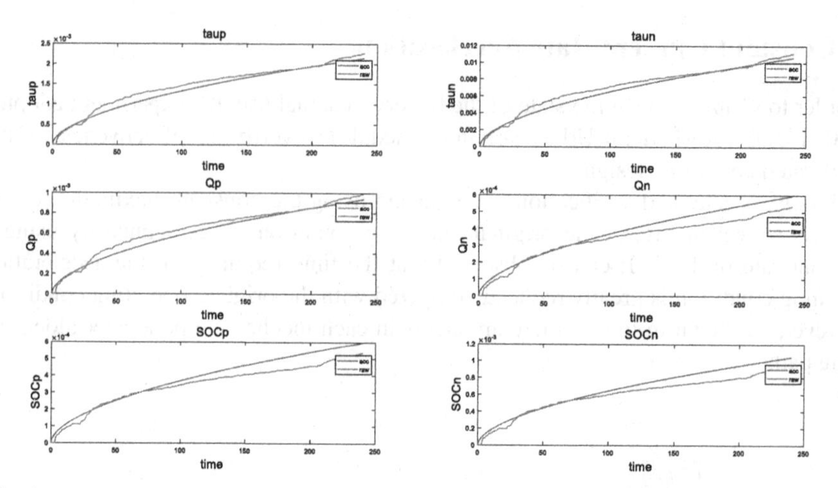

Fig. 3. Comparison of accelerated aging paths for each aging parameter at 1.2C constant rate

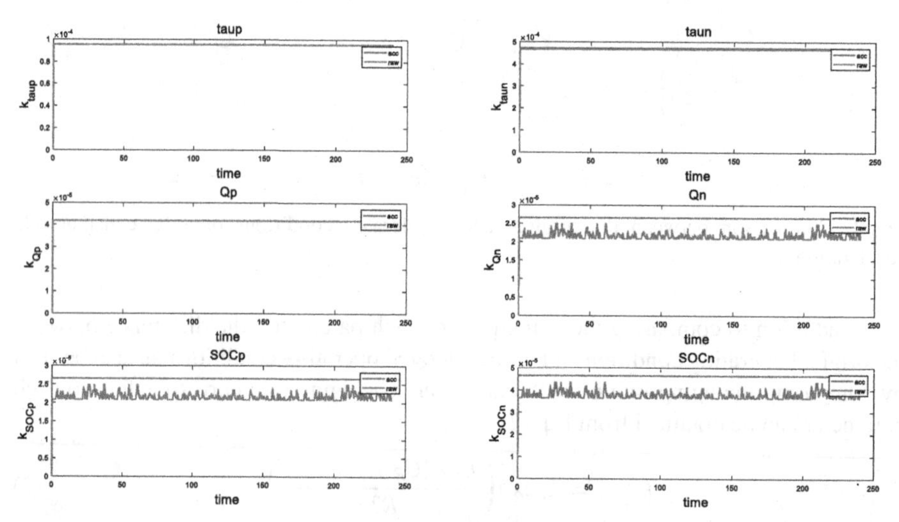

Fig. 4. Comparison of aging rate of each physical parameter accelerated by 1.2C constant rate

Therefore, under the premise of constant rate current acceleration, the following figure is obtained by optimizing the current rate. It can be seen that the error is minimized within the range of 0.47–0.48C. Considering the acceleration time, we choose the constant rate of 0.48C to accelerate the operating condition. The comparison between the original and accelerated conditions after acceleration is shown in Fig. 6, and the comparison of each aging parameter is shown in Fig. 7.

It can be seen that although the original operating condition of this acceleration method is close to the current operating condition at the beginning and end of the operating condition, the aging path of the accelerated operating condition in the intermediate stage is quite different from that of the original operating condition, and the accelerated

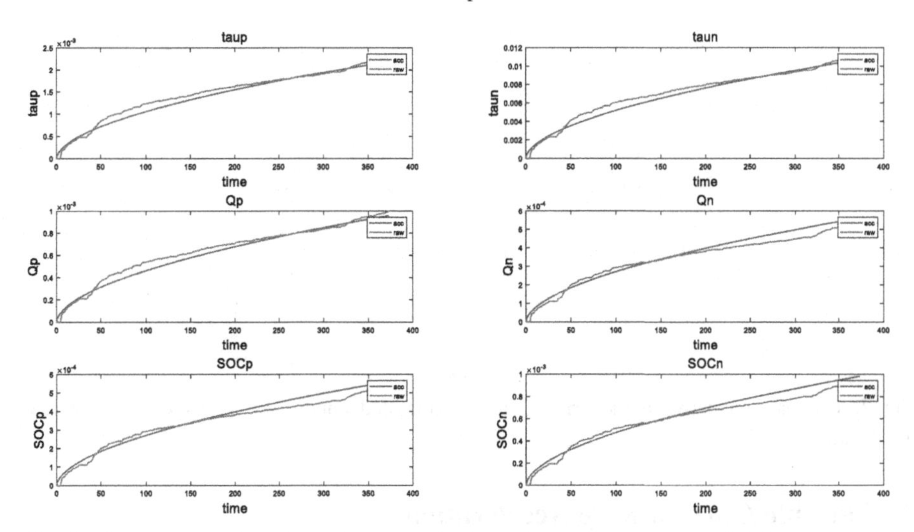

Fig. 5. Comparison of accelerated aging paths for each aging parameter at 0.8C constant rate

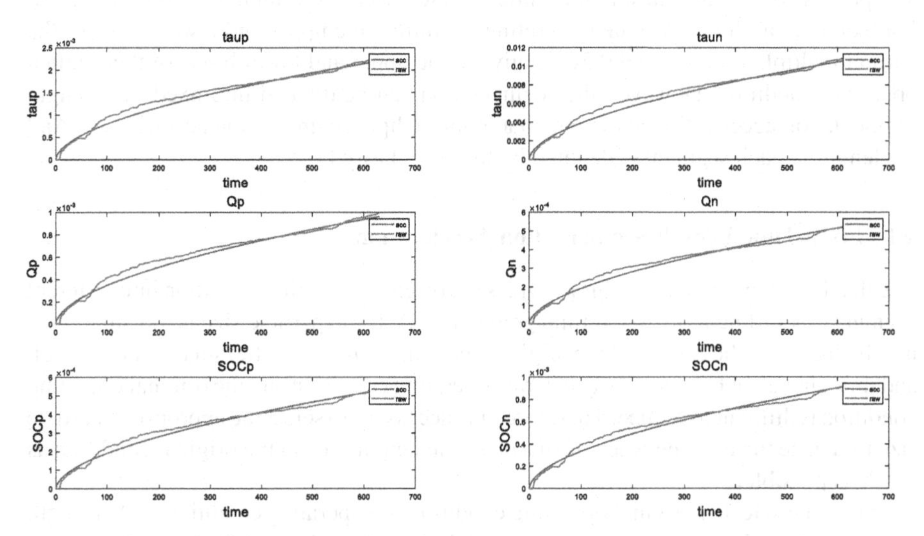

Fig. 6. Comparison of accelerated aging paths for each aging parameter at 0.8C constant rate

operating condition of the battery should follow the principle of the same acceleration path as far as possible. Therefore, the original method is optimized on this basis, that is, the original operating condition is segmented to conform to Eq. (4). The aging amount of the original condition corresponding to each segment on the accelerated condition should be as equal as possible to the original condition.

$$\Delta Q_{acc\_loss} = \Delta Q_{raw\_loss} \tag{4}$$

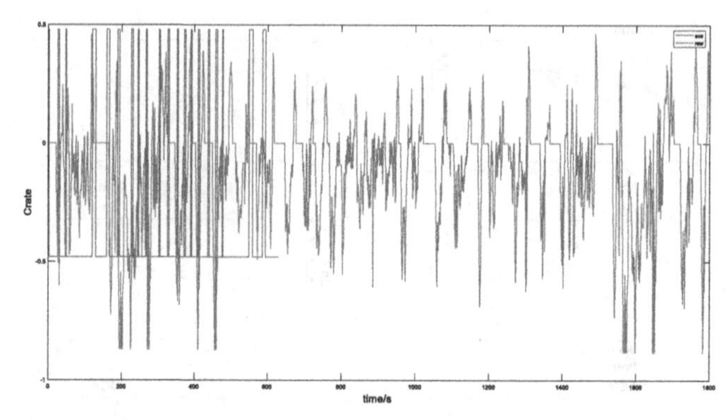

Fig. 7. Comparison between the original and accelerated conditions of 0.48C constant rate acceleration

4 Variable Current Rate Acceleration

The premise of variable current multiplier acceleration is to limit the current multiplier. For example, in this accelerated operating condition, the upper and lower limits of the current multiplier are set to ±0.2C relative to the upper and lower limits of the original operating condition. However, the segmentation can be divided into fixed time length segmentation acceleration and fixed charge throughput segmentation acceleration. And the length of each segment is denoted by the granularity D.

4.1 Fixed Time Length Segmentation Acceleration

The fixed time length segmentation divides the original operating condition into different segments according to the fixed time granularity D. If the particle size is too small, it is highly likely that the current of the original operating condition is too small in an interval, and the aging error between the accelerate operating condition and the original operating condition is difficult to reduce. Therefore, it is necessary to select the appropriate particle size to reduce the error between the acceleration condition and the original condition as much as possible.

Since the selected original operating condition is a operating condition with a length of 1800s, the selected optimization granularity is a factor of 1800, that is, from 90s, 100s. In addition, the MAPE value of the aging path curve of the original operating condition and the aging path curve of the accelerated operating condition were used as the objective function. The obtained results are shown in Fig. 8.

It can be seen that when the particle size is 180s, the MAPE of the two aging curves is the smallest, so this particle size is taken to generate the accelerated operating condition. After accelerated in this way, the FUDS operating condition of the 1800s was accelerated to 673s, the aging curve error between the original operating condition and the accelerated operating condition was 1.14%, and the overall acceleration rate was 2.67, which effectively realized the acceleration of the original operating condition (Fig. 9).

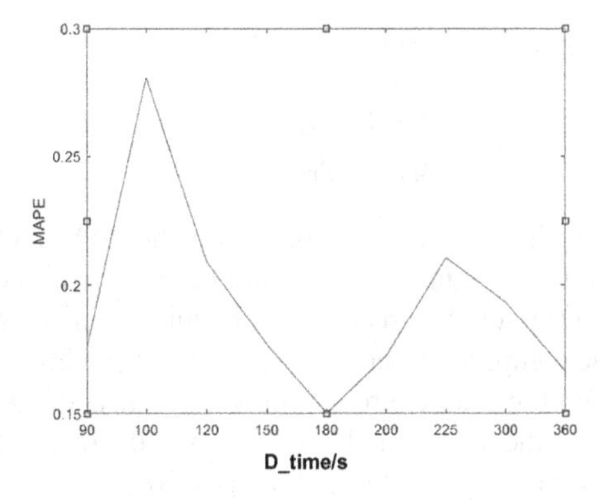

Fig. 8. Comparison of optimization results for different time lengths

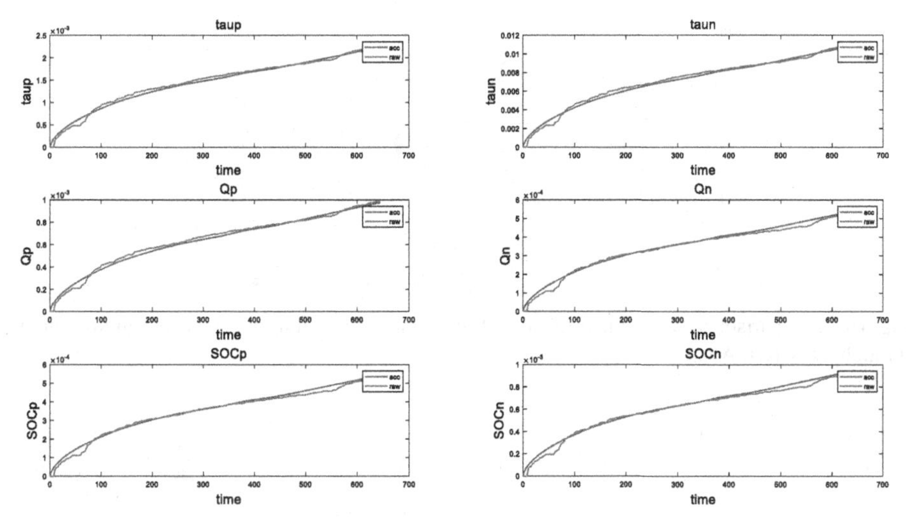

Fig. 9. The aging path comparison of each parameter between the accelerated operating condition and the original operating condition when the particle size D is 180s

4.2 The Granularity D is Optimized by the Charge Throughput Constraint

Although the fixed time length granularity D can meet the conditions for the generation of the accelerated operating condition, it is difficult to have a good mechanism explanation for why the fixed operating condition is selected. Therefore, this method tries to obtain the appropriate granularity D through certain mechanism constraints. Try to use charge throughput to constrain granularity, because the aging model adopted by charge in this study regards aging decay as a function of charge throughput, so using this method to constrain granularity can better reduce the error between the aging path under accelerated condition and the aging path under original condition.

Try to constrain the granularity D using the Eq. (5):

$$\begin{aligned} Q_D &\supseteq \{Q_C \cdot Q_{DS}\} \\ Q_D &\geq Q_{\min} \end{aligned} \tag{5}$$

That is, both charging and discharging conditions should be included in granularity D, and a minimum charge throughput Q_{min} should be specified. The charge throughput Q_D in each granularity D should be greater than the minimum charge throughput Q_{min}.

Figure 10 shows the comparison between the original condition and the accelerated condition when the minimum charge throughput is 1/10 Ah, and Fig. 11 shows the aging path comparison between the accelerated condition and the original condition for each parameter when the minimum charge throughput is 1/10 Ah.

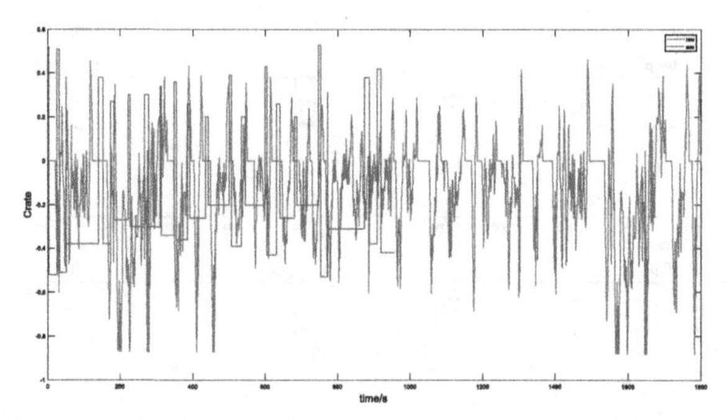

Fig. 10. Comparison between the original and accelerated conditions when the minimum charge throughput is 1/10 Ah

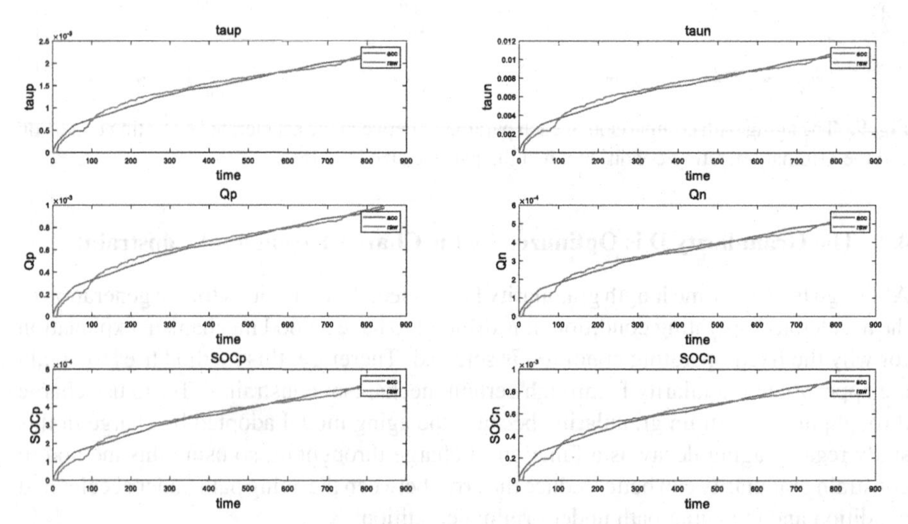

Fig. 11. Comparison of aging paths for each parameter between the accelerated and original conditions when the minimum charge throughput is 1/10 Ah

Taking the minimum charge throughput as the constraint, it is found that when the charge throughput constraint is 13/150 Ah, the aging path error between the original condition and the accelerated condition is the smallest (Fig. 12).

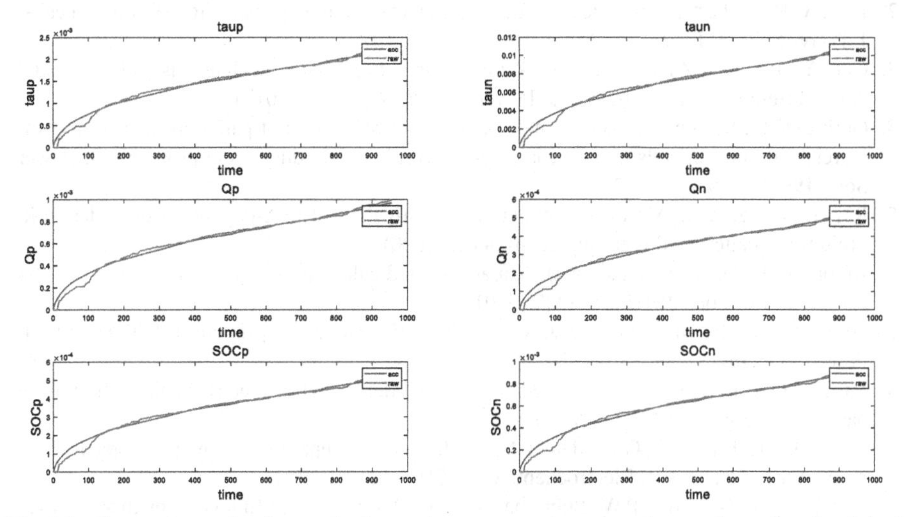

Fig. 12. Comparison of aging paths for each parameter between the accelerated and original conditions when the minimum charge throughput is 13/150 Ah

It can be seen that although the fixed charge throughput segmentation may result in larger deviations in some segments compared to the original condition, the overall error is still small, and the interpretability is better than the fixed time segmentation. After accelerated in this way, the FUDS operating condition of the 1800s was accelerated to 952s, the aging curve error between the original operating condition and the accelerated operating condition was 1.32%, and the overall acceleration rate was 1.89, which effectively realized the acceleration of the original operating condition with more clear mechanism explanation.

5 Conclusion

Using the acceleration method based on charge conservation to generate acceleration conditions can effectively accelerate the aging life experiment. In order to make the aging path of the lithium-ion battery consistent before and after acceleration, it is necessary to analyze the actual operating condition, and then the original operating condition is segmented and accelerated by using different current ratios. Two different segmentation methods have been proposed: the segmentation constrained by time and the segmentation constrained by charge throughput. The segmented approach with charge throughput as constraint is more reasonable in mechanism. And the segmented method with time constraint can make the acceleration rate larger. Both methods can effectively accelerate the operating condition, and the aging mechanism of the battery before and after acceleration is still the same as the original operating condition.

References

1. Han, X., Lu, L., Zheng, Y., et al.: A review on the key issues of the lithium ion battery degradation among the whole life cycle. eTransportation **1,** 100005 (2019)
2. Birkl, C.R., Roberts, M.R., McTurk, E., et al.: Degradation diagnostics for lithium ion cells. J. Power Sour. **341**, 373386 (2017)
3. Gao, Y., Jiang, J., Zhang, C., et al.: Lithium-ion battery aging mechanisms and life model under different charging stresses. J. Power Sour. **356**, 103114 (2017)
4. Gering, K.L., Sazhin, S.V., Jamison, D.K., et al.: Investigation of path dependence in commercial lithium-ion cells chosen for plug-in hybrid vehicle duty cycle protocols. J. Power Sour. **196**(7), 3395–3403 (2011)
5. Raj, T., Wang, A.A., Monroe, C.W., et al.: Investigation of path-dependent degradation in lithium-ion batteries. Batter. Supercaps **3**(12) (2020)
6. Bloom, I., Cole, B., Sohn, J., et al.: An accelerated calendar and cycle life study of Li-ion cells. J. Power Sour. **101**(2), 238247 (2001)
7. Wang, J., Liu, P., Hicks-Garner, J., et al.: Cycle-life model for graphite-LiFePO4 cells. J. Power Sour. **196**(8), 39423948 (2011)
8. Suri, G., Onori, S.: A control-oriented cycle-life model for hybrid electric vehicle lithium-ion batteries. Energy **96**, 644653 (2016)
9. Ramadass, P., Haran, B., Gomadam, P.M., et al.: Development of first principles capacity fade model for Li-ion cells. J. Electrochem. Soc. **151**(2), A196 (2004)
10. Lawder, M.T., Northrop, P.W., Subramanian, V.R.: Model-based SEI layer growth and capacity fade analysis for EV and PHEV batteries and drive cycles. J. Electrochem. Soc. **161**(14), A2099 (2014)
11. Atalay, S., Sheikh, M., Mariani, A., et al.: Theory of battery ageing in a lithium-ion battery: capacity fade, nonlinear ageing and lifetime prediction. J. Power Sour. **478**, 229026 (2020)
12. Guo, D., Yang, G., Feng, X., et al.: And aging of lithium ion battery accelerated life condition of path automatically generated method. J. Electr. Eng. Technol. (18), 4788–4797+4806 (2022). https://doi.org/10.19595/j.carolcarrollnki. 1000-6753. The tces. 210336. (in chinese)
13. Laidler, K.J.: The development of the Arrhenius equation. J. Chem. Educ. **61**(6), 494 (1984)
14. Guo, D., Yang, G., Feng, X., et al.: Physics-based fractional-order model with simplified solid phase diffusion of lithium-ion battery. J. Energy Storage **30**, 101404.1–101404.11 (2020). https://doi.org/10.1016/j.est.2020.101404
15. Ouyang, M., Feng, X., Han, X., et al.: A dynamic capacity degradation model and its applications considering varying load for a large format Li-ion battery. Appl. Energy **165**, 48–59 (2016)
16. USABC Electric Vehicle Battery Test Procedure Manual. Revision 2, DOE/ID-10479, January 1996

Adaptive Design of Uni-Variate Alarm Systems Based on Statistical Distance Measures

Mohsen Asaadi[1(✉)], Koorosh Aslansefat[2], Iman Izadi[3], and Fan Yang[1]

[1] Department of Automation, Tsinghua University, Haidian District, Beijing 100871, People's Republic of China
Asaadim10@mails.tsingua.edu.cn, yangfan@tsinghua.edu.cn
[2] School of Computer Science, University of Hull, Hull, UK
k.aslansefat@hull.ac.uk
[3] Isfahan University of Technology, Isfahan, Iran
iman.izadi@iut.ac.ir

Abstract. Alarm systems are utilized in the process industries to notify operators of abnormal process conditions or equipment faults. The alarm system must be appropriately constructed to maximize the possibility of safe and efficient operation. With the advancement of industrial and information technologies, real-time monitoring has shown to be an efficient method of ensuring operational safety and efficiency. Although the performance of alarm systems is an essential part of distributed control systems and has improved over the last decade, adaptive design of univariate alarm systems received remarkably little attention in research. We propose an adaptive designing method to evaluate and enhance the performance of an alarm system at runtime in this research. The approach is shown a flowchart, which enables adaptive design of alarm system based on the statistical characteristics of the process variable by considering both the performance deterioration of the alarm system itself, and the distributional shift of the process variable. As a result, the real-time adjustment of alarm system design parameters would be achievable, and at the same time the alarm system will operate more efficient. The proposed method is validated using a simulated example.

Keywords: Alarm System · Distributional Shift · Adaptive Design · Statistical Difference Measures · Performance Deterioration

1 Introduction

Nowadays, advanced industrial monitoring systems such as Supervisory Control and Data Acquisition (SCADA) and Centralized Monitoring System (CMS) are equipped with an enormous number of sensors thanks to the rapid growth in

This work was supported by the National Natural Science Foundation of China (No. 61873142) and the National Science and Technology Innovation 2030 Major Project (No. 2018AAA0101604) of the Ministry of Science and Technology of China.

B. Xin et al. (Eds.): IWACIII 2023, CCIS 1931, pp. 101–115, 2024.
https://doi.org/10.1007/978-981-99-7590-7_9

the sensing technologies. Those sensors can be used for condition monitoring purposes and by using pre-defined thresholds they can illustrate different normal and abnormal states of the system. For the abnormal states of the system that can be caused due to a component malfunctioning or fault occurrence, a monitoring and alarm system will generate alarms [1]. The generated alarms enable the operators to understand and solve the system's issues and plan for its maintenance. The cost and performance of operation and maintenance of industrial systems are highly associated with the performance of alarm systems. Thus, an alarm system with a low false detection and miss detection rate can have a huge economical benefit for a targeted industry, especially for the one that has accessibility limits like offshore wind farms [2]. Alarm management is challenging for any company, given the large number of possible alarms that might be generated at the same time [3]. For instance, operators faced difficult situations due to redundant and confusing information provided to them during the catastrophic accident at the nuclear power plant at Three Mile Island in 1979, the worst nuclear accident in US history. Much of the information collected was irrelevant and illusory during the accident [4].

To increase the safety and efficacy many actions, tools, evaluation metrics and policies have been created. For example, three specific indices have been used to analyse an alarm system's performance and safety in case of abrupt faults: Averaged Alarm Delay (AAD), Missed Alarm Rate or Probability (MAR/MAP), False Alarm Rate or Probability (FAR/FAP) by [5]. To design and optimize alarm systems for mixture processes and intermittent faults, [6] presented a time-variant finite mixture model to statistically model the behaviour of a process variable which is affected by an intermittent fault. The calculation methods for FAR and AAD are provided and a new time-variant missed alarm rate (MAR(t)) is introduced which reflects the missed alarm rate during the emerging stage of an intermittent fault. Process Variables are the parameters or quantities we wish to control at the correct limit.

[7] developed generalized delay timers, a novel way for improving traditional delay timer systems. In addition, a Markov model was used to calculate FAR, MAR, and Expected Detection Delay (EDD) in order to compare the performance of this technique to that of a standard delay timer. Regarding the performance evaluation of monitoring systems with adaptive and variable alarm thresholds, [8] has introduced a new method using combination semi-Markov process and temporal logic gates. [9] has proposed a new technique for getting optimum filter for alarms that incorporates plant and control system information while allowing the independence requirement to be relaxed.

By investigating through the existing literature for performance assessment and improvement of alarm systems, it is clear that there are few publications regrading adaptive design of alarm systems facing the concept drift or distributional shift in the upcoming data. To address the above-mentioned issues in the monitoring and alarm systems, in this paper, we proposed an adaptive approach which has the following contributions: A) Real time evaluation of the designed alarm system and B) Adjust the designing parameters of alarm system based on statistical difference measures.

2 Problem Formulation

2.1 Detecting Alarm States

The most common way of detecting an alarm state is to compare the value of a process variable to a constant high (low) alarm trip point, i.e.

$$x_a(t) = \begin{cases} 1 & x(t) > x_{htp} \quad \text{or} \quad x(t) < x_{ltp} \\ 0 & x_{ltp} \leq x(t) \leq x_{htp} \end{cases} \tag{1}$$

where $x(t)$ denotes the process variable, $x_a(t)$ denotes the alarm variable, and x_{htp} and x_{ltp} respectively denote the high and low trip points. A drum level is an example, which is related with the high alarm trip point 100 and the low alarm trip point -100 on a large scale thermal power plant. Figure 1(a) presents 1-hour samples of $x(t)$ with sampling period 1 s. A discrete-valued alarm variable $x_a(t)$ may be used to mathematically describe alarm states in alarm systems. In Fig. 1(b), the samples of two alarm variables connected with $x(t)$ are shown. That is, the high (low) alarm variable is assigned the value 1 if it is more (less than) 100(-100), and 0 otherwise [3]. The alarm occurrence and alarm clearance are defined as the change of alarm variables from 0 to 1 and from 1 to 0, respectively. Take note of the two rapid changes in the low alarm variable between 23:47:38 and 23:47:42, which are apparent in the magnified plot in Fig. 1.

2.2 Abrupt Faults

Assume that a process variable is in its normal state with distribution $p(x)$. An abrupt fault causes a change in the statistical properties of the variable, for instance, its mean. In this case, the PDF of the variable instantly changes to $q(x)$ in the faulty state. Figure 2 illustrates the distributions, $p(x)$ and $q(x)$ which indicate the variable distributions in normal and abnormal operation states, respectively. For this process variable, the well-known alarm performance indices are defined as [10]:

- **False Alarm Rate (FAR)**: In this work, the likelihood of an alarm occurring during the normal state is indicated by p_1. When the process variable is in its normal operating condition, the FAR index shows the potential of an alarm triggering. This indicates that the operator has received an incorrect alarm that does not need response. The FAR value is computed as follows in accordance with the process variable distribution (notice that due to the computational similarity of the indices for high and low trip points, the index is calculated exclusively for the high trip point): The probability of an alarm triggering is indicated by the above index; the process variable exceeds the high trip point. In other words, the alarm is raised when the process is operating normally.

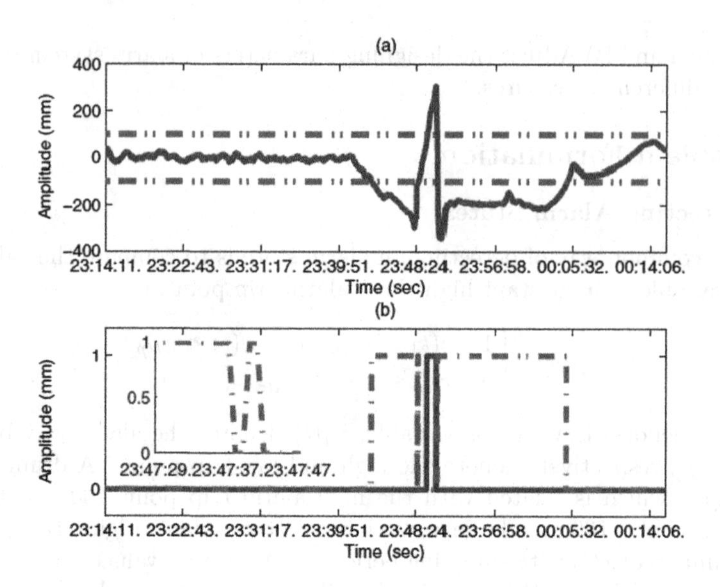

Fig. 1. (a) Samples of a process variable (solid) were collected in association with alarm trip points (dot-dash). (b), alarm variables $x_a(t)$ with high (solid) and low (dot-dash) alarm trip points [3]

- **Missed Alarm Rate (MAR)**: The likelihood of missing an alarm during a faulty or abnormal condition is given by q_2. The MAR index represents the likelihood of an alarm not being triggered when a process variable is operating in an abnormal manner [10]. This index is more critical than the FAR index since a false alarm may disturb the operator and waste his time, while a missed alarm may result in an incident while the operator is ignorant of its existence.

For the process variable in Fig. 2, we define the four parameters p_1, p_2, q_1, and q_2 as

$$\begin{aligned} p_1 &= \int_{x_{htp}}^{\infty} p(x)\mathrm{d}x, \quad p_2 = \int_{-\infty}^{x_{htp}} p(x)\mathrm{d}x \\ q_1 &= \int_{x_{htp}}^{\infty} q(x)\mathrm{d}x, \quad q_2 = \int_{-\infty}^{x_{htp}} q(x)\mathrm{d}x \end{aligned} \tag{2}$$

Then, it is straightforward to show that the performance indices are calculated as [10]:

$$\mathrm{FAR} = p_1, \mathrm{MAR} = q_2 \tag{3}$$

Here, x_{htp} is the alarm threshold, t_a is the time of the first alarm after the fault occurrence, t_f is the time of the fault occurrence. In Fig. 2, the adjusted threshold is indicated by a red vertical dash-line. In this figure, the overlapped region might lead to false detection since the designed alarm system is unable

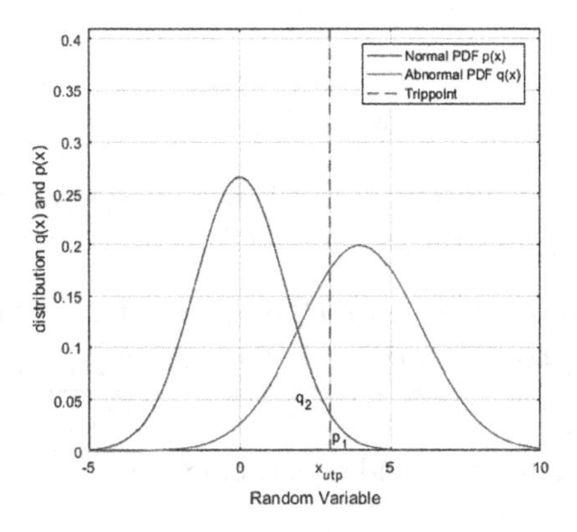

Fig. 2. Normal ($p(x)$) and abnormal ($q(x)$) PDFs

to determine which state of operation the process variable is associated with. In other words, the generated alarm might be a false alarm, or the non-alarm sample of alarm variable could be a missed alarm. The challenge of determining the best threshold is a classification problem, and the overlap region in Fig. 2 may lead to false detection if the designed alarm systems wrongly indicate the state which data belongs to. By considering this figure, one can see the region where two PDFs combine, as well as the potential for errors. We may determine the likelihood of an error [11]

$$P(error) = \int_{-\infty}^{\infty} P(error|x)P(x)dx \tag{4}$$

where, $P(error|x)$ can be calculated as the minimum of both probabilities of the normal and abnormal data sets as (5).

$$P(error|x) = \min_{x\in(-\infty,\infty)} [P(NS|x), P(ANS|x)] \tag{5}$$

NS and ANS denote the normal and abnormal states of operation, respectively, as shown in Fig. 2. The likelihood of error may be expressed in two parts by dividing the space into two regions denoted by the variables $R_1 = x \in R|x < X_{utp}$ and $R_2 = x \in R|x > X_{utp}$.

$$\begin{aligned} P(error) &= P(x \in R_1, NS) + P(x \in R_2, ANS) \\ &= \int_{R_1} P(x|NS)P(NS)dx + \\ &\int_{R_2} P(x|ANS)P(ANS)dx \end{aligned} \tag{6}$$

in other words, $P(x \in R_1, NS), P(x \in R_2, ANS)$ are FAR, and MAR indices, respectively. To ease the minimization problem, consider the following inequality rule [12].

$$\min[a,b] \leq a^{\gamma}b^{1-\gamma} \quad \text{where} \quad a,b \geq 0 \quad \text{and} \quad 0 \leq a \leq 1 \tag{7}$$

The equation (5) may be written as (8). When considering the worst-case situation or upper bound error, the '$\leq$' in (7) might be interpreted as '='.

$$\begin{aligned} P(error|x) &= min\left[P(NS|x),\ P(ANS|x)\right] = \\ min&\left[\frac{P(x|NS)\,P(NS)}{P(x)}, \frac{P(x|ANS)\,P(ANS)}{P(x)}\right] \end{aligned} \tag{8}$$

By using the inequality rule and equation (8), we have

$$\begin{aligned} P(error|x) \leq &\left(\frac{P(x|NS)\,P(NS)}{P(x)}\right)^{\gamma} \times \\ &\left(\frac{P(x|ANS)\,P(ANS)}{P(x)}\right)^{1-\gamma} \end{aligned} \tag{9}$$

The equation (10) can be driven through equations (4) to (9).

$$\begin{aligned} P(error) &\leq (P(NS))^{\gamma}\,(P(ANS))^{1-\gamma} \\ &\int_{-\infty}^{+\infty} (P(x|NS))^{\gamma}\,(P(x|ANS))^{1-\gamma}\,dx \end{aligned} \tag{10}$$

It is critical to evaluate the worst-case scenario in safety assurance, which might lead to (11), also known as the Chernoff upper bound of error [12].

$$\begin{aligned} P(error) &= P(NS)^{\gamma}\,P(ANS)^{1-\gamma} \\ &\int_{-\infty}^{+\infty} P(x|NS)^{\gamma}\,P(x|ANS)^{1-\gamma}\,dx \end{aligned} \tag{11}$$

The integral component of (11) could be solved using (12) [12] if the probability distributions of the classes follow normal or exponential distribution families.

$$\int_{-\infty}^{+\infty} P(x|NS)^{\gamma}\,P(x|ANS)^{1-\gamma}\,dx = e^{-\theta(\gamma)} \tag{12}$$

The $\theta(\gamma)$ can be calculated using (13) where μ and Σ are mean vector and variance matrix of each class respectively.

$$\begin{aligned} \theta(\gamma) = \frac{\gamma(1-\gamma)}{2}&\left[\mu_2-\mu_1\right]^T\left[\gamma\Sigma_1+(1-\gamma)\,\Sigma_2\right]^{-1} \times \\ &\left[\mu_2-\mu_1\right] + 0.5\,log\frac{|\gamma\Sigma_1+(1-\gamma)\,\Sigma_2|}{|\Sigma_1|^{\gamma}\,|\Sigma_2|^{(1-\gamma)}} \end{aligned} \tag{13}$$

The equation (13) basically becomes the Bhattacharyya distance when $\alpha = 0.5$. When $\Sigma_1 = \Sigma_2$, it can be demonstrated that this value is the optimal [12,13]. The Bhattacharyya distance will be utilized to show the technique in this research for simplicity. It should be noted that the estimated error bound could be larger than the true number in certain circumstances. This is permissible, however, since an overestimation of the classifier error would not pose a safety risk (although it may impact performance). The probability of obtaining a proper classification may be computed using (14) since the $P\,(error)$ and $P\,(correct)$ are complimentary.

$$P\,(correct) = 1 - \sqrt{P\,(NS)\,P\,(ANS)}\;e^{-\theta(\gamma)} \tag{14}$$

In most cases, the Chernoff upper limit of error is used to determine the separability of two classes of data, however in this case, equation (14) is used to determine the similarity of two classes. In other words, if you compare a class's $P\,(error)$ to itself in an optimized environment, the answer should be one, whereas $P\,(correct)$ should be zero. The obvious reason is to see whether the data distribution during training matches the data distribution seen in the field (or not).

The integral component of $P\,(error)$ may be transformed to the cumulative distribution function as (15) if $P\,(NS) = P\,(ANS)$.

$$\begin{aligned} P\,(error) &= \left(\int_{-\infty}^{T} P_{NS}\,(x)dx + \int_{T}^{+\infty} P_{ANS}\,(x)dx\right) \\ &= 1 - (F_{ANS}\,(T) - F_{NS}\,(T)) \end{aligned} \tag{15}$$

In addition, Equation (15) illustrates the link between likelihood of error (and also accuracy) and statistical difference between two Cumulative Distribution Functions (CDF) of two states. ECDF-based statistical measures such as the Kolmogorov-Smirnov distance (KSD) (Eq. 16) and similar distance measures can be used to predict the error at runtime [14,15].

$$P(error) \approx 1 - KSD = 1 - \sup_{x}\,(F_{ANS}\,(x) - F_{NS}\,(x)) \tag{16}$$

It should be noted that not all ECDF-based distances are constrained between zero and one, and may need the adjustment of a coefficient as a measure of precision estimate in certain circumstances. The relationship between ECDF-based distance and accuracy will be examined in Sect. 4.

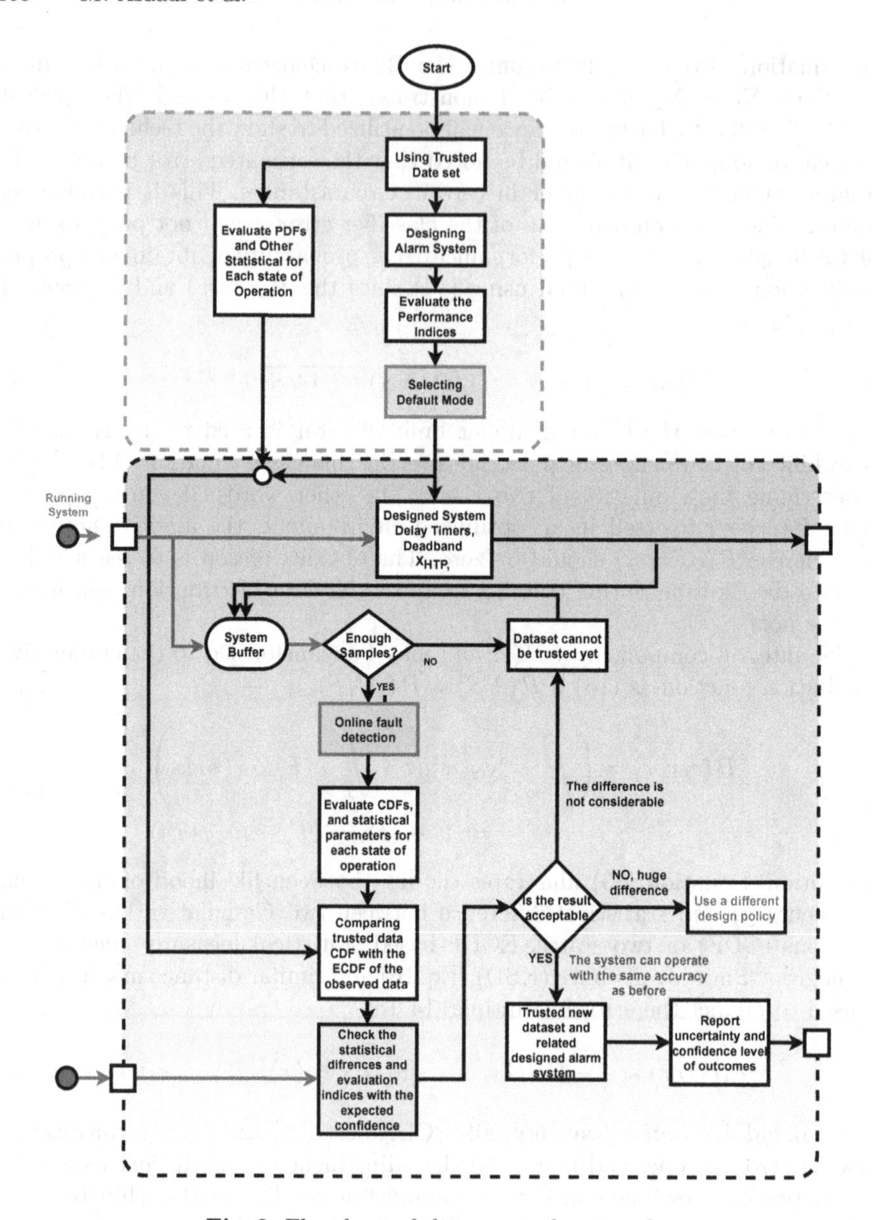

Fig. 3. Flowchart of the proposed approach

3 Safe Designed Alarm System

First and foremost, it is important to stress that the emphasis of this research is on the design of alarm systems for abrupt faults. The flowchart in Fig. 3 demonstrates how we see the idea being used in practice. The designing phase and the application phase are the two main sections of this flowchart. I) The designing phase is an offline approach that uses historical data from a given process variable to design an alarm system based on alarm system performance indices such as missed alarm rate, false alarm rate, and alarm average delay. In order to construct the performance assessment indices, this phase contains the change detection method for the process variable. In the second phase, all indices of the ideal design would be saved for future comparison. II) The second or the application phase is an adaptive approach in which real time data is provided to the system; in this stage, it is not known anything about the statistical characteristics of the real-time data. For example, consider an alarm system designed to monitor the pressure of the main steam driving the power turbine for a thermal power plant. The design policy supposed to make the alarm system able to trigger an alarm as soon as an abrupt is occurred by the least amount of time. In the application phase, it is important to keep in mind that the incoming data isn't classified as faulty or non-faulty. As a result, it is impossible to predict if the designed alarm system will operate as well as it did during the designing phase. The PDF and statistical parameters of each class could be estimated as input samples are gathered. Because the system needs a sufficient number of samples to correctly detect the statistical difference, a buffer of samples may also be required before proceeding. Using the modified Chernoff error bound presented in [11], the statistical difference of each state of operation in the designing phase and application phase is compared. If the statistical difference is very low, the designed alarm system results and accuracy could be trusted. In the power turbine's example, the alarm system would continue its operation in this case by holding the designed policies considered in the designing stage. Conversely, if the statistical difference is greater, the findings and accuracy of the designed alarm system are no longer regarded acceptable (because to the huge disparity between the trusted and observed data). In this case, the system should use an alternative design policy or notify a human operator. In the above example, the alarm system could ask the operator to justify the designing parameters of the alarm system.

4 Statistical Difference Values

In this section, the statistical distances values used in the application phase of the flowchart in comparison stage of the statistical values indicated by the yellow box in Fig. 3 are proposed. There would be a buffer in the application phase to gather enough samples. An expert should determine the buffer size at design time so that the gathered data contains the statistical properties of the operation state. It is worth noting that the future data is not considered to belong to a specific operation state. After collecting sufficient samples, the designed alarm system from the previous step will be used to clarify the operation state based on the generated alarms.

The statistical properties of buffered data are gathered and compared to the initial data set using ECDF-based statistical distance measures such as Kolmogorov-Smirnov (KS), Kuiper (K), Anderson-Darling (AD), Cramer-Von

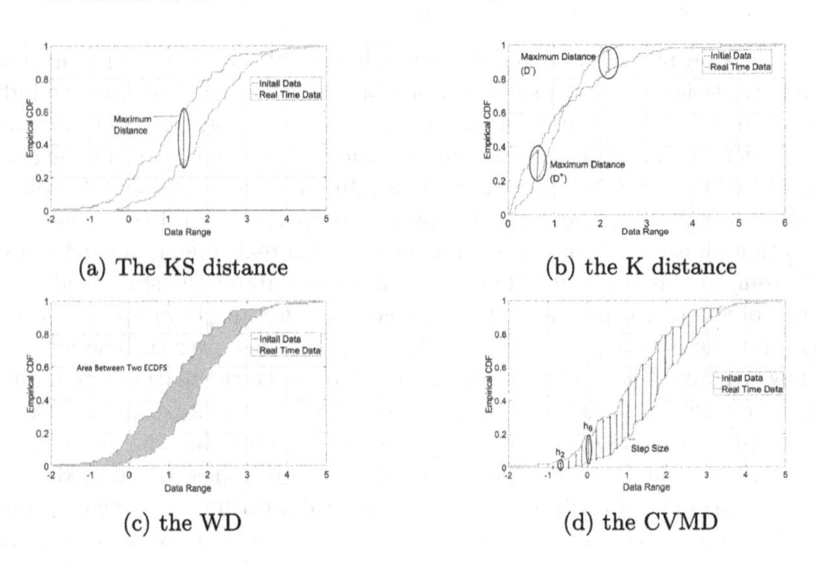

(a) The KS distance

(b) the K distance

(c) the WD

(d) the CVMD

Fig. 4. Different statistical distance measures.

Mises (CVM), and Wasserstein (W) [14]. Additionally, throughout the design phase, an expected confidence level for each statistical distance measures should be determined. The confidence level will be determined using the comparison described before and will be compared to the predicted confidence threshold once again. Three distinct possibilities were examined: 1) when the confidence is slightly lower than the threshold, the system should collect additional data; 2) when the confidence is significantly higher than the predefined threshold, it is assumed that the upcoming data have not been seen by the designed alarm system previously and a human-in-the-loop procedure should be considered; and 3) when the confidence is higher than the predefined threshold, the designed alarm system's results will be accepted and a report of the system's findings will be stored. To illustrate, consider a case in which a process variable is impacted by natural noise, which alters the process variable's statistical behavior. As a consequence, the number of alarms generated varies, and the safe design algorithm warns the operator. The operator will determine whether or not the process variable (say, a chemical process) is running properly. If the process is running properly, the alarm system must be redesigned to include the newly buffered data. Otherwise, the alarm system successfully identified the anomalous condition. This algorithm notifies the operator simply by comparing the statistical difference values and also the estimated FAR and MAR of the buffered data and compare it with those of initial data. In Fig. 4, different statistical measures and their differences are shown; since, this work is generally inspired by the [16], we use the same explanation used to explain the statistical difference measures applications in comparing the ECDFs of the trusted (initial) data and the real time data. As can be seen, the KS distance between two ECDFs quantifies their maximum value. The KS distance is incapable of determining which ECDF has a greater value, however the Kuiper distance can quantify two maximums up and down. When two sets have the same mean but distinct variances, the Kuiper distance provides a more accurate metric than the KS distance. As shown in Fig. 4(c), the WD may compute the area between two ECDFs in some way. As a result, the WD will be more sensitive to changes in the distributions' shape. The CVM distance is comparable to the WD distance, except it is quicker. When the step size of the CVM algorithm is reduced, the results approach those of the WD. [16] provides further insight on ECDF distance measures. Based on the specific attributions of the above statistical distances, we can change the policy of design, and have a more detailed view on the monitored data. This helps us adjust the designing parameter of the alarm system in a real time way and enhance the performance of the designed alarm system.

5 Simulated Example

In this section we brought an example to show how the flowchart is working. Figure 5 depicts the statistical deference measures in relation to accuracy measures for the basic classification approach, which is a simple linear classifier. In the flowchart explored for this study, at the evaluation stage, statistical difference measures are used to compare real-time data to initial data. Figure 5 (a-b) show the accuracy changes respect to WD and KS measurements. As can be seen, there is a predictable manner of how the values are changing. The Fig. 5 (c-d) show the WD and KS changes with respect to the different values of variance. In the following example, we used the Monte-Carlo simulation in order to know whether there is a predictable manner of changing or not. And also, instead of accuracy values we used the well-known performance indices of alarm system.

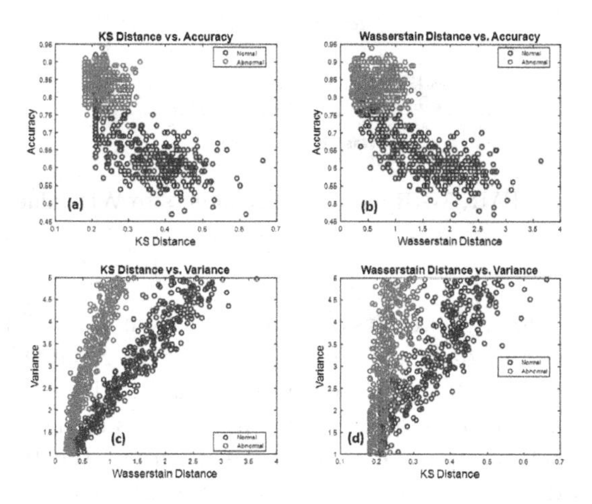

Fig. 5. Increasing the variance of test data (from 1 to 5) for both normal and abnormal classes: a-b) Accuracy in relation to WD and KSD c-d) Variance values in relation to WSD and KSD [16]

At this point, we are aware there is an orderly fashion relation between WD measures and the indices. Based on this, we can provide strategies for designing alarm systems that are based on real-time WD measurements of the data and predicted values of MAR and FAR. We just take into account the ROC curve threshold optimization for the alarm system in this report.

Example 1: In this example we calculated the optimum threshold for process variable $x \sim N(2,1)$ as its normal operation distribution, and $x \sim N(4,1)$ for abnormal operation distribution. The optimum threshold is considered as the tagged one to the knee point of the ROC curve. Which in here the optimum threshold is 3.25 (the optimization method is the same as it is in [5]). We do the same calculation for different distributional shifts in terms of variance values. Based on the flowchart, only some of the WD values are accepted, and this happens through a comparison of WD values to a predefined threshold (in here, $D_{th} = 0.5$). In other words, some WD values correspond to data shifts which do not make any obvious changes in the statistical behavior of the data. This threshold is adjusted based on the importance level of the process variable, in terms of safety and security. We also applied different data shift equivalent to 1:0.05:5 on the variance value of both normal and abnormal data set individually and predict the FAR and MAR indices based on the Monte-Carlo simulation.

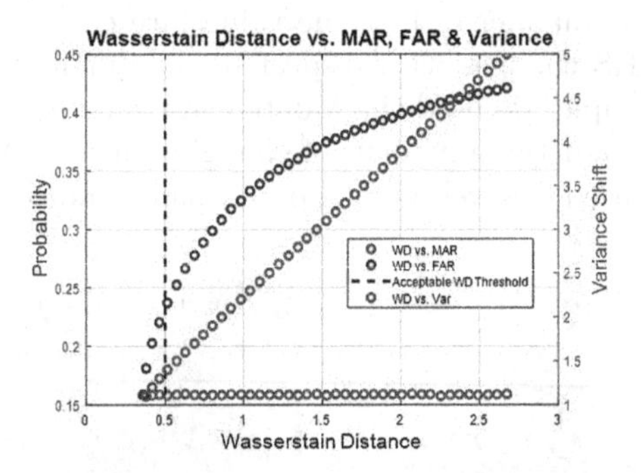

Fig. 6. Corresponding FAR, MAR and Variance indices to WD values - Variance Shift for normal data

Figures 6, and 7 show the results of Monte Carlo simulation for variance shifts on normal and abnormal data sets, respectively. For the Monte Carlo simulation, data are generated 5×10^5 times for both normal and abnormal conditions. On the basis of false negative and false positive arrays of the confusion matrix constructed for each observation to evaluate accuracy, the frequency of false alarms and missed alarms (1 as positive and 0 as negative) is calculated. For normal operating state, the average number of alarm occurrences is determined by averaging the false positive occurrence numbers, and for abnormal operation state, the average number of missed alarms is determined using the same method

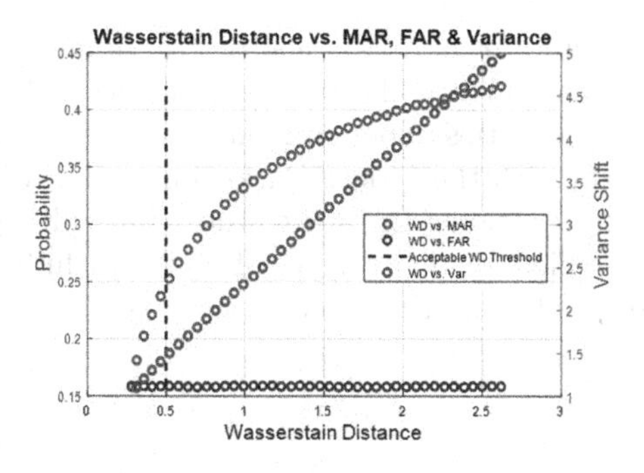

Fig. 7. Corresponding FAR, MAR and Variance indices to WD values - Variance Shift for abnormal data

as for normal operation state. It can be assumed, based on the Monte-Carlo results, that the shift in variance causes predictable changes in MAR and FAR. Consider the situation where the WD statistical difference is 0.70, the predicted MAR index is 0.28, and the pre-adjusted threshold is 3. Since the change only applied to the abnormal state, the FAR index is the same as when the alarm system was initially designed. By applying the ROC curve, the alarm system's threshold is redesigned. Figure 8 illustrates the ROC curve used to determine the optimal threshold in light of the new data shift. The new MAR is 0.21 and the new FAR is 0.2 in accordance with the optimal threshold of 0.28.

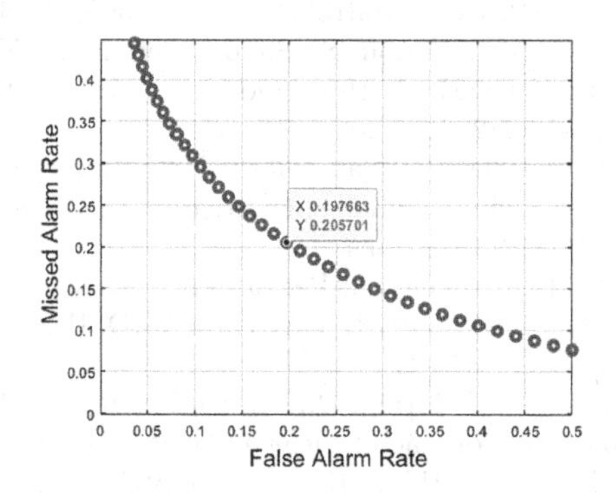

Fig. 8. Used ROC curve to set the threshold

6 Conclusion

For the first time, we attempted to suggest a adaptive designing method of an alarm system in the presented work. We evaluated the degree of dissimilarity between the real-time process variable and the data for the process variable used to design the alarm system using statistical deference measures. We illustrated our work using a flowchart, which clarifies the sequence of the method's various stages and the operators' duties depending on the method's output. At last, we validate the method through the Monte-Carlo simulation, which the results are consistent with the expectations. In future study, we will expand the approach such that it may be applied for many types of faults, including intermittent and incipient faults, and also for multi-variate alarm systems.

References

1. Izadi, I., Shah, S.L., Shook, D.S., Chen, T.: An introduction to alarm analysis and design. IFAC Proc. Vol. **42**(8), 645–650 (2009). Elsevier
2. May, A., McMillan, D.: Condition based maintenance for offshore wind turbines: the effects of false alarms from condition monitoring systems. In: ESREL (2013)
3. Wang, J., Yang, F., Chen, T., Shah, S.L.: An overview of industrial alarm systems: main causes for alarm overloading, research status, and open problems. IEEE Trans. Automat. Sci. Eng. **13**(2), 1045–1061 (2016). IEEE
4. Zang, H., Yang, F., Huang, D.: Design and analysis of improved alarm delay-timers. IFAC-PapersOnLine **48**(8), 669–674 (2015). Elsevier
5. Xu, J., Wang, J., Izadi, I., Chen, T.: Performance assessment and design for univariate alarm systems based on FAR, MAR, and AAD. IEEE Trans. Automat. Sci. Eng. **9**(2), 296–307 (2011). IEEE
6. Asaadi, M., Izadi, I., Hassanzadeh, A., Yang, F.: Assessment of alarm systems for mixture processes and intermittent faults. J. Process Control **114**, 120–130 (2022)
7. Adnan, N.A., Cheng, Y., Izadi, I., Chen, T.: Study of generalized delay-timers in alarm configuration. J. Process Control **23**(3), 382–395 (2013). Elsevier
8. Aslansefat, K., Gogani, M.B., Kabir, S., Shoorehdeli, M.A., Yari, M.: Performance evaluation and design for variable threshold alarm systems through semi-Markov process. ISA Trans. **97**, 282–295 (2020). Elsevier
9. Roohi, M.H., Chen, T., Guan, Z., Yamamoto, T.: A new approach to design alarm filters using the plant and controller knowledge. Indust. Eng. Chem. Res. **60**(9), 3648–3657 (2021). ACS Publications
10. Xu, J., Wang, J., Izadi, I., Chen, T.: Performance assessment and design for univariate alarm systems based on FAR, MAR, and AAD. IEEE Trans. Automat. Sci. Eng. **9**(2), 296–307 (2012). IEEE
11. Aslansefat, K., Sorokos, I., Whiting, D., Tavakoli Kolagari, R., Papadopoulos, Y.: SafeML: safety monitoring of machine learning classifiers through statistical difference measures. In: Zeller, M., Höfig, K. (eds.) IMBSA 2020. LNCS, vol. 12297, pp. 197–211. Springer, Cham (2020). https://doi.org/10.1007/978-3-030-58920-2_13
12. Fukunaga, K.: Introduction to Statistical Pattern Recognition. Elsevier (2013)
13. Nielsen, F.: 2018 IEEE International Conference on Acoustics, Speech and Signal Processing (ICASSP), The Chord Gap Divergence and a Generalization of the Bhattacharyya Distance, pp. 2276–2280 (2018)
14. Deza, M.M., Deza, E.: Distances in probability theory. In: Encyclopedia of Distances, pp. 257–272. Springer, Heidelberg (2014). https://doi.org/10.1007/978-3-662-44342-2_14
15. Mathias, R.: Empirical behaviour of tests for the beta distribution and their application in environmental research. Stochast. Environ. Res. Risk Assessm. **25**, 79–89 (2011)

16. Aslansefat, K., Kabir, S., Abdullatif, A., Vasudevan, V., Papadopoulos, Y.: Toward improving confidence in autonomous vehicle software: a study on traffic sign recognition systems. Computer **54**(8), 66–76 (2021). IEEE
17. Naghoosi, E., Izadi, I., Chen, T.: A study on the relation between alarm deadbands and optimal alarm limits. In: Proceedings of the 2011 American Control Conference, pp. 3627–3632 (2011). IEEE
18. Izadi, I., Shah, S.L., Shook, D.S., Kondaveeti, S.R., Chen, T.: A framework for optimal design of alarm systems. IFAC Proc. Vol. **42**(8), 651–656 (2009)

Correlation Analysis Between Insomnia Severity and Depressive Symptoms of College Students Based on Pseudo-Siamese Network

Ya-fei Wang[1], Yan-ling Zhu[1], Peng Wu[1], Meng Liu[1,2,3], and Hui Gao[3](✉)

[1] Department of Nursing,
Anhui Medical College, Hefei 230601, Anhui, People's Republic of China
zhuyanling@ahyz.edu.cn

[2] Affiliated Maternal and Child Care Hospital of Nantong University, Nantong 226000, Jiangsu, People's Republic of China

[3] Department of Pediatrics, The First Affiliated Hospital of Anhui Medical University, Hefei 230022, Anhui, China
gh20190130@163.com

Abstract. To explore the correlation between emotional mood and sleep quality in a college student population, we propose a new method based on pseudo-siamese network, which can quickly diagnose the causes of depression. This paper investigated the association between distinct sleep levels and depressive mood, with the aim of enhancing sleep quality and treatment approaches. The meta-test was conducted to examine the influencing factors and the results showed that the *OR* of the association between insomnia severity and depression in university students was 2.128, with a 95% confidence interval of 1.603–2.824 and a Z of 5.23, $P < 0.05$. Students with low sleep levels had a 2.128 times higher risk of experiencing depression compared to those with high sleep quality. Furthermore, a significant correlation was observed between low sleep levels and heightened depressive mood among students. The association model, employing a pseudo-siamese network, was developed by controlling for variables including weight, diet, age, and gender. The independent variable of sleep quality was entered into the model to determine the level of depressed mood among university students. The findings hold significance in analyzing the relationship between insomnia severity and depressed mood among university students, as well as in improving the overall sleep quality among this population.

Keywords: Pseudo-siamese Network · Sleep Quality · Depressed Symptoms

1 Introduction

During their time at university, students undergo rapid psychological development and face many challenges. These include navigating group dynamics, coping with academic responsibilities, managing employment prospects and coping with emotional stress. These factors can contribute to the emergence of detrimental emotions such as anxiety

B. Xin et al. (Eds.): IWACIII 2023, CCIS 1931, pp. 116–127, 2024.
https://doi.org/10.1007/978-981-99-7590-7_10

and depression, subsequently exerting a negative impact on their sleep quality [1]. On the contrary, insomnia has a propensity to induce or exacerbate symptoms of depression and anxiety [2]. Currently, the mental well-being of university students has become a widespread concern, with depression emerging as a predominant issue among graduates. Depression is characterized by an abnormally low, distressing, and negative emotional state in which individuals experience feelings of self-criticism and self-blame. This can lead to self-harming tendencies and suicidal thoughts, significantly impacting both individuals' daily lives and society as a whole [3]. According to studies conducted in China, a considerable percentage of university students, ranging from 25.7% to 31.2%, are reported to experience depression. Moreover, the detection rate of depression among university students is observed to be progressively increasing year after year [4]. The sleep quality of university pupils is intricately linked to the physical health, as a good quality of sleep plays a vital role in promoting their overall physical and psychological well-being. Notably, a scholarly survey revealed that among high-level university students, approximately 2.13 out of every 10 individuals are at risk of developing depression. Sleep disturbances are a prominent clinical manifestation of depression, and extensive research has demonstrated substantial changes in signaling pathways among individuals with depression [5, 6]. Research has indicated that inadequate sleep can readily shift the central nervous system from an aroused to an inhibited state, resulting in various negative consequences [7].

At present, pseudo-siamese networks has become an important research direction of computer vision and sound. It is a computer-aided algorithm that can assist doctors in quickly diagnosing the causes of depression in university students and it is able to achieve rapid diagnosis by doctors by docking the neural signals in university students' brains to build a pseudo-siamese model. Therefore, the use of pseudo-siamese networks to study the causes of depression and its mechanisms affecting university students can help reduce and prevent its onset and make it better. Computer models and machine algorithms can be employed to analyze the relationship between insomnia severity and depression among university students. This analytical approach can assist healthcare professionals in efficiently identifying and diagnosing depression in university students. Several researchers have utilized data mining algorithms to construct predictive models concerning the link between insomnia severity and stress levels in students. Ultimately, regression prediction analysis is employed to derive the outcomes regarding the relationship between sleep quality and depression in university students [8, 9].

The main contributions of this paper are as follows. Firstly, it employs a pseudo-siamese network model to assess the insomnia severity and levels of depression among college students. Secondly, it investigates the present state of sleep and mode among college pupils, along with associated factors, to establish a correlation between sleep status and depression. Lastly, it proposes a depression intervention as a means to enhance the sleep quality and mental health of college students, thereby addressing the identified issues effectively.

2 Methodology

2.1 Data

In this study, 2050 volunteers were recruited from three universities in Nanjing and Hefei cities between September and December 2022. After the data collection process, 1850 valid responses were obtained, resulting in a return rate of 90.24%. The study employed a school-based survey methodology aimed at assessing behaviors and mental health among Chinese university students. Among the volunteers, 505 were first-year students, 520 were second-year students, 440 were third-year students, and 385 were fourth-year students. All surveys were administered on-site within the schools, where trained investigators provided guidance to the volunteers, facilitating the completion of the written questionnaires.

2.2 Evaluation Methodology

Health Questionnaire-9 (PHQ-9). The PHQ-9 is a self-rating scale consisting of nine items designed to screen for depressive symptoms. It has demonstrated robust psychometric properties in various primary care settings, indicating its effectiveness as a screening tool [10]. As shown in Table 1, it assesses the presence and severity of depressive symptoms experienced within the past two weeks and a higher score on the scale indicates a greater severity of depressive symptoms [11].

Table 1. Contents of PHQ-9 scale.

Answer code	Experience of depressive symptoms	Degree of depression
0	None	without
1	several days	mild
2	More than half the days	moderate
3	Almost every day	serious

Insomnia Severity Index (ISI). The ISI (Insomnia Severity Index) is a tool used to assess the severity of insomnia. It consists of seven items, and volunteers respond to each item using a scale of 0, 1, 2, 3, or 4. These responses correspond to the options "not at all," "mild," "moderate," "severe," and "extremely severe," respectively. The total score on the ISI ranges from 0 to 28. Higher scores indicate more severe insomnia. Cronbach's alpha is 0.873[12].

2.3 Statistical Analysis of Correlation Model Based on Pseudo-Siamese Network

If the confidence interval of a mediating (indirect) effect includes 0, it indicates that the effect is not statistically significant at the 5% significance level. This means that there is no significant evidence to support the presence of a mediating effect. Regarding the algorithm based on a pseudo-siamese network, it has demonstrated favorable performance in image classification tasks. The structural form of the pseudo-siamese network consists of a twin layer, the core part of which is the twin layer, whose main function is the extraction and output of sleep features [13]. The computational process that is referring to involves the utilization of multiple twin kernels in a given scenario. Its objective is to decrease the image size and mitigate overfitting while preserving the original features. The fully connected layer serves to transform the extracted features from the preceding layer into one-dimensional features. Lastly, the attrition layer, which is the final component in the network, enables a comparison between the model's results and the actual data. A well-performing classification model aims to minimize the error in these comparison results. The structure consists of multiple layers of neurons, which are interconnected based on the input and output. These layers include one or more hidden layers positioned between the input and output layers. Within each hidden layer, every neuron is connected to neurons in the previous (m-1st) hidden layer and the subsequent (m+1st) hidden layer. This interconnected structure enables information flow and computation throughout the network. Pseudo-siamesening neural networks have a number of characteristics [14]. Firstly, neural networks possess a remarkable parallel data processing capability. This means that multiple computations can be performed simultaneously, leveraging the power of parallel processing. Additionally, the use of multiple neurons allows for the execution of multiple operations in parallel. Consequently, a neural network can efficiently handle large-scale processing by distributing the workload across multiple neurons. Secondly, neural networks excel in their ability to combine and integrate information. Through neuron-to-neuron associative weighting, the network efficiently processes diverse and uncertain information, enabling effective analysis and decision-making. Thirdly, neural networks possess the capability of self-education. The associative capacity of neurons is determined by their weights, which undergo adjustments during the network's evolution. This dynamic nature allows the system to engage in continuous learning and upgrading. In a neural network, each neuron represents a specific type of information. The collaborative operation of these neurons determines the overall performance of the network in processing information. Consequently, even if a single neuron experiences an error or failure, it does not have a detrimental impact on the network as a whole or its ability to process information effectively. The distributed nature of information processing ensures robustness and fault tolerance within the network. One of the prominent forms of deep learning is the pseudo-siamese network, which leverages twin operations to capture the individual characteristics of neurons. This algorithm belongs to the family of forward neural networks and exhibits a deep neural network structure. It utilizes twin operations to process input data and extract meaningful features for learning and inference tasks. The pseudo-siamese network is widely employed in various domains due to its ability to handle complex patterns and facilitate efficient information processing [14, 15]. The pseudo-siamese network includes a pseudo-siamese layer, a pooling layer and a fully connected layer. The specific operation of the pseudo-siamese layer method involves multiplying each corresponding unit by a circular center using a pseudo-siamese kernel of a specific size (see Fig. 1).

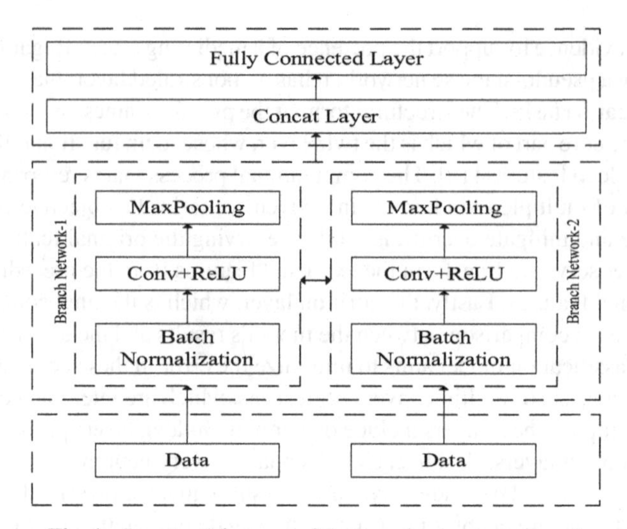

Fig. 1. Structural features of the pseudo-siamese network

2.4 Data Processing

To perform text retrieval, a pseudo-siamese network is utilized after filtering users. The process involves segregating the identified volunteers with depression from those with normal emotions. By logging in with cookies, the original microblogs of the selected volunteers are mined in a specific time sequence. Subsequently, the collected data is filtered and organized. Following the data crawling process, a total of 382 volunteers exhibiting depressive symptoms and 372 normal controls are obtained. The text information from the microblogs is saved on a per-user basis, serving as the fundamental unit for further analysis and processing. This approach allows for the efficient extraction and organization of relevant textual data for subsequent investigations.

$$L_j = \frac{\sum_{i=a} \lambda x^3 e^x}{\sqrt{\eta \ln x^3}} \tag{1}$$

$$L_i = \iint_{i=a} \ln x^a \varepsilon \vec{x} dxdy \tag{2}$$

$$K_a = \overline{\overline{\sin x^2 \cdot \ln x^x}} \tag{3}$$

$$K_b = \coprod_{i=a}^{n} \ln x^a \cdot \sin x^3 \tag{4}$$

$$V = \frac{\left\| \sin x + \cos x^2 \right\|}{2\mu} \tag{5}$$

$$R_a = \begin{vmatrix} 0 & 1 \\ 1\ x & 0 \\ x\ 0 & 0 \end{vmatrix} \tag{6}$$

where L_j denotes the number of iterations, L_i denotes the loss energy, K_a denotes the data similarity, K_b denotes the convolution step, V denotes the random gradient and R_a denotes the output data feature set [15].

2.5 Correlation Model Establishment and Test Plan

This experiment used AMOS17.0 software to construct seven different structural equations. To analyze the factors influencing the quality evaluation score, survey grade, publication time, and sampling method, meta-regression was conducted. The results revealed that these factors collectively explain 88.6% of the observed differences. Notably, the regression coefficient obtained in this analysis was 0.0958 lower compared to the regression coefficient obtained using the random effect model in the meta-analysis. This indicates that research quality, survey grade, publication time, and sampling method significantly contribute to the observed variations in the quality evaluation score. The meta-regression provides valuable insights into the factors influencing the overall quality assessment of the studies analyzed. The Beg rank correlating method was used to test the insomnia situation deviation: $Z = 0.94 < 1.96$, $P > 0.05$, no significant insomnia situation deviation was found. By applying the Rosenthal method, a failure safety degree of 821 is obtained, indicating that it satisfies the specified requirements. This means that among the 821 papers analyzed on the correlation between depression and insomnia, there is substantial evidence supporting the claim that poor sleep quality is indeed a risk factor for depression. The robustness of the findings and the large number of papers supporting this relationship provide strong support for the hypothesis. As shown in Table 2.

Table 2. Test of sleep quality deviation by Beg rank correlation method.

Independent variable	Case	Maximum value	Average value	Deviation
Study load	1850	5	2.56	0.642
Self-assessment of health status	1850	5	2.2	0.790
Self assessment of family economic conditions	1850	1	1.67	0.373
One child	1850	2	1.67	0.486

3 Results

3.1 General Demographic Characteristics

The detection rates of mild, moderate, and severe depressive symptoms among 1850 college students were 25.4%, 5.2%, and 2.5%, respectively. Except for different nationalities, the detection rates of depressive symptoms among medical students with different demographic characteristics, such as gender, education level, family residence, and so on, were statistically significant ($P < 0.05$), as seen in supplementary material.

3.2 Mediation Effect Analysis

In the analysis conducted using a random effects model, the results revealed an odds ratio (*OR*) of 2.128, indicating that 95% of university students experienced sleep disturbances. This finding was statistically significant with a *Z*-score of 5.23 and a *P*-value less than 0.05. Additionally, it was observed that 5% of university students reported poor sleep quality, which had a significant impact on changes in their depressed mood.

To further investigate the data, a subgroup analysis was performed based on different types of research data, publication time, and sampling methods. This analysis revealed significant differences in sleep quality, activity time, survey year, and sampling method, as depicted in Fig. 2. These findings highlight the variations in sleep patterns among university students across different research studies, time periods, and sampling techniques, emphasizing the importance of considering these factors when examining the relationship between sleep quality and depressed mood.

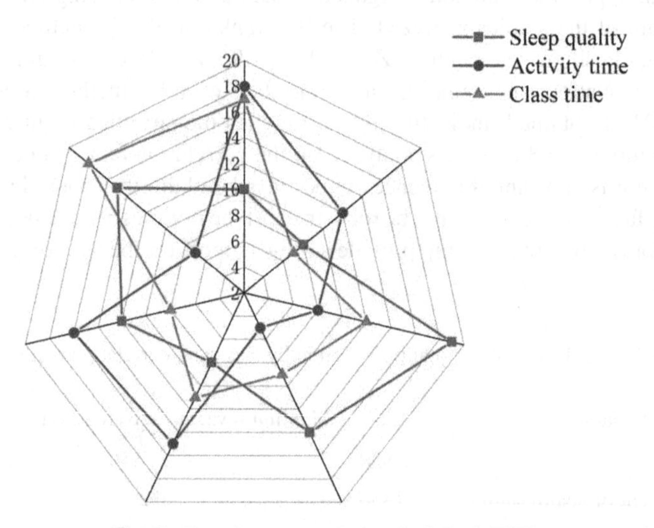

Fig. 2. Data heterogeneity analysis by META

Based on the subgroup meta-analysis results, a significant correlation between poor sleep quality and depression was observed across most subgroups. However, it is worth noting that in the randomly selected classes from the sample of grades 1 to 4, there was no significant difference detected ($Q = 68.39$, $P > 0.05$). This finding suggests that the relationship between insomnia level and depression may not be consistent across all academic years. It is possible that other factors, such as academic workload or social dynamics, may play a more influential role in these specific grade levels, masking the direct association between poor sleep quality and depression. Meta-regression analysis was calculated as follows.

$$Z_1 = \sum_{a=1} \ln x^2 \beta \sin y^3 \tag{7}$$

$$Z_2 = \frac{\sqrt{V(x) \sum_{i=a} \ln x^a}}{2\varepsilon} \tag{8}$$

$$Y_1 = \eta \sin x^2 \iint_{a=1} \ln x^a dxdy \tag{9}$$

$$Y_2 = [Z_1(x) + \sum \ln x^{2y} V(x)] \tag{10}$$

$$U = \sum_{i=a}^{n} \begin{vmatrix} x & 1 & 1 \\ 0 & x & 0 \\ 0 & 1 & 1 \end{vmatrix} \tag{11}$$

$$I = \int_{\mathrm{i}=a} \sum_{n=1} \sin x^2 \prod_{i=a} \eta \tan x^a \tag{12}$$

where Z_1 denotes regression coefficient, Z_2 denotes reliability coefficient, Y_1denotes validity coefficient, Y_2 denotes sleep score, U denotes ISI score and I denotes significance.

The model used in this study demonstrated a good fit, as indicated by the fit indices: CFI (Comparative Fit Index) of 0.950, TLI (Tucker-Lewis Index) of 0.988, RMSEA (Root Mean Square Error of Approximation) of 0.007, and SRMR (Standardized Root Mean Square Residual) of 0.004. These values suggest that the model adequately represents the data and fits the observed relationships well, as depicted in Fig. 3. The pseudo-siamese network regression analysis conducted on the relationship between T1 sleep levels and T2 sleep duration revealed a significant effect of T1 depression on T2 sleep ($Beta = 0.18, P < 0.01$). This finding suggests that initial levels of depression have an impact on subsequent sleep quality. However, when T1 depression was controlled for, the expected effect of T1 sleep levels on T2 depression was not significant ($P > 0.05$), indicating that sleep levels alone may not have a direct influence on subsequent depression levels.

3.3 Physical Activity Impact

This study provides evidence for the presence of a mediating factor in the relationship between subjective sleep levels and depression among university students. The findings indicate that subjective insomnia level scores have a positive effect on depression ($P < 0.001$), implying that poorer sleep level is associated with higher levels of depression. Additionally, subjective insomnia level scores are positively correlated with negative values ($P < 0.001$), indicating that lower subjective sleep quality is associated with a more negative emotional state. Moreover, when considering the prediction of depression, the use of subjective sleep quality scores with the awareness reassessment method has a greater impact on depression ($P < 0.001$). This suggests that cognitive reassessment plays a moderating role in the relationship between subjective insomnia and depression in university students. In the pseudo-siamese network model, the maximum effect size

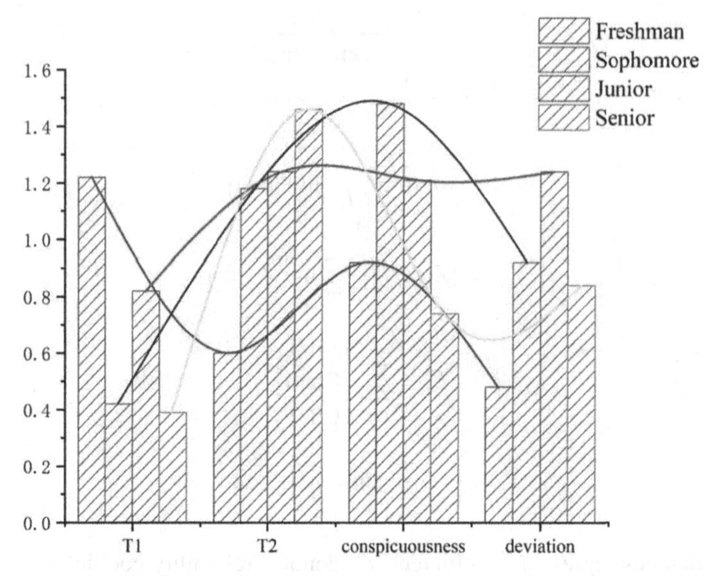

Fig. 3. Pseudo-siamese Network regression analysis

with cognitive reassessment is calculated as −0.21 * 0.29 = 0.0609, while subjective sleep level has a maximum effect size of 0.33, corresponding to a coefficient of 0.391 and an intermediate effect size of 16.38%. These findings highlight the significant association between sleep quality and depressed mood in university students, as indicated by the neural data established by the neurons in the pseudo-siamese network model. Furthermore, the time spent in physical activity is identified as a significant mediating variable.

4 Discussion

Among university students, anxiety and depression are the main mental health issues. Senior students show a higher tendency towards insomnia, and there is a higher proportion of university students who habitually stay up late, have poor academic performance, irregular breakfast habits, and engage in late-night snacking. The detection rates of mild, moderate, and severe depressive symptoms among 1850 college students were 25.4%, 5.2%, and 2.5%, respectively. A study carried out showed that depression amongst medical students is a global issue [16]. Generally speaking, sleep problems may be a predictive factor for depression, as the authors of studies found: students reported somatic symptoms rather than psychological problems [17]. Additionally, low family income, the habit of staying up late, poor academic performance, irregular breakfast habits, and lack of exercise were identified as significant influencing factors for depression among university students. There is also a significant association between sleep patterns and depressive symptoms, which is consistent with previous results [6, 18].

To investigate the correlation between college students' insomnia situation and depressive mood, this study establishes a pseudo-siamese network model space and conducts correlation analyses using meta-neural data nodes. In the diagnosis and identification of depression, a classification model is typically constructed, and relevant rules are extracted to facilitate screening, identification, and diagnosis of depression. Hence, it is crucial to assess the accuracy of the depression classification model in order to make correct predictions and diagnoses, ultimately improving clinical outcomes. Given that the collected information often contains redundant data that can hinder prediction accuracy, it becomes essential to extract relevant information that aids in prediction through feature information processing and feature selection. Feature extraction and engineering play a significant role in accurately forecasting pseudo-siamese network models. With the abundance of collected information, it becomes necessary to extract meaningful features that contribute to the prediction task. This involves processing the information to identify and select the most relevant features for accurate forecasting. The success of the pseudo-siamese network model relies heavily on the effective extraction and matching of features, as they directly impact the model's predictive capabilities in the context of depression diagnosis and prognosis [14]. On this basis, the machine learning method based on neural network will be used to construct depression risk factors, biomarkers, and so on. Standardized data will be obtained through pre-processing, and divided into different training sets and test sets. After training the training sets, the model will be finally evaluated, and it will be continuously improved at the evaluation stage. Based on the prediction of depression, machine learning can reflect the factors of depression and various disease indicators, such as influencing factors, symptoms and physiological characteristics.

The results demonstrate that college students with insufficient sleep time and poor insomnia situation exhibit greater fluctuations in the twin space model of mental mood, whereas those with sufficient sleep time and higher sleep quality display fewer fluctuations in their mental mood test. These findings indicate that college students with poor sleep quality are more likely to experience depressive symptoms compared to those with better sleep habits. This study found the relationships between insomnia situation, depression, cognitive reassessment, and physical activity time. Overall, this study emphasizes the importance of considering subjective sleep and its impact on mental health outcomes, particularly in the context of university students. The findings provide insights into potential intervention strategies that target sleep quality and cognitive reassessment to improve mental well-being in this population. A meta-analysis on sleep quality showed that insomnia situation affected 65% of pupils in Europe [19]. Poor sleep quality is associated with lower academic performance. It is also correlated with the onset of depression and stress [20]. This study found the relationships between sleep quality, depression, cognitive reassessment, and physical activity time. Overall, this study emphasizes the importance of considering subjective sleep quality and its impact on mental health outcomes, particularly in the context of university students. The findings provide insights into potential intervention strategies that target sleep quality and cognitive reassessment to improve mental well-being in this population.

In future meta-analyses, it is recommended to incorporate prospective findings such as cohort studies to further explore the association between insomnia levels and depression among university students. Additionally, efforts should be made to include subjects with high-quality sleep scores in meta-analyses to further examine the relationship with depression. The results underscore the connection between college students' sleep quality and their risk of depression, emphasizing that those with poor sleep quality are at higher risk.

Acknowledgements. This study was supported by Key projects of universities in Anhui Province of China (2022AH052325).

Author Contributions. YFW and YLZ designed the study. PW, LM, and HG performed the survey research, YFW and HG analyzed the data. All authors read and approved the final version.

References

1. Wang, F., Bíró, É.: Determinants of sleep quality in college students: a literature review. Explore **17**(2), 170–177 (2021). https://doi.org/10.1016/j.explore.2020.11.003
2. Kaya, F., Bostanci Daştan, N., Durar, E.: Smart phone usage, sleep quality and depression in university students. Int. J. Soc. Psychiatry **67**(5), 407–414 (2021). https://doi.org/10.1177/002076402096020
3. Christian, L.M., Carroll, J.E., Porter, K., Hall, M.H.: Sleep quality across pregnancy and postpartum: effects of parity and race. Sleep Health **5**(4), 327–334 (2019). https://doi.org/10.1016/j.sleh.2019.03.005
4. Gao, L., Xie, Y., Jia, C., Wang, W.: Prevalence of depression among Chinese university students: a systematic review and meta-analysis. Sci. Rep. **10**(1), 15897 (2020). https://doi.org/10.1038/s41598-020-72998-1
5. Downing, M.J., Millar, B.M., Hirshfield, S.: Changes in sleep quality and associated health outcomes among gay and bisexual men living with HIV. Behav. Sleep Med. **18**(3), 406–419 (2020). https://doi.org/10.1080/15402002.2019.1604344
6. Dudo, K., Ehring, E., Fuchs, S., Herget, S., Watzke, S., Unverzagt, S., Frese, T.: The association of sleep patterns and depressive symptoms in medical students: a cross-sectional study. BMC Res. Notes **15**(1), 1–6 (2022). https://doi.org/10.1186/s13104-022-05975-8
7. Rahmani, M., Rahmani, F., Rezaei, N.: The brain-derived neurotrophic factor: missing link between sleep deprivation, insomnia, and depression. Neurochem. Res. **45**(2), 221–231 (2020). https://doi.org/10.1007/s11064-019-02914-1
8. Lueke, N.A., Assar, A.: Poor sleep quality and reduced immune function among college students: perceived stress and depression as mediators. J. Am. Coll. Health, 1–8 (2022). https://doi.org/10.1080/07448481.2022.2068350
9. Gomes, S.R.B.S., von Schantz, M., Leocadio-Miguel, M.: Predicting depressive symptoms in middle-aged and elderly adults using sleep data and clinical health markers: a machine learning approach. Sleep Med. **102**, 123–131 (2023). https://doi.org/10.1016/j.sleep.2023.01.002
10. Wang, W., Bian, Q., Zhao, Y., Li, X., Wang, W., Du, J., et al.: Reliability and validity of the Chinese version of the patient health questionnaire (PHQ-9) in the general population. Gener. Hosp. Psychiatry **36**(5), 539–544 (2014). https://doi.org/10.1016/j.genhosppsych.2014.05.021
11. Löwe, B., Kroenke, K., Herzog, W., Gräfe, K.: Measuring depression outcome with a brief self-report instrument: sensitivity to change of the patient health questionnaire (PHQ-9). J. Affect. Disord. **81**(1), 61–66 (2004). https://doi.org/10.1016/S0165-0327(03)00198-8

12. Liu, Z., Liu, R., Zhang, Y., Zhang, R., Liang, L., Wang, Y., et al.: Association between perceived stress and depression among medical students during the outbreak of COVID-19: the mediating role of insomnia. J. Affect. Disord. **292**, 89–94 (2021). https://doi.org/10.1016/j.jad.2021.05.028
13. Parkin, S.R.: Practical hints and tips for solution of pseudo-merohedric twins: three case studies. Acta Crystallogr. E Crystallogr. Commun. **77**(5), 452–465 (2021). https://doi.org/10.1107/S205698902100342X
14. Zhang, Y., Tiňo, P., Leonardis, A., Tang, K.: A survey on neural network interpretability. IEEE Trans. Emerg. Top. Comput. Intell. **5**(5), 726–742 (2021)
15. Wang, H., Wu, Y., Min, G., Miao, W.: A graph neural network-based digital twin for network slicing management. IEEE Trans. Ind. Inform. **18**(2), 1367–1376 (2020)
16. Rotenstein, L.S., et al.: Prevalence of depression, depressive symptoms, and suicidal ideation among medical students: a systematic review and Meta-analysis. JAMA **316**(21), 2214–2236 (2016). https://doi.org/10.1001/jama.2016.17324
17. Abdelaziz, A.M.Y., Alotaibi, K.T., Alhurayyis, J.H., Alqahtani, T.A., Alghamlas, A.M., Algahtani, H.M., et al.: The association between physical symptoms and depression among medical students in Bahrain. Int. J. Med. Educ. **8**, 423–427 (2017). https://doi.org/10.5116/ijme.5a2d.16a3
18. Wu, M., Liu, X., Han, J., Shao, T., Wang, Y.: Association between sleep quality, mood status, and ocular surface characteristics in volunteers with dry eye disease. Cornea **38**(3), 311–317 (2019). https://doi.org/10.1097/ICO.0000000000001854
19. Rao, W.W., et al.: Sleep quality in medical students: a comprehensive meta-analysis of observational studies. Sleep Breath. **24**(3), 1151–1165 (2020). https://doi.org/10.1007/s11325-020-02020-5
20. Almojali, A.I., Almalki, S.A., Alothman, A.S., Masuadi, E.M., Alaqeel, M.K.: The prevalence and association of stress with sleep quality among medical students. J. Epidemiol. Glob. Health **7**(3), 169–174 (2017). https://doi.org/10.1016/j.jegh.2017.04.005

Construction and Research of Pediatric Pulmonary Disease Diagnosis and Treatment Experience Knowledge Graph Based on Professor Wang Lie's Experience

Qingyu Xie and Wei Su(✉)

School of Medical Information, Changchun University of Chinese Medicine, Changchun 130117, Jilin, China
suwei@ccucm.edu.cn

Abstract. To construct a knowledge graph of Professor Wang Lie, a Master of Traditional Chinese Medicine(TCM), on the diagnostics and treatment of pediatric pulmonary diseases, by providing a foundation for the inheritance of his academic thoughts and clinical experience in TCM. In this study, we focused on Professor Wang's diagnostics and treatment experience in pediatric pulmonary diseases. By utilizing unstructured text data and integrating his academic thoughts and clinical experience, a knowledge graph was built using the Neo4j graph database. Unstructured text experience data meeting the requirements was selected, entered into the Excel® table to establish the Professor Wang's database for the diagnosis and treatment of pediatric pulmonary diseases, and standardized processing of data was performed. In the Neo4j graph database, the schema-level graph comprised 20 entity concept labels, 29 entity nodes, 28 entity relationships, and 22 types of entity relationships. The data-level graph included 20 entity concept labels, 870 entity nodes, 22 types of entity relationships, and 1469 entity relationships. This enabled the visualization of Professor Wang's diagnostics, treatment principles, prescription strategies, and medication patterns for pediatric pulmonary diseases. By constructing a knowledge graph of Professor's diagnostics and treatment for pediatric pulmonary diseases based on the Neo4j graph database, Knowledge extraction, integration, and representation were attained. This work also lays the foundation for optimizing TCM treatment plans for pediatric pulmonary diseases and building intelligent diagnosis and expert systems for TCM-based treatment in the pediatric population.

Keywords: Pediatric pulmonary diseases · Knowledge graph · Neo4j graph database · Professor Wang Lie

1 Introduction

Pediatric lungs are delicate and not fully developed, making children susceptible to various pulmonary diseases when attacked by external pathogenic factors. The lungs function like a canopy and shelter the internal organs and are the first line of defense

B. Xin et al. (Eds.): IWACIII 2023, CCIS 1931, pp. 128–138, 2024.
https://doi.org/10.1007/978-981-99-7590-7_11

against invading pathogens. The lungs govern the respiration, have an opening in the nose, and manifest through body hair. Pathogenic influences can easily invade the lungs without obstruction. External pathogenic factors, known as the "six excesses", primarily affect the lungs and result in, common pulmonary conditions in children, such as cold, cough, bronchitis, recurrent respiratory infections, and asthma. Therefore, pulmonary diseases are very widespread conditions in the field of pediatrics [1]. Professor Wang Lie, a Master of Traditional Chinese Medicine (TCM), has dedicated his entire career to the research of pediatric lung diseases and respiratory disorders. He possesses unique perspectives and a wealth of experience in diagnosing, treating, and prescribing herbal formulas; in addition to utilizing medicinal herbs to treat pediatric pulmonary diseases [2]. Over the years, Professor Wang has accumulated a wealth of clinical data and academic experience in the field of pediatric TCM treatment. In light of the need to inherit and innovate upon the academic thoughts and clinical experience of esteemed TCM practitioners, the introduction of modern information technology becomes crucial to achieve the digitization, intelligence, and transmission of this valuable TCM diagnostic and treatment experience.

The concept of knowledge graph was introduced by Google in 2012. It is a structured representation method that describes concepts, entities, and their relationships in the objective world in a structured form. Knowledge graphs utilize graph structures to model, identify, and infer complex associations between entities. They have found extensive applications in various domains such as intelligent question answering and big data analytics [3–6]. In the research field, knowledge graphs have gained increasing attention among researchers. Yuanyuan Cao [7] constructed an intelligent question answering system based on a knowledge graph for campus network services. The intelligent question answering system realized the convenience and intelligentization of network information services for teachers and students on campus, and improved the service efficiency of campus informatization. Jinlei Liu [8] utilized the Neo4j graph database to build a knowledge graph for coronary heart disease, applying knowledge graph technology to visualize the process of syndrome differentiation and treatment, TCM diagnosis, and medication prescription. This approach allows TCM practitioners to intuitively represent the relationship between diagnosis, treatment processes, and data, thereby providing methods for standardizing and formalizing the diagnosis and treatment of coronary heart disease in TCM.

The exploration and inheritance of the academic thoughts and clinical experience of esteemed TCM practitioners have always been highly regarded in the field of TCM. Professor Wang has accumulated an impressive amount of clinical data and academic experience in the diagnosis and treatment of pediatric pulmonary diseases, particularly by establishing a comprehensive system of principles, methods, formulas, and herbal medicines. In this study, we have utilized Professor Wang's diagnostic and treatment knowledge for pediatric pulmonary diseases as the data source. We designed and constructed the pattern layer and data layer of a knowledge graph, by utilizing the Neo4j graph database to store the knowledge graph. This enables the dynamic visualization display and semantic retrieval functionalities, hence assisting young TCM practitioners to learn Professor Wang's diagnostic and treatment experience for pediatric pulmonary diseases. It supports the inheritance of esteemed TCM practitioners academic thoughts and clinical knowledge, provides new tools to understand the diagnostic characteristics of renowned TCM practitioners and for better clinical decision-making, and lays the foundation for the deep integration of artificial intelligence and TCM.

2 Materials and Methods

2.1 Data Sources

The data in this study were unstructured text materials that have been publicly published. Relevant literature was electronically retrieved from databases such as Chinese Journal Net, Wanfang Data and CNKI Biological and Medical Literature. Literature and works that could not be obtained through databases were manually retrieved from the library of Changchun University of Chinese Medicine. Literature papers with Professor Wang's clinical experience as the theme were screened, with "Professor Wang Lie" as the keyword or subject retrieval term, and related literature materials were retrieved manually. As a result, 860 papers were included, and repeated or unqualified papers were excluded, 152 papers were reserved.

2.2 Inclusion Criteria

1) Research literature published from January 2003 to May 2023, focusing on the clinical diagnostic and treatment experience of renowned TCM master Professor Wang.
2) Clinical experience literature specifically related to pediatric pulmonary diseases.
3) The clinical experience literature should include comprehensive records, such as, disease name, symptoms, etiology and pathogenesis, disease nature and location, tongue appearance, pulse condition, treatment methods, and prescribed medications.

2.3 Exclusion Criteria

1) Diseases with diagnoses that combine other visceral diseases (pulmonary diseases combine heart and liver diseases)
2) Review articles, conference reports, personal reports.

2.4 Standardized Processing of Data

Referring to the National Prescribed Textbook of Chinese Medical Industry in the "13th Five-Year Plan" Chinese Medicines [9], the Chinese medicine names, flavors, and meridians that appear in the prescriptions were standardized in a unified manner, such as specifying "Fritillaria" as "Bulbus Fritillaria cirrhosa D.Don", "Perilla" as "Perilla frutescens", and "Alkekengi officinarum var. Franchetii (Mast.) R.J.Wang" as "Calyx seu Fructus Physalis".

2.5 Knowledge Extraction

The process of knowledge extraction includes entity extraction, relation extraction and attribute extraction, and knowledge extraction is presented in the form of SPO triples [10]. For example, what is the best way to extract knowledge regarding "Pericarpium Citri Reticulatae treats cough". "Pericarpium Citri Reticulatae" is the name of a Chinese medicine, which can be regarded as an independent entity, and the attribute is Chinese medicine. "Cough" can also be regarded as an entity, and the attribute is a disease. Then the relationship between Pericarpium Citri Reticulatae and cough is the treatment relationship. The representation form of SPO triples is [Pericarpium Citri Reticulatae] <treatment relationship> [cough]. Therefore, in this study, the extraction of entity concepts starts with manual annotation. Relevant entities, relationship attributes, and their relationships were extracted from the clinical experience of Professor Wang in diagnosing and treating pediatric pulmonary diseases. The standards set by National Administration of Traditional Chinese Medicine for terminologies in the TCM industry were used for references in extraction of entity concepts. It combines Professor Wang's academic thoughts and diagnostic-treatment pathways for pediatric pulmonary diseases. After discussions with experts from Professor Wang's research team, a total of 20 entity concept labels were determined, including people, book titles, theories, diseases, contents, symptoms, stages, types, causes, pathogenesis, disease nature, disease location, therapeutic methods, prescriptions, Chinese herbal medicines, effects, alternative names, characteristics, medicinal properties, and meridians. Furthermore, 22 semantic relationship types were identified, namely writing, founding, guiding, content is, symptom is, including, stage is, has, cause is, pathogenesis is, disease nature is, disease location is, therapeutic method is, treats, consists of, effect is, classified as, derived from, alternative name is, characteristic is, medicinal property is, and meridian affiliation is. The relationships between these entity concepts are specified in Table 1.

Table 1. Professor Wang Lie's knowledge graph relationship summary table

Head	Tail	Relation
people	book_title	writes
people	theory	founds
theory	disease	guides
theory	content	content_is
disease	disease	includes
disease	stage	stage_is
disease	type	has
disease	symptom	symptom_is

(continued)

Table 1. (*continued*)

Head	Tail	Relation
disease	cause	cause_is
disease	pathogenesis	pathogenesis_is
disease	disease_nature	disease_nature_is
disease	disease_location	disease_location_is
disease	therapetic_method	therapeutic_method_is
type	causes	cause_is
type	symptom	symptom_is
prescription	disease	treats
prescription	herbal	consists_of
herbal	type	treats
herbal	disease	treats
herbal	effect	effect_is
herbal	herbal	classifies
herbal	book_title	derived_from
herbal	alternative_name	alternative_name_is
disease	alternative_name	alternative_name_is
herbal	characteristic	characteristic_is
herbal	medicinal_property	medicinal_property _is
herbal	meridian	meridian_affiliation_is

2.6 Knowledge Graph Construction Method

Neo4j is an open-source, high-performance, graph database based on Java. Unlike traditional relational data model graph databases, it stores structured text data in a graph network and visually displays the relationship associations between complex data [11, 12]. The main method is to use nodes and edges to correlate the relationships between complex data. The nodes in each graph network represent entities or concepts, and edges represent the relationships among entities and between entities and concepts [13, 14]. Therefore, this study used the Neo4j graph database to store the knowledge graph. Based on the extracted data, it was converted into a CSV file, The code has been written in Python3.6 Py2neo to import the database to draw the knowledge graph.

3 Results

3.1 Pattern Layer Graph

After referring to the TCM terminology standards established by National Administration of Traditional Chinese Medicine and consulting experts from the Professor Wang's research team, a total of 20 entity concept labels, 29 entity nodes, 28 entity relationships, and 22 entity relationship types were finalized in the schema-level graph construction. Based on the predefined entity concepts and semantic relationships, the knowledge graph schema layer for Professor Wang's expertise in pediatric pulmonary diseases was constructed (Fig. 1).

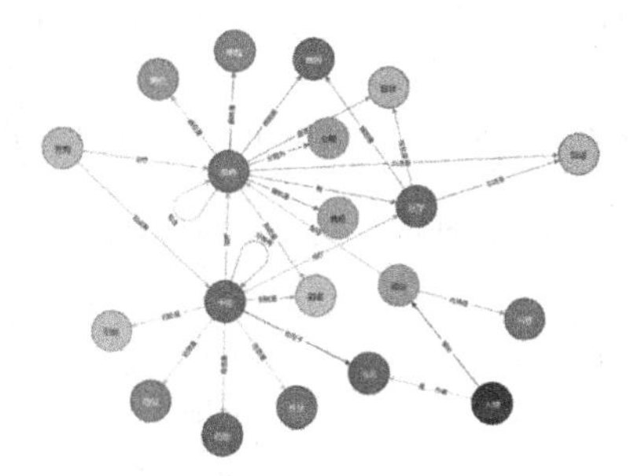

Fig. 1. Knowledge Graph Model Layer of Professor Wang Lie's Diagnosis and Treatment of Pediatric Pulmonary Diseases

3.2 Data Layer Graph

A total of 1,470 clinical experience and treatment data regarding Professor Wang's expertise in pediatric pulmonary diseases were collected for this study. These data were transformed into a CSV file and imported into the Neo4j graph database using Python. The construction was performed based on the 20 entity concept labels and 22 entity relationship types, resulting in 870 entity nodes and 1,469 entity relationships (Table 2). By executing the Cypher query "MATCH (n:*) RETURN n," the knowledge graph representing Professor Wang's experience in diagnosing and treating pediatric pulmonary diseases could be obtained (Fig. 2).

4 Application of Professor Wang's Knowledge Graph for the Diagnosis and Treatment of Pediatric Pulmonary Diseases

Cypher language is a powerful query language of Neo4j [14]. With the help of the match statement to query-specified nodes or relationships and the where statement to set query conditions, the entities and relationships between entities in diagnostic experience can be queried [15–17], providing references for the utilization of Professor Wang's knowledge resources for the diagnosis and treatment of pediatric pulmonary diseases.

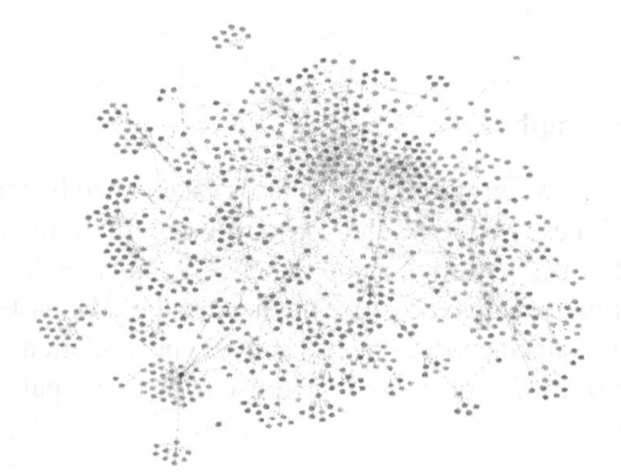

Fig. 2. Professor Wang's Experience Knowledge Graph on the Treatment of Pediatric Pulmonary Diseases

Table 2. Label and Relationship Type Count Table

Label	Nodes Number	Relationship Type	Count
people	1	writes	17
book_title	18	founds	26
theory	28	guides	12
disease	119	content_is	14
content	14	symptom_is	271
symptom	202	includes	29
stage	11	Stage_is	19
type	47	has	43
cause	35	classifies	4
pathogenesis	16	causeis	35
disease_nature	1	Pathogenesis_is	16
disease_location	4	disease_nature_is	1
therapeutic_method	63	disease_location_is	16
prescription	56	therapeutic_method_is	70
herbal	177	treats	230
effect	97	consists_of	494
alternative_name	18	effect_is	120
characteristic	1	derived_from	1
medicinal_property	11	aiternative_name	18
meridian	10	characteristic_is	1
		medicinal_property_is	13
		meridian_affiliation_is	29

Through the use of Cypher language's MATCH query, it is possible to visualize the semantic relationships [18] within the graph. To provide a clearer and more intuitive representation of the query graph, each entity concept label is represented by nodes with different colors. For example, to understand Professor Wang's insights, treatments, and herbal formulas for treating "Pediatric Asthma in Remission," the following query can be executed: MATCH (n: Disease) where n. name = "Pediatric Asthma in Remission" RETURN n. This query will return a graph, as shown in Fig. 3. From the graph, it can be observed that "Pediatric Asthma in Remission" falls under the category of pediatric asthma. The symptoms associated with this condition include productive cough, shortness of breath, and fatigue. The pathogenesis involves lung-spleen-kidney deficiency and phlegm stagnation, with the affected areas being the lungs, spleen, and kidneys. The treatment approach focuses on clearing heat, purging the lungs, stopping the cough, resolving phlegm, tonifying the spleen, and nourishing the kidneys. Within the context of pediatric asthma in remission, there are two pattern types identified: namely "Pediatric Asthma with Spleen Deficiency and Phlegm Accumulation" and "Pediatric Asthma with Lung Heat Stagnation." According to Professor Wang's academic thoughts and treatment system developed over many years, during the remission stage of an asthma attack, when the pathogenic factors have subsided and the body is in a state of deficiency, qi deficiency and internal dampness may lead to the accumulation of phlegm, resulting in the primary symptom of excessive cough with phlegm. Treatment in this stage primarily focuses on supplementing deficiencies and resolving phlegm. Therefore, for the pattern of lung heat stagnation, the prescription "Xie Fei Fang" (Clearing Heat and Purging the Lungs Formula) is used, which consists of Scutellaria baicalensis Georgi, Calyx seu Fructus Physalis, Chelidonium majus, Fritillaria cirrhosa D.Don, Eriobotrya japonica Thunb, Stemona japonica, Pinellia rhizome praeparatum cum alumine, and Fructus trichosanthis. For the pattern of spleen deficiency with phlegm accumulation, the prescription "Hua Tan Fang" (Resolving Phlegm Formula) is used, which consists of Euryale ferox, Pinellia rhizome praeparatum cum alumine, Exocarpium Citri Rubrum, Wolfiporia cocos, Platycodon grandiflorus, Fritillaria cirrhosa D.Don, Fructus trichosanthis, Radix Glehniae. With the constructed knowledge graph, the corresponding information about the disease etiology, pathogenesis, treatment methods, and prescriptions can be obtained by executing the appropriate query statements include the disease name.

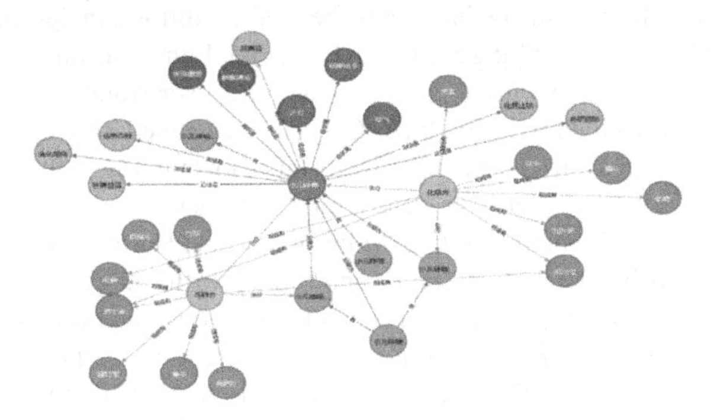

Fig. 3. Knowledge Graph of Asthma Remission Period in Children

By executing the query "MATCH (n:Symptom) where n.name = "cough" RETURN n", a graph with represents the symptom group related to the symptom "cough" can be obtained as shown in Fig. 4. The associated symptoms include cough in relation to pediatric asthma, pediatric bronchitis, pediatric pneumonia, recurrent respiratory infections in children, pediatric nasal congestion, pertussis in children, common cold in children, and pediatric cough. Based on the formed knowledge graph by prior queries, more relationships can be revealed and explored by clicking on any entity label in the graph. For example, clicking on the entity label for pediatric nasal congestion will lead to a visual representation centered around pediatric nasal congestion in the graph. The symptoms manifested in this case include sudden or recurrent sneezing, nasal itching, nasal congestion, clear and watery nasal discharge, coughing, wheezing, eye-related symptoms, mouth-breathing, and associated sleep snoring. Clicking further on the entity label for the remission period of pediatric nasal congestion will provide a visual display of related information such as etiology, pathogenesis, treatment methods, prescriptions, and composition of herbal formulas specific to the remission period of pediatric nasal congestion.

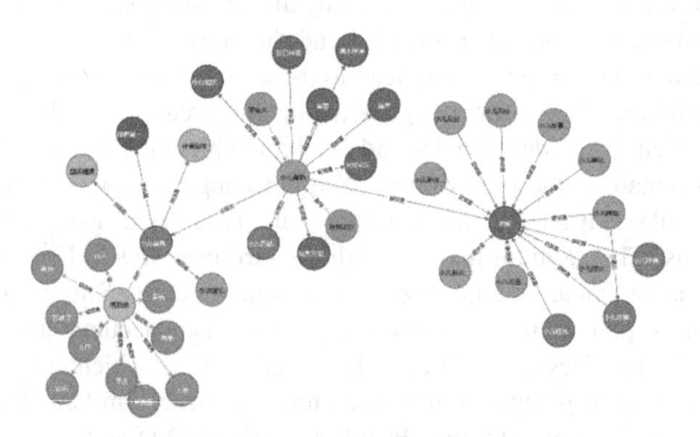

Fig. 4. Knowledge Graph of Cough

Professor Wang is skilled in treatment of various diseases using Chelidonium majus, such as fever, cough, wheezing, and asthma in children. He has developed numerous renowned formulas for treating pediatric lung-related conditions, including Xie Fei Fang (Clearing Lung Formula) and Ping Fei Fang (Balancing Lung Formula), in which Chelidonium majus was commonly used. To understand Professor Wang's perception of Chelidonium majus, the following query "MATCH can be executed: (n:Herb) where n.name = " Chelidonium majus " RETURN n". The results shown in Fig. 5 indicate that Chelidonium majus belongs to the Papaveraceae family and has various folk names, including Picrorhizae rhizoma, Rhizoma Coptidis, Munronia henryi Harms, and Gelsemium elegans. It has a slightly bitter and warm nature and belongs to the lung and spleen meridians. Chelidonium majus is known for its effects in relieving cough and wheezing, alleviating pain, promoting diuresis, killing bacteria, and resolving toxins. Chelidonium majus is included in formulas such as Xie Fei Fang, Ping Fei Fang, Huan Xiao Fang, Xiao Er Xiao Ke Chuan Fang, Bi Xiao Fang, and Xiao Er Bai Bei Zhi Ke Ling. Professor Wang pioneered the use of Chelidonium majus in treating coughs, starting with its application for pertussis (whooping cough), which yielded good therapeutic effects. Later, Chelidonium majus was also used to treat wheezing. Since Professor Wang first introduced it to pediatric clinical practice, it has been effectively used in the treatment of pediatric pertussis, bronchitis, asthma, pneumonia, and other conditions.

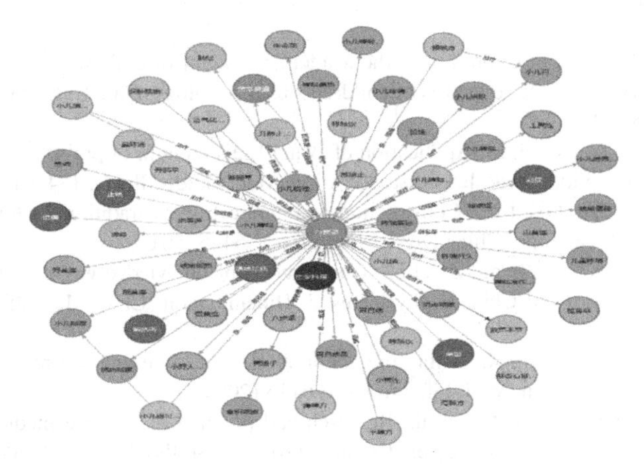

Fig. 5. Chelidonium majus Knowledge Graph

5 Conclusion

Professor Wang is a third-generation national master of TCM and and he has been a doctor of TCM for pediatric disease more than 60 years. He has extensive clinical experience, particularly in the prevention and treatment of pediatric lung-related diseases. Professor Wang has proposed numerous famous theories, including theories on cough, three-stage treatment of asthma, among others, totaling over 10 kinds of theories. Unprecedented in the history of TCM pediatrics are specific TCM medicines for pediatric asthma, such as Xiao Er Xiao Ke Chuan, Xiao Er Fei Re Ping, and Xiao Er Yi Qi Gu Ben Jiao Nang, and et al. The wealth of academic ideas and clinical knowledge of Professor Wang is mostly in the form of unstructured text, comprising scattered, unordered, and independent concepts or viewpoints. Therefore, there is an urgent need to integrate and summarize this knowledge into a structured format, to provide a pathway for inheriting and promoting the academic ideas and clinical experience of esteemed senior TCM practitioners.

This study focuses on utilizing Professor Wang's diagnostic and treatment experience in pediatric lung-related diseases as the data source, integrating his academic ideas and clinical experience into a knowledge graph. It explores the principles, evidence, formulas, and herbs used by Professor Wang in the diagnosis and treatment of pediatric lung-related diseases. This will lay the foundation for optimizing TCM treatment plans for pediatric lung-related diseases and the development of intelligent diagnostic and expert treatment systems in TCM for pediatrics illnesses.

Acknowledgment. Key research and development projects of Jilin Province Science and Technology Development Plan (20210204120YY).

References

1. Bai, N.: Preliminary exploration of the academic thought of children's pharmacology and cough theory instraight deduction of children's pharmacology. Clin. Res. Tradit. Chin. Med. **13**(33), 4–5 (2021)
2. Guo, T., Sun, L., Wang, L.: Application of Wang Lie's thought of prevention and treatment of diseases in children with lung diseases. Chin. Med. Guide **18**(17), 134–6+45 (2021)
3. Yu, T., Li, J., Yu, Q., et al.: Construction and application of traditional Chinese medicine health knowledge graph. China Digit. Med. **12**(12), 64–6 (2017)
4. Deng, Y., Zhou, W., Zhang, Z., et al.: Construction of knowledge graph based on medical records of famous old Chinese physicians. J. Hunan Tradit. Chin. Med. **35**(07), 186–187 (2019)
5. Yang, Y., Li, Y., Zhong, X.: A link prediction model for traditional Chinese medicine case knowledge graph. Chin. Med. Inf. **39**(03), 1–6+15 (2022)
6. Niu, Z., Zhang, X., Xu, W., et al.: Research and practice of Chinese medical knowledge learning based on knowledge graph. Comput. Program. Skills Maint. (02), 20–23 (2023)
7. Cao, Y., Su, W.: Research on knowledge graph construction and Q&A application of campus network operation and maintenance service based on Neo4j. Adv. Educ. Humanit. Soc. Sci. Res. **1**(3), 360–366 (2023)
8. Liu, J., Hui, X., Zhang, Z., et al.: Construction of coronary heart disease knowledge graph based on TCM diagnosis and treatment guidelines. Chin. J. Exp. Tradit. Med. Formulae **29**(07), 208–215 (2023)
9. Zhao, Y., Li, Y., Zhang, J.: Construction of traditional Chinese medicine knowledge graph for influenza. Chin. J. Med. Libr. Inf. Sci. **30**(05), 24–30 (2021)
10. Zhong, G.: Chinese Materia Medica, 4th edn. China Press of Traditional Chinese Medicine, Beijing (2016)
11. Cao, H., Xu, J., Dou, F.: Research progress of graph neural network in knowledge graph construction and application. Comput. Era **6**, 35–38 (2020)
12. Wu, D., Zhou, Z., Shang, H.: Analysis of research hotspots and trends of TCM health examination based on knowledge graph. Comput. Era (02), 59–61+5 (2022)
13. Chen, S., Xia, S., Deng, W., et al.: Research on coronary heart disease TCM syndrome differentiation and treatment knowledge graph based on Neo4j. China Med. Guide **18**(21), 138–141 (2021)
14. Tong, Y., Chen, L.: Analysis of hotspots and trends in TCM health examination based on knowledge graph. Chin. Med. Manag. Mag. **31**(08), 194–196 (2023)
15. Li, L., Li, Y., Xu, H., et al.: Research and practice of building intelligent Q&A APP for traditional chinese medicine health preservation based on knowledge graph. J. Med. Inf. **43**(07), 50–54 (2022)
16. Xiao, F., Zhang, S., Hu, Z.: Research on disease prevention and control knowledge graph based on Neo4j. Electron. Technol. Softw. Eng. **22**, 180–182 (2021)
17. Yin, D., Zhou, L., Zhou, Y., et al.: Research on "graph search mode" design of traditional Chinese medicine prescription knowledge graph. Chin. J. Tradit. Chin. Med. Inf. **26**(08), 94–98 (2019)
18. Yang, Y., Li, Y., Shuai, Y., et al.: Construction of knowledge graph based on traditional Chinese medicine case records. J. Med. Inf. **43**(10), 50–54 (2022)

A Novel SEIAISRD Model to Evaluate Pandemic Spreading

Hui Wei[1,2] and Chunyan Zhang[1,2(✉)]

[1] Department of Automation, College of Artificial Intelligence, Nankai University, Tianjin 300071, China
[2] Tianjin Key Laboratory of Intelligent Robotics, Nankai University, Tianjin 300071, China
zhcy@nankai.edu.cn

Abstract. Starting in late March 2021, several countries have been massively promoting vaccination, sparking a global debate on whether containment or suppression strategies response to COVID-19 remain necessary. To evaluate the potential impact of SARS-CoV-2 vaccines and control strategies, a data-driven, discrete-time compartmental model is developed, incorporating the unparalleled characteristics of the epidemic. This model is then calibrated to country-level reported active cases, accumulative recovered and deceased cases, as well as daily new cases. Indeed, this paper applies the calibrated model to reconstruct the transmission dynamics of the outbreak over time, especially after the emergence of various vaccines. The results, combined, demonstrate that the current vaccines appear to be insufficient to eliminate this disease, despite the increasing number of vaccinations, highlighting that such control strategies or restrictions are not supposed to be completely lifted across the world. Finally, there is no denying that the proposed model and framework can be readily extended to other countries, states, regions, or organizations.

Keywords: COVID-19 · Data-driven approach · Epidemiological model · Decision-making

1 Introduction

The ongoing COVID-19 epidemic, an international public health emergency caused by the novel coronavirus [18], has been posing unprecedented challenges to social economics, humanitarian rights, and healthcare systems around the world [4]. COVID-19 was officially declared a global pandemic by the World Health Organization on 11 March 2020 [21]. This epidemic, as of late October 2021, has spread rapidly to over 220 countries, causing more than 240 million accumulative confirmed cases and 4.9 million deaths, which thus be regarded as a more terrible pandemic compared to the Spanish flu. Among these countries, the most representative examples, affected by the epidemic, include such as

Supported by the National Natural Science Foundation of China under Grants 62073174, 62073175 and 91848203.

B. Xin et al. (Eds.): IWACIII 2023, CCIS 1931, pp. 139–152, 2024.
https://doi.org/10.1007/978-981-99-7590-7_12

America with over 46 million cases and 763 thousand deaths, India with over 34 million cases and 450 thousand deaths, Brazil with over 21 million cases and 600 thousand deaths, England with over 8.9 million cases and 140 thousand deaths, as well as Russia with over 8.3 million cases and 235 thousand deaths, etc.

China is the first batch of countries to experience the challenge associated with the COVID-19 epidemic. From 23 January to 8 April 2020, the Chines authority, by imposing a series of stringent lockdown measures or non-pharmaceutical interventions, successfully contained the large-scale outbreak of COVID-19 in Wuhan, the capital of Hubei province [15]. This control strategy, to date, results in less than 100 thousand confirmed cases and 4.7 thousand deaths for mainland China. Since then, these strategies such as enhanced testing, contact tracing, household quarantine, mask-wearing, and social distancing, etc., were successively followed by South Korea, Australia, Singapore, Denmark, and other countries [13]. Despite such rigorous requirements, those countries have successfully given evidence to the possibility of the aforementioned interventions for alleviating the spread of the outbreak. Such extreme strategies, nevertheless, may be able to interrupt temporarily the disease transmission, yet come at the price of tremendous economic losses or social destruction. In the countries that have curbed the initial outbreak of COVID-19, there is still a continued controversy over when to relax, where to be implemented, and how to reopen economic and social activities [6].

To help policy-makers to formulate optimal policies for restarting economic activity and normal social functioning, while avoiding a future resurgence of the outbreak, mathematical models are proved as a unique, yet efficient tool for this aim. Specifically, mathematical models can contribute profound insight for them into the potential impact of existing control strategies, as well as into the evaluation of possible governmental policies [1]. A typical example of the latter, for instance, is that some countries have experienced a thorough shift in defending the so-called "herd immunity" policies even without any interventions. This, to a large degree, can be contributed to the model developed by the Imperial College London that predicts extensive death tolls before achieving this intention. More incisive insights into the transmission dynamics of such large-scale pandemics can be benefited by using SIR- or SEIR-type compartmental models [22]. For the COVID-19 epidemic, several compartmental models have been flourished, considering the unparalleled characteristics of this epidemic such as asymptomatic infection, longer incubation period, and super-spreader event [10],?, etc., even incorporating post-epidemic interventions (e.g., distancing-like measures or quarantine scenarios) [11]. Those compartmental models have indicated that the disease transmission, through rapidly testing, quarantining and contact tracing, can be significantly interrupted [2]. Alternatively, most of them were established with incomplete data, affecting the confidence in produced results for driving public health policy [5]. Validating COVID-19 modelling predictions is thus of great importance.

So far, for public health policies imposed in severely affected countries, it has been characterized by insufficient adherence to social distancing, logistic challenges for implementing large-scale contact-tracing, along with deficient detecting kits and relevant supplies [9]. Recent researches reveal that increased detecting and contact-tracing efficiency may contribute to reducing the significantly increasing number of accumulative confirmed cases and deaths in America, England, Italy, and other countries [20]. Several modelling researches have looked at the spatial dynamics of the outbreak and its evolution in the region at the state level [17] but, to our knowledge, limited work has focused on the functions of diverse containment and suppression strategies [19]. Comprehending the degree to which these strategies, such as enhanced detecting, contact tracing and quarantining, local lockdown, as well as social distancing, etc., influence locally the disease transmission is fundamental for predicting and preventing second or even multiple resurgences of the epidemic [8]. This, in return, will promote the optimization of these strategies to alleviate the harmful economic burden caused by COVID-19, while developing energetically effective vaccines and associated therapeutics. Most modelling researches, however, were gained under a common assumption, that is, without an effective vaccine. Among those researches, few took into account the impact of vaccines and their relevant vaccination information. This has inevitably sparked a debate for us on whether such control strategies are necessary, and on how to release them while escaping from future outbreaks, after the emergency of various vaccines.

In this work, our primary purpose is to further explore the evolution of the COVID-19 epidemic over time in each stage but, more importantly, after the emergency of various vaccines. Specifically, a data-driven, discrete-time compartmental model, named SEIAISRD, is developed, incorporating the aforementioned characteristics of COVID-19, even combining the effectiveness of various vaccines and relevant vaccination information. The model parameters, using the extended Kalman filter algorithm, are practically calibrated to country-level reported active cases, accumulative recovered and deceased cases, as well as daily new cases from 24 January 2020 to 22 October 2021. Additionally, this calibrated model is then applied to evaluate the potential impact of current SARS-CoV-2 vaccines and control strategies implemented in each country, presenting a data-fitting comparison between the estimated and reported cases. Finally, this model, by just recalibrating and recalculating the assumed values for model parameterization, can be extended or readily applicable to reevaluate the current situation responding to the epidemic for those representative countries, such as America, India, England, China, and Brazil, etc.

2 Methods

2.1 The SEIAISRD Model

This paper has modified the generalized SIR or $SEIR$ model to capture the evolution of COVID-19 over time in each stage, especially after the emergence of various vaccines. The SEIAISRD model, combining the unprecedented characteristics of this epidemic, assumes that susceptible individuals (S), to a certain potential, can become exposed individuals (E), through any contact with the

SARS-COV-2 virus. After passing for an incubation period, E, who is in latent yet non-infectious stage, can either be transmitted to asymptomatic infected (I_A) or symptomatic infected (I_S). Finally, the infected individuals will be transmitted to the removed compartments, identifying recovered (R) or deceased (D) individuals. Note that, I_A will be recovered following hospitalization, whereas I_S, especially those with severe symptoms, are faced with the risk of death in the initial outbreak. The flow diagram of this model is schematically described in Fig. 1. To balance fidelity and identifiability of the calibrated model, it is reasonable for us to assume that all infected cases are effectively tested or isolated, and thus unavailable for transmitting the virus.

The proposed model, describing, respectively, the dynamics of the following six compartments (e.g., S, E, I_A, I_S, R and D), is given by:

$$\frac{dS}{dt} = -\frac{(1-\alpha)\beta S(I_A + I_S)}{N}, \tag{1}$$

$$\frac{dE}{dt} = \frac{(1-\alpha)\beta S(I_A + I_S)}{N} - \epsilon E, \tag{2}$$

$$\frac{dI_A}{dt} = \rho_1 \epsilon E - \gamma_1 I_A, \tag{3}$$

$$\frac{dI_S}{dt} = \rho_2 \epsilon E - (\gamma_2 + \mu) I_S, \tag{4}$$

$$\frac{dR}{dt} = \gamma_1 I_A + \gamma_2 I_S, \tag{5}$$

$$\frac{dD}{dt} = \mu I_S, \tag{6}$$

where β and γ denote, respectively, the infection rate and the recovery rate. Among which, γ_1 is the recovery rate of asymptomatic infected individuals (I_A), yet γ_2 is that of symptomatic infected individuals (I_S). $\alpha \in [0,1]$is a parameter modelling the effectiveness of SARS-COV-2 vaccines. The remaining parameters include such as the mean exposed period ϵ (day^{-1}), the case fatality rate mu, as well as the proportion of exposed individuals (E) proceeding to asymptomatic (or symptomatic) infections ρ_1 (or ρ_2), obviously, $\rho_1 + \rho_2 = 1$. Remark that since this epidemic is transmitted by a person-to-person pattern, and no evidence showing that parasite vector or environmental parameters have immensely influenced its infection rate, it thus is assumed that β is the same for all countries. Besides, the actual population of the country N can be divided into the aforementioned six compartments and satisfies

$$N = S + E + I_A + I_S + R + D. \tag{7}$$

Of note, the actual epidemic data of COVID-19 for each country [12], collected by the Johns Hopkins CSSE Repository, includes the following reported cases, such as the active cases, accumulative recovered and deceased cases, as well as daily new confirmed cases. To fit accurately the proposed model to these data, a discrete-time augmented SEIAISRD model, by augmenting the infection rate β and daily new confirmed individuals (C) as additional state variables, can be obtained as follows:

$$S(k+1) = S(k) - \frac{(1-\alpha)\beta(k)S(k)(I_A(k)+I_S(k))\Delta t}{N} + \omega_1(k), \tag{8}$$

$$E(k+1) = E(k) + \frac{(1-\alpha)\beta(k)S(k)(I_A(k)+I_S(k))\Delta t}{N} - \epsilon E(k)\Delta t + \omega_2(k), \tag{9}$$

$$I_A(k+1) = I_A(k) + \rho_1\epsilon E(k)\Delta t - \gamma_1 I_A(k)\Delta t + \omega_3(k), \tag{10}$$

$$I_S(k+1) = I_S(k) + \rho_2\epsilon E(k)\Delta t - (\gamma_2+\mu)I_S(k)\Delta t + \omega_4(k), \tag{11}$$

$$R(k+1) = R(k) + \gamma_1 I_A(k)\Delta t + \gamma_2 I_S(k)\Delta t + \omega_5(k), \tag{12}$$

$$D(k+1) = D(k) + \mu I_S(k)\Delta t + \omega_6(k), \tag{13}$$

$$C(k+1) = C(k) + (\gamma_1+\gamma_2+\mu)(I_A(k)+I_S(k))\Delta t - C(k)\Delta t + \omega_7(k), \tag{14}$$

$$\beta(k+1) = \beta(k) + \omega_8(k), \tag{15}$$

where Δt is the simulation time step (this paper sets $\Delta t = [0.1, 0.001]$). The White Gaussian noise $\omega(k) = [\omega_1(k) \quad \omega_2(k) \quad \cdots \quad \omega_8(k)]^T$ is utilized to express model uncertainty and is assumed to be uncorrelated. Extending a parameter to a new state variable, note that, is an ordinary measure when estimating model parameters using the extended Kalman filter (EKF) (more details are illustrated on page 422 of [16]).

2.2 Estimating the Effective Reproduction Number

The effective reproduction number, R_t, is defined by the average number of second-generation infected cases (e.g., asymptomatic or symptomatic infected cases) transmitted from a single infected individual at a certain time t [14]. It is still unclear whether the individuals recovered from the COVID-19 epidemic will be re-infected, whereas initial evidence reveals that this is little possibility [23]. This paper thus assumes that only a single infected case relevant to this epidemic may occur to any single individual. As reminded above, R_t, as a qualitative index, is often used to capture the transmission dynamics of the epidemic. For example, as $R_t > 1$, the epidemic will spread rapidly among the population, whereas the epidemic will gradually disappear for $R_t < 1$. Accordingly, it presents a quantitative tool of whether further control efforts or interventions are necessary to curtail the spread of COVID-19. Based on the series of daily new confirmed cases, R_t is considered here to be the source of the general update equation during a birth process, and defined as

$$R_t = \frac{S(t)}{N}\beta(1-\alpha)\left(\frac{\rho_1}{\gamma_1} + \frac{\rho_2}{\gamma_2+\mu}\right), \tag{16}$$

where the term $S(t)/N$ is utilized to compensate the reduction in susceptible individuals, and then, multiplies by the infected risk of each contact for infected individuals and their untraced contacts, $\beta(1-\alpha)$, and their mean infection period, $\left(\frac{\rho_1}{\gamma_1}+\frac{\rho_2}{\gamma_2+\mu}\right)$. This may be, respectively, explained by the contribution of asymptomatic infected (I_A), $\frac{S(t)}{N}\beta(1-\alpha)\frac{\rho_1}{\gamma_1}$, and symptomatic infected (I_S), $\frac{S(t)}{N}\beta(1-\alpha)\frac{\rho_2}{\gamma_2+\mu}$.

In this section, the extended Kalman filter (EKF) is adopted by us to estimate dynamically the effective reproduction number. For brevity of the reading pubic, an augmented state vector is defined as $x(k+1) = [S(k+1) \quad E(k+1) \quad I_a(k+1) \quad I_s(k+1) \quad R(k+1) \quad D(k+1) \quad C(k+1) \quad \beta(k+1)]^T$, such that the discrete-time augmented SEIAISRD model (8)-(15) can be modified as $x(k+1) = f(x(k)) + \omega(k)$. Since then, let us regard $\hat{x}(k)$ as the estimated vector of $x(k)$ from the EKF. $f(x(k)) = f(\hat{x}(k)) + J_f(\hat{x}(k))(x(k) - \hat{x}(k))$ is obtained by applying first-order Taylor series expansion to f at $\hat{x}(k)$, among which, the Jacobian matrix is $J_f(\hat{x}(k))$ described by

$$J_f(\hat{x}(k)) = \begin{bmatrix} J_{11} & 0 & J_{13} & J_{14} & 0 & 0 & 0 & J_{18} \\ J_{21} & J_{22} & J_{23} & J_{24} & 0 & 0 & 0 & J_{28} \\ 0 & \rho_1\varepsilon\Delta t & 1-\gamma_1\Delta t & 0 & 0 & 0 & 0 & 0 \\ 0 & \rho_2\varepsilon\Delta t & 0 & J_{44} & 0 & 0 & 0 & 0 \\ 0 & 0 & \gamma_1\Delta t & \gamma_2\Delta t & 1 & 0 & 0 & 0 \\ 0 & 0 & 0 & \mu\Delta t & 0 & 1 & 0 & 0 \\ 0 & 0 & J_{73} & J_{74} & 0 & 0 & 1-\Delta t & 0 \\ 0 & 0 & 0 & 0 & 0 & 0 & 0 & 1 \end{bmatrix}, \tag{17}$$

where

$$J_{11}(\hat{x}(k)) = 1 - \frac{(1-\alpha)\beta(k)(I_A(k)+I_S(k))\Delta t}{N}, \tag{18}$$

$$J_{13}(\hat{x}(k)) = J_{14}(\hat{x}(k)) = -\frac{(1-\alpha)\beta(k)S(k)\Delta t}{N}, \tag{19}$$

$$J_{18}(\hat{x}(k)) = -\frac{(1-\alpha)S(k)(I_A(k)+I_S(k))\Delta t}{N}, \tag{20}$$

$$J_{21}(\hat{x}(k)) = \frac{(1-\alpha)\beta(k)(I_A(k)+I_S(k))\Delta t}{N}, \tag{21}$$

$$J_{22}(\hat{x}(k)) = 1 - \epsilon\Delta t, \tag{22}$$

$$J_{23}(\hat{x}(k)) = J_{24}(\hat{x}(k)) = \frac{(1-\alpha)\beta(k)S(k)\Delta t}{N}, \tag{23}$$

$$J_{28}(\hat{x}(k)) = \frac{(1-\alpha)S(k)(I_A(k)+I_S(k))\Delta t}{N}, \tag{24}$$

$$J_{44}(\hat{x}(k)) = 1 - (\gamma_2+\mu)\Delta t, \tag{25}$$

$$J_{73}(\hat{x}(k)) = J_{74}(\hat{x}(k)) = (\gamma_1+\gamma_2+\mu)\Delta t. \tag{26}$$

Further details of the extended Kalman filter applied to the SIR or SEIR-type model are illustrated in [16].

2.3 Data-Fitting and Sensitivity Analysis

Real-world reported cases, originating from 24 January 2020 to 22 October 2021, have been used to fit the SEIAISRD model. All data-fitting analyses have been performed with a MATLAB optimization toolbox using the extended Kalman filter. This optimization algorithm, given all inputs that vary within the ranges found from several researches, is verified as the optimal solution. Infected initial inputs have been determined from reported active cases, accumulative recovered, and deceased cases, whereas the remaining initial inputs could vary during the optimization process. To describe the inherent uncertainty of COVID-19, consequently to demonstrate the impact of combined control strategies and vaccines, the aforementioned results are the average outputs by running over 50 independent repetitive simulations. Specifically, the model parameters α, β, ϵ, γ_1, γ_2, μ together with the initial inputs have been chosen, incorporating a $\pm 20\%$ maximum variation from their normal values. It is suggested that the proposed model and algorithm are robust to several parameter variations, demonstrating the feasibility to curtail this epidemic and expandability to be extended to other countries across the world. Estimated 95% confidence intervals have taken the 1% perturbations of uniform distribution into account, to assess the corresponding confidence in the results, and finally to compare the fitting results with actual reported cases, such as active cases, accumulative recovered and deceased cases, as well as daily new confirmed cases. COVID-19 active cases, accumulative recovered and deceased cases, daily new confirmed cases, as well as vaccination information for mainland China, are obtained by the National Health Commission of the People's Republic of China [7] and are available at http://www.nhc.gov.cn/. Whereas for the representative countries (e.g., America, India, Brazil, Russia, France, Turkey, England, Argentina, Italy, and Colombia), collected by Johns Hopkins CSSE Repository, can be available at https://github.com/CSSEGISandData/COVID-19.

3 Results

3.1 Model Formulation and Validation

To capture the evolution of the ongoing COVID-19 epidemic over time in each stage, especially after the emergence of various vaccines, a data-driven, discrete-time compartmental SEIAISRD model is performed to evaluate the impact of current SARS-CoV-2 vaccines and control strategies for several representative countries (such as China, America, India, Brazil, Russia, France, Turkey, England, Argentina, Italy, and Colombia). The SEIAISRD model, combining the unprecedented characteristics of COVID-19, main includes the following compartments: susceptible, S; exposed, E; asymptomatic infected, I_A; symptomatic infected, I_S; recovered, R; and deceased, D; described schematically in Fig. 1. This is an epidemiological modelling tool that can reconstruct the transmission dynamics of COVID-19 within a population. As shown in Fig. 1, this model assumes that susceptible individuals (S) become exposed individuals (E), to a certain probability, through contact with any of the infected compartments (e.g., asymptomatic infected (I_A) or symptomatic infected (I_S), which means that the exposed individuals (E) are in latent, yet non-infectious stage. After passing for a mean exposed period, the exposed individuals (E) are then transitioning to asymptomatic infected (I_A) or symptomatic infected (I_S). Finally, the infected individuals will be transitioning to the removed compartments, identifying recovered (R) or deceased (D) individuals.

Table 1. Calibrated model parameters for the SEIAISRD model.

Parameter	Definition
N	Population size of the corresponding country
α	The effectiveness of SARS-COV-2 vaccine
β	Transmission rate
ϵ	Mean exposed period
ρ_1	Proportion of exposed transitioning to asymptomatic infected (I_A)
ρ_2	Proportion of exposed transitioning to symptomatic infected (I_S)
γ_1	Recovery rate of asymptomatic infected (I_A)
γ_2	Recovery rate of symptomatic infected (I_S)
μ	The case fatality rate

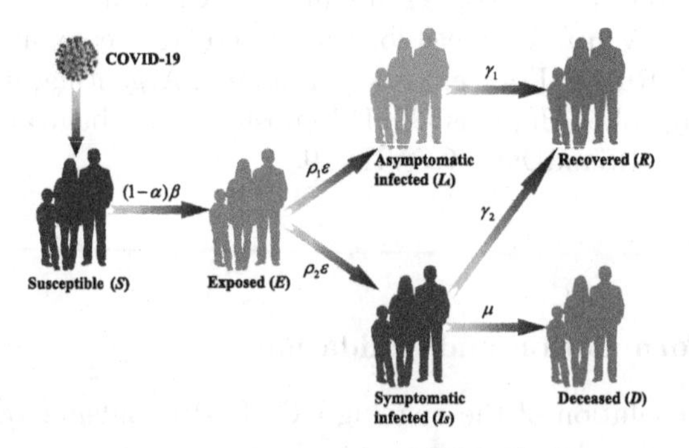

Fig. 1. Schematic diagram of the SEIAISRD model structure. The SEIAISRD, a date-driven, discrete-time compartmental model, is presented to reconstruct the transmission dynamics of the ongoing COVID-19 epidemic after the emergency of various vaccines. This model, combining the unprecedented characteristics of COVID-19, includes mainly six compartments, such as susceptible (S), exposed (E), asymptomatic infected (I_A), symptomatic infected (I_S), recovered (R), and deceased (D).

As regards validation, the model parameters, using the extended Kalman filter algorithm, are calibrated to country-level reported active cases, recovered cases, deceased cases, and daily new cases from 24 January 2020 to 22

October 2021. Estimating all the parameters regarding each country allows us to not only precisely fit the reported COVID-19 cases, but also to describe the diverse country situations and the disparate impacts of government policies aiming to contain this epidemic spread for those selected countries right now. Calibrated model parameters are highlighted in Table 1. A detailed description of the SEIAISRD model consideration, parameterization, and sensitivity analysis are further explained in Methods. Besides, further details on the real-world reported cases and vaccination information for each country can be found at https://github.com/CSSEGISandData/COVID-19, as collected by Johns Hopkins CSSE Repository [12].

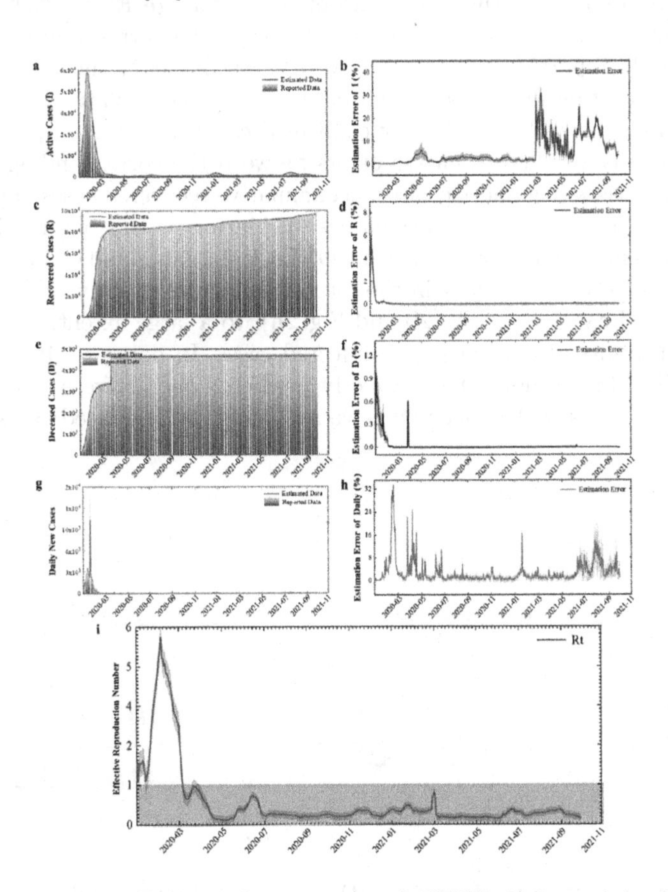

Fig. 2. Real-time modelling and fitting results of COVID-19 for mainland China: (a) Active cases; (b) Estimation error of active cases; (c) Accumulative recovered cases; (d) Estimation error of accumulative recovered cases; (e) Accumulative deceased cases; (f) Estimation error of accumulative deceased cases; (g) Daily new confirmed cases; (h) Estimation error of Daily new confirmed cases. (i) Effective reproduction number R_t.

3.2 Case Studies for the Representative Countries

As all we know, China is the first batch of the countries to experience the challenge caused by COVID-19, a case study on transmission dynamics of the epidemic for mainland China is thus investigated. Unlike most previous researches, this paper underscores the evolution of this epidemic over time in each stage, especially after the emergence of various vaccines. To this aim, the model parameters are calibrated to match the crucial characteristics of the reported COVID-19 cases. The results, as described in Fig. 2, show numerical simulations of real-time modelling and data fitting of COVID-19 for mainland China, including a comparative analysis of the relative Root Mean Square Errors (for brevity, it is next referred to as estimation errors) between the corresponding estimated and real-world reported cases. Among which, Fig. 2a, c, e, and g show, respectively, the evolution of the estimated active cases, accumulative recovered and deceased cases, as well as daily new confirmed cases. Figure 2b, d, f, and h represent, respectively, the estimation errors between the corresponding estimated and reported cases (e.g., active cases, accumulative recovered cases and deceased cases, as well as daily new confirmed cases), hinting at this observation is in a great agreement with recent reports of COVID-19 for mainland China. It is worth stressing that, the fluctuations in the estimation errors of active cases and daily new confirmed cases (e.g., Fig. 2b and Fig. 2h), to a large extent, are related to the abroad inputs of asymptomatic infected cases. In doing so, it is confirmed that the SEIAISRD model has correctly predicted the active cases, daily new confirmed cases, as well as accumulative recovered and deceased cases.

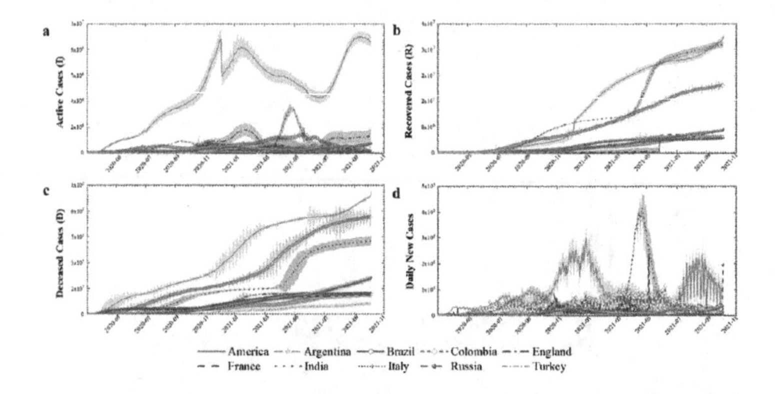

Fig. 3. Real-time fitting results of COVID-19 for several representative countries: (a) Active cases; (b) Accumulative recovered cases; (c) Accumulative deceased cases; (d) Daily new confirmed cases. Those selected countries are chosen from the top ten countries regarding accumulative confirmed cases, such as America, India, Brazil, Russia, France, Turkey, England, Argentina, Italy, and Colombia in order. The shaded error bands denote, respectively, 95% confidence intervals of the corresponding mean by running over 50 independent repetitive experiments. For easy comparison, raw data, represented by this plot, are selected from 1 March 2020 to 22 October 2021.

Furthermore, the time-varying effective reproduction number for mainland China, presented in Fig. 2i, is estimated by using the SEIAISRD model together with the extended Kalman filter, indicating that this epidemic keeps to a typical trajectory. It is well known that the effective reproduction number, R_t, is the average number of second-generation infected cases (e.g., asymptomatic or symptomatic infected cases) transmitted from a single infected individual at a certain time t [14]. Generally speaking, R_t, as a qualitative index, is often used to describe the real-time transmission dynamics of epidemic disease. For example, as $R_t > 1$, the epidemic will spread rapidly among the population, whereas the epidemic will gradually disappear for $R_t < 1$. It is worth stressing that the SEIAISRD model would be readily extended to other regions, states or countries. To this end, the representative examples of the countries, which belong to the top ten countries regarding accumulative confirmed cases (e.g., America, India, Brazil, Russia, France, Turkey, England, Argentina, Italy, and Colombia), are chosen to demonstrate the applicability of the proposed model. After which, most model parameters remain constant except for the effectiveness of the SARS-COV-2 vaccine α. The results, as shown in Figs. 3, 4, display the evolution of COVID-19 after the emergence of various vaccines for those selected countries, incorporating a comparative analysis of the estimation errors between the corresponding estimated and real-world reported cases.

Figure 3 implies the real-time data-fitting results of several representative countries, such as America, India, Brazil, Russia, France, Turkey, England, Argentina, Italy, and Colombia. Of which, Fig. 3a, b, c, and d demonstrate that the proposed model can predict, respectively, the estimated active cases, accumulative recovered cases, accumulative deceased cases, and daily new confirmed cases for those selected countries, with real-world reported cases falling within its 95% confidence intervals. The results expose that, under current non-pharmaceutical interventions and SARS-COV-2 vaccines, this disease can not be thoroughly eliminated. Even though the increasing contact tracing and social distancing, they are still experiencing an increase in reported active, accumulative deceased, and daily new confirmed cases. France, Italy, and Brazil, for instance, may experience a continued increase in active cases within the following periods (Fig. 3a). America, Brazil, and India, on average, result in a sustained increase in accumulative deceased cases since the outbreak of this epidemic, whereas India, especially, has experienced an exponential increase since 1 April 2021 (Fig. 3c).

Bringing down and keeping up $R_t < 1$ is essential to mitigate the spread of the ongoing COVID-19 epidemic. The results, obtained systematically from Fig. 4, evaluate the potential of preserving $R_t < 1$ for those representative countries under current non-pharmaceutical interventions and SARS-COV-2 vaccines. It is therefore predicted that for the remaining countries (e.g., America, India, Brazil, Russia, France, Turkey, England, and Colombia) except Italy and Argentina, under the existing interventions and vaccinations coverage rate, curbing this epidemic appears to be impossible without additional control efforts, as the effective reproduction numbers are still insufficient to reduce and preserve $R_t < 1$. The Indian authority, for example, has ignored the warnings of scientists and relaxed a series of quarantine measures starting in middle March 2021, causing people not strictly comply with quarantine rules, such as mask-wearing and social distancing [3]. This, on average, results in the peaks of 3,737,715 active cases (95% CI: 3,735,337-3,740,093), 453,704 accumulative deceased cases (95% CI: 453,641-453,767), and 406441 daily new confirmed cases (95% CI: 405,405–407,477) as of 22 October 2021. The aforementioned results, additionally, should not be regarded as the optimized interventions combining all possible model parameters, but rather as a distinct assessment of current situations for curtailing this epidemic around the world.

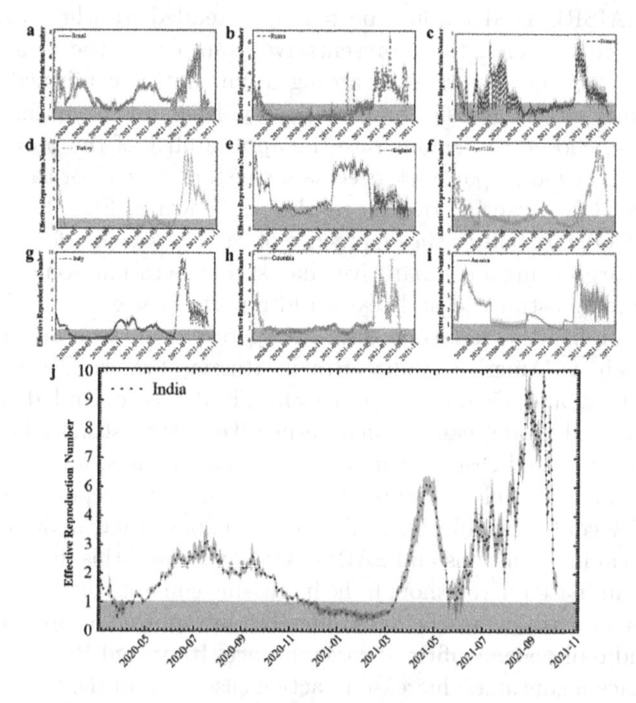

Fig. 4. Time-varying effective reproduction number Rt for several representative countries: (a) Brazil; (b) Russia; (c) France; (d) Turkey; (e) England; (f) Argentina; (g) Italy; (h) Colombia; (i) America; (j) India. Those selected countries are chosen from the top ten countries regarding accumulative confirmed cases, such as America, India, Brazil, Russia, France, Turkey, England, Argentina, Italy, and Colombia in order. The green rectangular region indicates $R_t < 1$. The shaded error bands denote, respectively, 95% confidence intervals of the corresponding mean by running over 50 independent repetitive experiments. For easy comparison, raw data, represented by this plot, are selected from 1 March 2020 to 22 October 2021. (Color figure online)

4 Conclusion

This paper has reconstructed the transmission dynamics of COVID-19 over time in each stage for mainland China. The mathematical model, SEIAISRD, has been calibrated to real-world epidemic reported cases, accounting for non-detected infections, symptomatic-asymptomatic infections distinction, variant-dependent transmission rates, the effectiveness of various vaccines, as well as the remaining epidemiological parameters. A considerable calibration result is the proportion of asymptomatic infections found to be 1.2% of that of the reported infected cases, yet its recovery rate found to be twice as high among symptomatic infections. Alternatively, in countries faced with advanced over-dispersion and super-spreader characteristics during the ongoing COVID-19 outbreak, the effective reproduction number, to a large degree, will be experiencing fluctuations as the asymptomatic/symptomatic infected cases decreases. This is more possible to help explain the case for countries with lower posterior values related to dispersion parameters, such as Brazil, Russia, France, England, America, Italy, as well as India, etc. It should, however, be worth stressing that the estimation errors for India continue to be fluctuating, compared to the remaining nine countries. It is well known that India is a developing country with a large population, and its healthcare resources, including the number of vaccinations, all fall behind those of developed countries. This phenomenon is thus extremely unreasonable. To sum up, it is imperative for us to conclude that there are some uncertainties in the actual epidemic data released by India. Future researches tend to address further aspects as, yet may not be limited to, exploring how the heterogeneity of the social contact network can affect the transmission dynamics of the outbreak, as well as declining uncertainty around the modelling approaches in different countries, regions, or crowds.

References

1. Aleta, A., et al.: Modelling the impact of testing, contact tracing and household quarantine on second waves of Covid-19. Nature Hum. Behav. **4**(9), 964–971 (2020)
2. Anderson, R.M., Heesterbeek, H., Klinkenberg, D., Hollingsworth, T.D.: How will country-based mitigation measures influence the course of the Covid-19 epidemic? Lancet **395**(10228), 931–934 (2020)
3. Asrani, P., Eapen, M.S., Hassan, M.I., Sohal, S.S.: Implications of the second wave of Covid-19 in India. Lancet Respir. Med. **9**(9), e93–e94 (2021)
4. Bastani, H., et al.: Efficient and targeted Covid-19 border testing via reinforcement learning. Nature **599**(7883), 108–113 (2021)
5. Block, P., et al.: Social network-based distancing strategies to flatten the Covid-19 curve in a post-lockdown world. Nat. Hum. Behav. **4**(6), 588–596 (2020)
6. Brauner, J.M., et al.: Inferring the effectiveness of government interventions against Covid-19. Science **371**(6531) (2021)
7. N.H.C.: Covid-19 prevention and control: vaccination information, People's Republic of China. http://www.nhc.gov.cn/xcs/yqjzqk/list_gzbd.shtml (2021)
8. Chiu, W.A., Fischer, R., Ndeffo-Mbah, M.L.: State-level needs for social distancing and contact tracing to contain Covid-19 in the United States. Nat. Hum. Behav. **4**(10), 1080–1090 (2020)

9. Collins, F.S.: Covid-19 lessons for research (2021)
10. Contreras, S., et al.: The challenges of containing SARS-CoV-2 via test-trace-and-isolate. Nat. Commun. **12**(1), 1–13 (2021)
11. Della Rossa, F., et al.: A network model of Italy shows that intermittent regional strategies can alleviate the Covid-19 epidemic. Nat. Commun. **11**(1), 1–9 (2020)
12. Dong, E., Du, H., Gardner, L.: An interactive web-based dashboard to track Covid-19 in real time. Lancet. Infect. Dis **20**(5), 533–534 (2020)
13. Gibney, E.: Whose coronavirus strategy worked best? scientists hunt most effective policies. Nature **581**(7806), 15–17 (2020)
14. Guerra, F.M., et al.: The basic reproduction number (r0) of measles: a systematic review. Lancet. Infect. Dis **17**(12), e420–e428 (2017)
15. Hao, X., Cheng, S., Wu, D., Wu, T., Lin, X., Wang, C.: Reconstruction of the full transmission dynamics of Covid-19 in Wuhan. Nature **584**(7821), 420–424 (2020)
16. Hasan, A., Putri, E.R., Susanto, H., Nuraini, N.: Data-driven modeling and forecasting of Covid-19 outbreak for public policy making. ISA transactions (2021)
17. Hoertel, N., et al.: A stochastic agent-based model of the SARS-CoV-2 epidemic in France. Nat. Med. **26**(9), 1417–1421 (2020)
18. Hsiang, S., et al.: The effect of large-scale anti-contagion policies on the Covid-19 pandemic. Nature **584**(7820), 262–267 (2020)
19. Li, Z., et al.: Active case finding with case management: the key to tackling the Covid-19 pandemic. Lancet **396**(10243), 63–70 (2020)
20. Monti, M., Torbica, A., Mossialos, E., McKee, M.: A new strategy for health and sustainable development in the light of the Covid-19 pandemic. Lancet **398**(10305), 1029–1031 (2021)
21. Organization, W.H.: Coronavirus disease 2019 (Covid-19) situation report-71 (world health organization, accessed 31 Mar 2020). http://www.who.int/docs/default-source/coronaviruse/situation-reports/20200331-sitrep-71-covid-19.pdf?sfvrsn=4360e92b_8 (2020)
22. Rockx, B., et al.: Comparative pathogenesis of Covid-19, MERS, and SARS in a nonhuman primate model. Science **368**(6494), 1012–1015 (2020)
23. Stokel-Walker, C.: What we know about Covid-19 reinfection so far. BMJ 372 (2021)

Keyword-based Research Field Discovery with External Knowledge Aware Hierarchical Co-clustering

Kai Sugahara and Kazushi Okamoto(✉)

The University of Electro-Communications,
1-5-1 Chofugaoka, Chofu, Tokyo 182-8585, Japan
{ksugahara,kazushi}@uec.ac.jp

Abstract. Helping researchers in understanding the current position and significance of their research field benefits both the individual and the development of science and technology. Clustering techniques are traditional approaches based on text mining of papers and their metadata, but suffer from inconsistent representations of words in metadata uniquely assigned by authors, resulting in low accuracy. To address this issue, we propose the application of HICCAM, a hierarchical co-clustering method that exploits auxiliary knowledge of clustered objects to discover research fields. Our first step involved constructing augmented matrices representing paper abstracts and research keywords from external domains. We then cooperatively and accurately clustered a relational matrix of these objects using these augmented matrices. To validate the effectiveness of our framework, we conducted a case study using conference papers published in the field of computer science and various auxiliary knowledge. The comparative analysis identified the domain of auxiliary knowledge that contributes to the research field detection, and the visualization results showed it effective for field discovery in terms of easy interpretation and scalability brought by the hierarchical algorithm.

Keywords: Hierarchical co-clustering · Research field discovery · Unsupervised transfer learning · Relational data analysis

1 Introduction

Numerous research papers in various research fields are published each year. As new research fields emerge and existing ones integrate, assisting researchers in comprehending the position and significance of their research fields becomes crucial. This effort stimulates ideas and inspiration for researchers, driving their projects forward, and also facilitates the progress of applied technologies. Additionally, organizing research domains aids the funding agencies in streamlining the evaluation process for grants, thereby enhancing the efficiency of grant reviews [12].

Extensive studies on research field discovery, closely related to community detection, have been carried out using various methodologies. Typically, detection approaches can be broadly categorized into topology-based and topic-based methods [5]. Topology-based methods construct networks where nodes

B. Xin et al. (Eds.): IWACIII 2023, CCIS 1931, pp. 153–166, 2024.
https://doi.org/10.1007/978-981-99-7590-7_13

are papers/researchers and edges represent relationships between papers or researchers by co-authorship, citation relationships, and publication destinations. Then, graph clustering methods including modularity-based [3] and spectral clustering [2,14] are used to identify the research communities. Although graph-based detection is consistent with our intuitive understanding, it may not provide satisfactory results if the target graph contains biases. For example, the presence of papers with few or an overwhelming number of citations [7].

In contrast, topic-based approaches do not focus directly on the connections between nodes. A straightforward and interpretable approach for filed discovery is to leverage text clustering by utilizing words present in papers and their metadata. Typically, keywords that characterize the papers are first extracted and preprocessed, such as stemming and stopword removal, and then existing clustering algorithms, such as k-means, are applied [6,8]. In [13], first a singular value decomposition was performed to deal with the high dimensionality in the paper-term matrix, and then topic detection was addressed by clusters generated by k-means. Other approaches pre-transform scientific documents into embeddings based on deep learning methods, which are then clustered using state-of-the-art clustering methods [11]. However, existing approaches lack several perspectives, as follows:

1. The phrasing of a paper largely relies on the authors. Therefore, the usage of words with the same connotation may be inconsistent, even in the same field.
2. Large corpora, where knowledge is implicitly built up by many people, may allow us to extract similarities between words. In other words, such words with the same meaning can be integrated by clustering with this implicit knowledge.
3. A given research area may be further divided into sub-fields, and the granularity of clustering should be carefully determined.

Based on these insights, this study investigates a framework employing the Hierarchical Co-clustering with Augmented Matrices (HICCAM) [10] for robustly detecting research fields against differences in the representation of keywords. HICCAM clusters two object sets simultaneously while exploiting their auxiliary knowledge in an end-to-end manner. We attempted to cluster paper-by-keyword matrices, usually prone to sparsity, by taking advantage of the auxiliary classification knowledge built in an external domain and an unsupervised transfer learning framework. Additionally, HICCAM employs a hierarchical co-clustering framework that can capture both the macro-level and micro-level structure of research communities. This hierarchical structure enables multiresolution analysis and allows researchers to explore communities at different levels of granularity.

We conducted a two-stage experiment to evaluate the effectiveness of HICCAM in research field discovery. This experiment consisted of a parameter study and a case study. First, auxiliary knowledge about the papers and keywords was prepared from several domains. Next, we discussed the effective domains, and the necessary extent of knowledge utilization for research community detection. Finally, we visualized the relationships among multiple objects/clusters and qualitatively evaluated the detection performance to determine whether it aligns with our intuition.

Our contributions are summarized follows:

- We propose a robust framework for discovering research fields using knowledge of words and sentences constructed from external domains, and for dealing with lexical variation.
- We use HICCAM, a co-clustering method, to generate paper and word clusters with high accuracy through natural end-to-end integration of auxiliary knowledge.
- Our experiment applied several types of auxiliary knowledge to identify potentially useful knowledge domains in research field discovery, and further, as a case study, demonstrated the utility of using HICCAM.

The rest of this paper is organized as follows. Section 2 introduces co-clustering and HICCAM. Section 3 describes an approach for discovering research fields using HICCAM. Section 4 summarizes and discusses the experimental results.

2 Background

2.1 Co-clustering

Co-clustering is an unsupervised learning technique that simultaneously clusters row objects X and column objects Y of relational matrix F. In general, co-clustering is superior to one-way clustering, which classifies only row objects, in achieving robustness and overcoming high dimensionality. Information-theoretic co-clustering (ITCC) [4] is widely used and performs as,

$$\min_{\hat{X},\hat{Y}} \left\{ I(X,Y) - I(\hat{X},\hat{Y}) \right\} \quad (1)$$

where $\hat{X}$ and $\hat{Y}$ denote the row and column clusters, respectively, and $I(X,Y)$ and $I(\hat{X},\hat{Y})$ are the mutual information in the normalized matrices of $F(X,Y)$ and $F(\hat{X},\hat{Y})$, respectively.

2.2 HICCAM

Sugahara and Okamoto proposed HICCAM [10] and demonstrated an improved classification accuracy in document clustering using an unsupervised transfer learning framework that introduced auxiliary knowledge regarding row and column objects. They first constructed two augmented matrices, G and H, representing the auxiliary knowledge of the 2-sided object X and Y of the relation matrix F, respectively, and then sequentially split the row and column objects based on the information theory.

In particular, the approach is based on the objective function,

$$q(\hat{X},\hat{Y}) = \frac{1}{1+\alpha+\beta}\left\{\frac{I_F(\hat{X},\hat{Y})}{I_F(X,Y)} + \alpha\frac{I_G(\hat{X},A)}{I_G(X,A)} + \beta\frac{I_H(\hat{Y},B)}{I_H(Y,B)}\right\} \in [0,1] \quad (2)$$

where A and B denote the feature objects of matrices G and H, respectively, and I_F, I_G, I_H are the mutual information in the normalized matrices of F, G, H respectively. Note that α and β are hyperparameters representing the influence of classification knowledge in augmented matrices G and H respectively. Subsequently, a cluster that most increases the value of the objective function q is selected and divided into two clusters by using a greedy algorithm. Therefore, HICCAM can generate dendrograms of row and column clusters, and provide interpretable clustering by tracking the splitting process. For details about the splitting algorithm, please refer to [10].

3 Method

This section describes our approach to discovering research fields, which consists of several components, as shown in Fig. 1.

3.1 Dataset Preparation

To discover research fields, we focused on papers related to "recommender systems" as it falls within our area of expertise. We independently collected metadata of papers published in KDD (The annual ACM SIGKDD Conference), SIGIR (International ACM SIGIR Conference on Research and Development in Information Retrieval), and RecSys (The ACM Conference on Recommender Systems) from 2018 to 2022. Papers lacking research keywords or abstracts were excluded, resulting in 2,471 papers for this study.

We also used 4,651 keywords assigned by the authors, all converted to lower case, as clustering features. We then constructed a relational matrix, $F(X,Y)$, to represent the relationship between the papers X and keywords Y. Each element, F_{ij}, of the matrix was set to 1 if keyword Y_j was associated with paper X_i, and to 0 otherwise. The resulting matrix F had a density of only 0.08%, indicating that it is highly sparse.

3.2 Auxiliary Knowledge Preparation

HICCAM is expected to achieve high classification accuracy by utilizing auxiliary vectors with the classification knowledge of papers and keywords. In this study, we generated two augmented matrices, $G(X,A), H(Y,B)$, by embedding the abstracts of the papers and the keywords into 350 dimensions. Specifically, $G_{i:}$ represents the abstract vector of the paper X_i, whereas $H_{j:}$ is the word vector of keyword Y_j. Additionally, we employed the following types of embedding domains to compare and identify the most useful domains for the research field discovery.

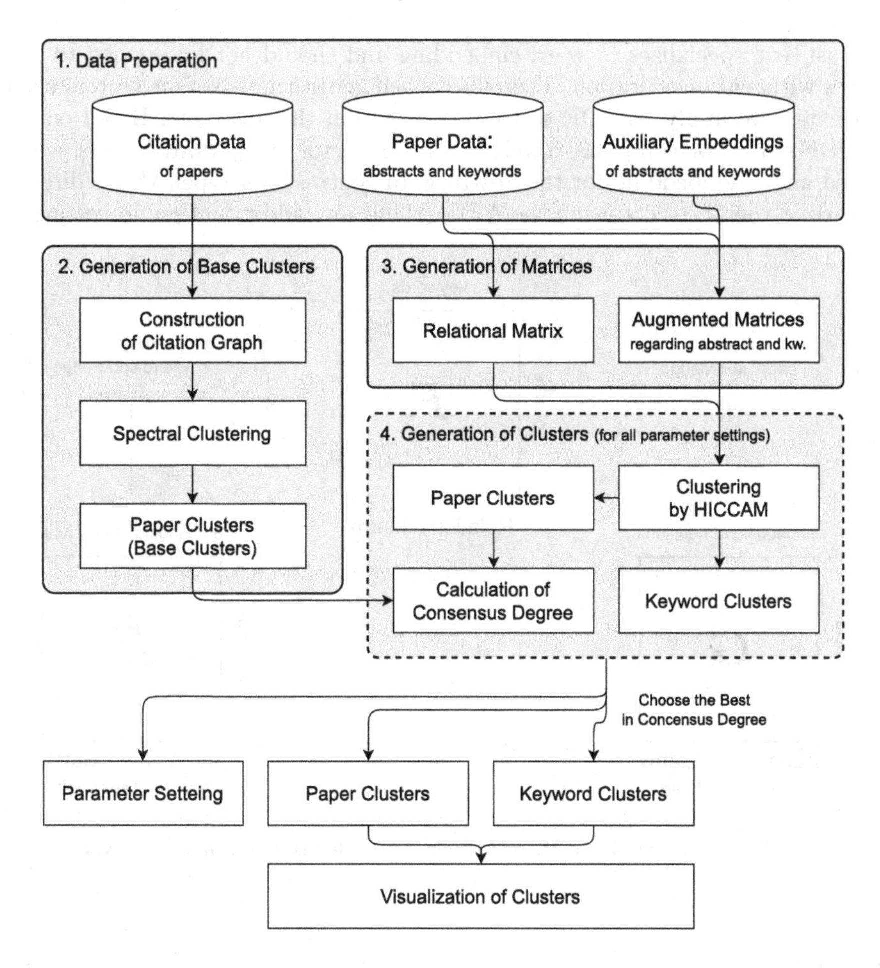

Fig. 1. Structure diagram of our approach for research field discovery

openai. We retrieved the abstract and keyword embeddings using OpenAI Embeddings [9]. We used the API[1] on June 19, 2023, and the resulting 1536-dimensional vector was reduced to 350 dimensions by PCA (Principal Component Analysis).

fasttext-wikipedia. Firstly, we downloaded Wikipedia Dumps[2] and extracted the text using WikiExtractor[3] to create a training corpus. Next, we trained the fastText model [1] on the corpus in 350 dimensions.

fasttext-arxiv. We collected the metadata of papers in the computer science category on arXiv to construct a training corpus from their abstracts. We then trained the fastText model [1] on the corpus in 350 dimensions.

[1] https://platform.openai.com/docs/guides/embeddings.
[2] https://dumps.wikimedia.org/enwiki/latest/enwiki-latest-pages-articles.xml.bz2.
[3] https://github.com/attardi/wikiextractor.

FastText specializes in word embedding and should not be applied to sentences without consideration. Therefore, when generating abstract vectors using fasttext-*, we apply TF-IDF to the words within the abstracts. Based on the TF-IDF scores, we computed a weighted mean vector of the word vectors, which served as the embedding for the abstract. In contrast, for OpenAI, we directly vectorized the abstracts using the API without any additional preprocessing.

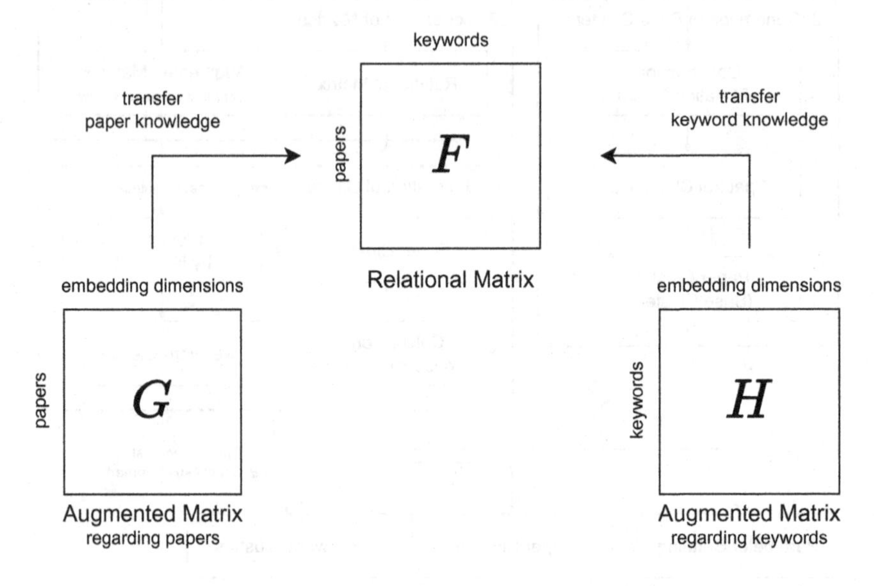

Fig. 2. Schematic of the HICCAM clustering for research field discovery

3.3 Clustering

Matrices F, G, H constructed in Sects. 3.1 and 3.2 were used as inputs for HICCAM to generate the paper and keyword clusters, as illustrated in Fig. 2. In addition to the influence parameters α, β mentioned in Sect. 2.2, HICCAM requires specifying a threshold θ for the objective function value q to determine when the splitting process should be terminated. To the best of our knowledge, there is no known method for determining these parameter values, hence this study used the method described in Sect. 3.4 to determine them. After tuning the hyperparameters, we employed HICCAM on the entire dataset and observed the clustering results.

3.4 Parameter Tuning

Since the dataset did not contain true labels for research field clustering, expert labeling is generally required for evaluation; however, this procedure was omitted in this study due to cost and time constraints. Instead, we generated paper clusters using a citation-based approach as the base labels in the following steps.

1. The references R_i cited in each paper X_i in the dataset were collected using the Crossref API[4]
2. Papers with fewer than three citation co-occurrences with any other paper were removed to generate a dense citation network
3. For each pair of papers (X_i, X_j), the Simpson coefficient was computed as the reference co-occurrence score given by, $score(X_i, X_j) = |R_i \cap R_j| / \min(|R_i|, |R_j|)$
4. A graph was constructed with the remaining papers as nodes and the calculated scores as edge weights
5. With 150 clusters, spectral clustering was applied to the graph to obtain the paper clusters as base labels

The large clusters that were generated, in which a few papers were commonly cited, were excluded. As a result, we extracted 149 base labels for 554 papers. After preparing the base labels, the following hyperparameters were considered as candidates for HICCAM:

- Abstract knowledge influence $\alpha \in \{0, 2^{-1}, 2^0, 2^1, 2^2, 2^3, 2^4\}$
- Keyword knowledge influence $\beta \in \{0, 2^{-1}, 2^0, 2^1, 2^2, 2^3, 2^4\}$.

HICCAM was then performed five times on the subsets of the dataset corresponding to the remaining 554 papers for each hyperparameter setting. To identify the best parameter settings, we used a commonly used clustering accuracy measure:

$$NMI = \frac{\sum_i \sum_j n_{ij} \log \frac{n \times n_{ij}}{n_i \times n_j}}{\sqrt{\left(\sum_i n_i \log \frac{n_i}{n}\right)\left(\sum_j n_j \log \frac{n_j}{n}\right)}} \in [0, 1]$$

where $n_i = |C_i|, n_j = |E_j|, n_{ij} = |C_i \cap E_j|$, and C_i, E_j denote the i-th base-cluster and j-th predictive cluster in HICCAM, respectively.

4 Results and Discussion

4.1 Parameter Study

First, we discuss the domains of auxiliary knowledge useful for research field discovery and their influences. We employed various combinations of embeddings for the augmented matrices G and H, including openai, fasttext-wikipedia, and fasttext-arxiv. Additionally, we performed a comprehensive analysis to examine variations in the "degree of consensus" by manipulating the influence parameter. Figure 3 shows the average NMI between the base clusters and the predicted paper clusters in HICCAM based on five trials. It is important to note that the

[4] https://www.crossref.org/documentation/retrieve-metadata/rest-api/.

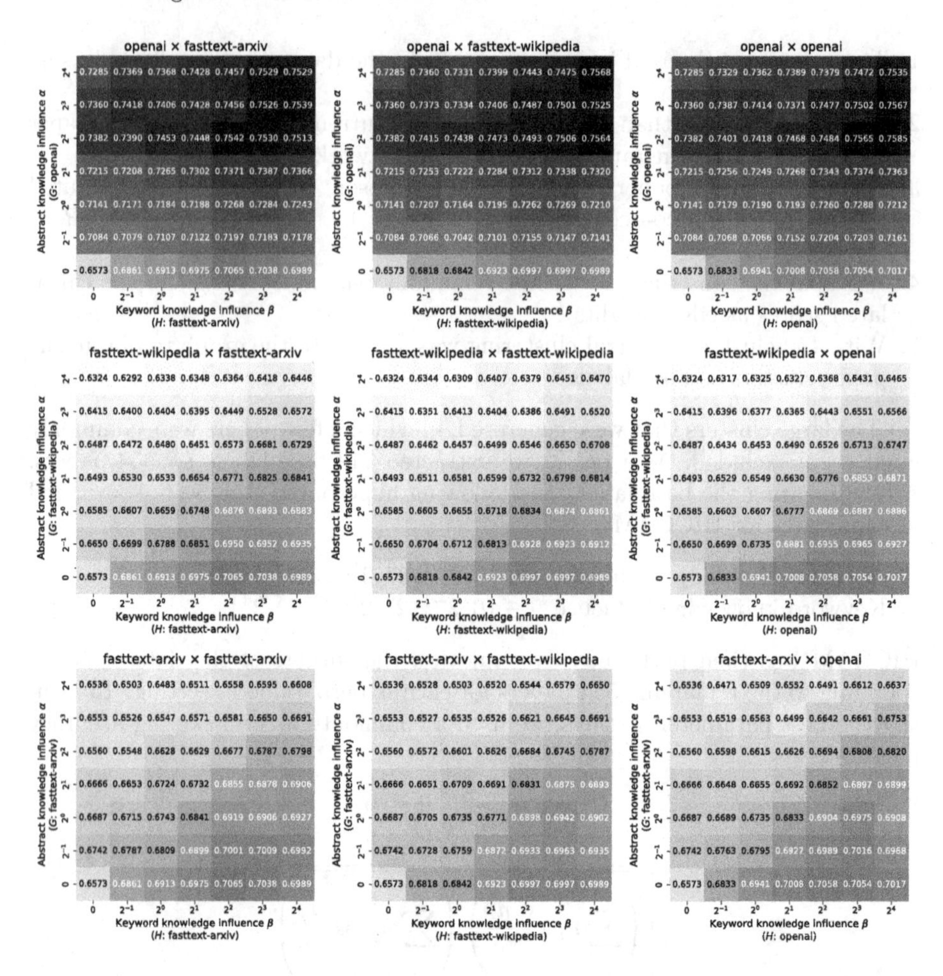

Fig. 3. Change in the average NMI of HICCAM with different influence parameters and embedding types

NMI value in this experiment does not reflect the reproducibility of the true paper clusters, but rather it indicates agreement with another citation-based approach.

Comparing the nine heatmaps, we can reveal the domain with the most effective auxiliary knowledge. For augmented matrix G (i.e., abstract embeddings), knowledge from openai consistently contributed the most because the embedding of openai was based on Transformer [9], which is superior in understanding the context of a sentence, and therefore was effective for abstract sentences, outperforming fasttext-*, which simply computes the average vector of words. Moreover, fasttext-arxiv's contribution to G was slightly higher but comparable to that of fasttext-wikipedia, suggesting that auxiliary knowledge is more effective in the research domain. A row-by-row horizontal comparison of the heatmaps showed that the augmented matrices H for all domains tended to contribute a similar amount, suggesting the absence of differences between the domains in terms of lexical variation.

Finally, we focused on the openai×openai heatmap, which tended to be high overall. By increasing the classification knowledge of the abstracts, i.e., by increasing α, the NMI grew, but dropped back above 2^3. This suggests that the inadequacy of clustering based on abstract knowledge alone and the requirement of appropriate use of knowledge. The same was true for the influence of auxiliary knowledge of keywords. Otherwise, the NMI improved as the viewpoint shifted to the upper right, suggesting the effectiveness of the coordinated use of abstract and keyword knowledge.

4.2 Case Study

In this section, we present some visualization results and discuss the research fields that were effectively detected based on the clustering outcomes with $\alpha = 2^2$ (using openai) and $\beta = 2^4$ (using openai). These parameter settings exhibited the highest NMI score in our parameter study.

Cluster Relationships. Because HICCAM clusters both the articles and keywords present in the relational matrix F, it inherently captures the relationships between these two types of clusters. To quantitatively represent the relationship between clusters, we introduce the relational score rel for paper cluster $\hat{X}_i$ and keyword cluster $\hat{Y}_j$ using the following definition:

$$rel(\hat{X}_i, \hat{Y}_j) = \frac{1}{|\hat{X}_i|} \sum_{x \in \hat{X}_i} \max_{y \in \hat{Y}_j} F(x, y) \in [0, 1]. \tag{3}$$

The score signifies the degree of relevance between the paper cluster $\hat{X}_i$ and topic (keyword cluster) $\hat{Y}_j$. It is important to note that HICCAM aims to mitigate lexical variations in this experiment; thus, we utilized the max operation to handle the included words.

Owing to the excessive number of clusters generated as a result of parameter tuning in the visualization, we defined a threshold θ for the objective function value q and stopped partitioning in the middle of the process. Figures 4, 5 show rel score heatmaps of clusters at $\theta = 0.15, 0.20$ respectively, and we defined the most frequent keyword as the label for each keyword cluster. Figure 4 shows the presence of various topics in the dataset and their corresponding paper clusters at the macro level. "Machine learning," "transformer," "reinforcement," and "deep learning" were relevant across all communities (paper clusters), while "recommender systems," "collaborative filtering," and "graph neural networks" exhibited bias towards specific communities. Figure 5 provides a more detailed view of the communities and topics. Notably, the similarities between topics offer interesting insights, such as the proximity between "fairness" and "explainability," as well as "question answering" and "information retrieval," highlight closely related topics.

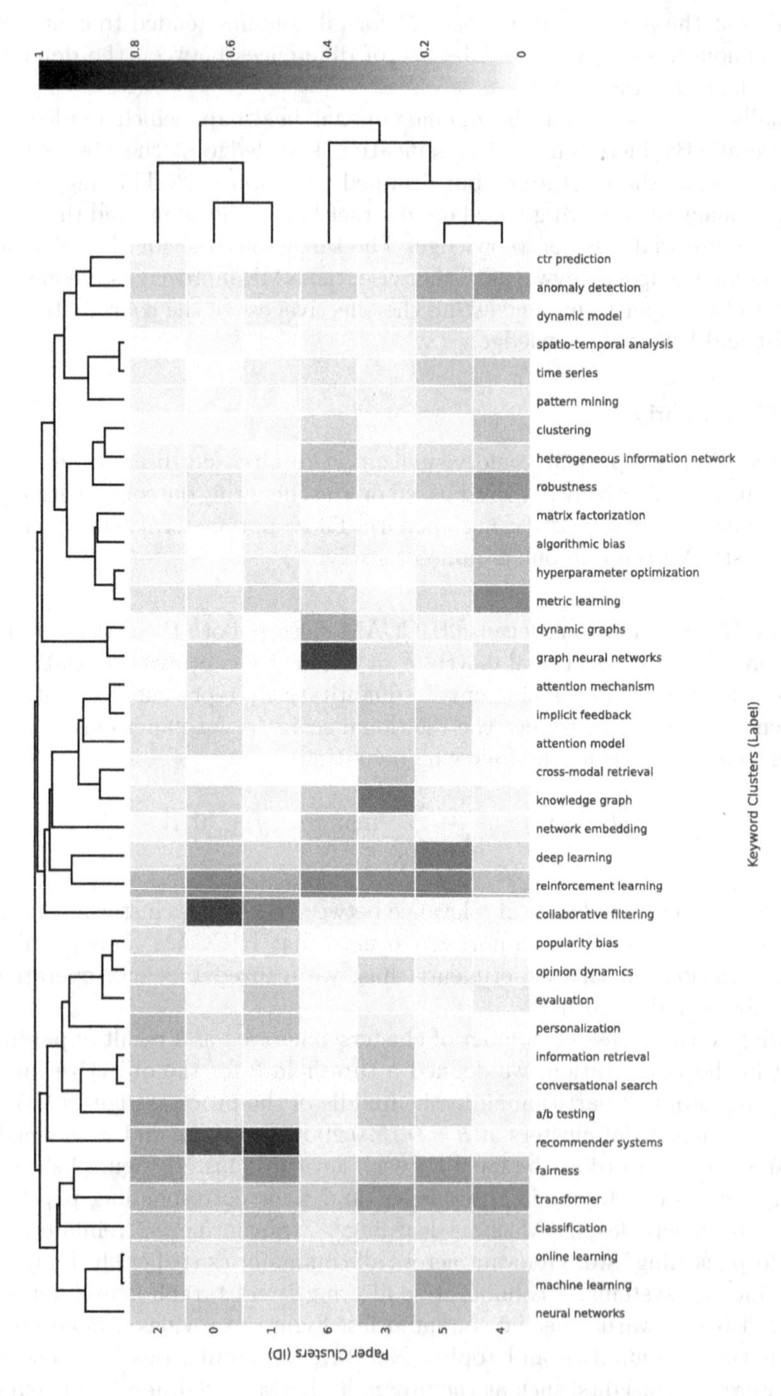

Fig. 4. Cluster relationships when θ is set to 0.15

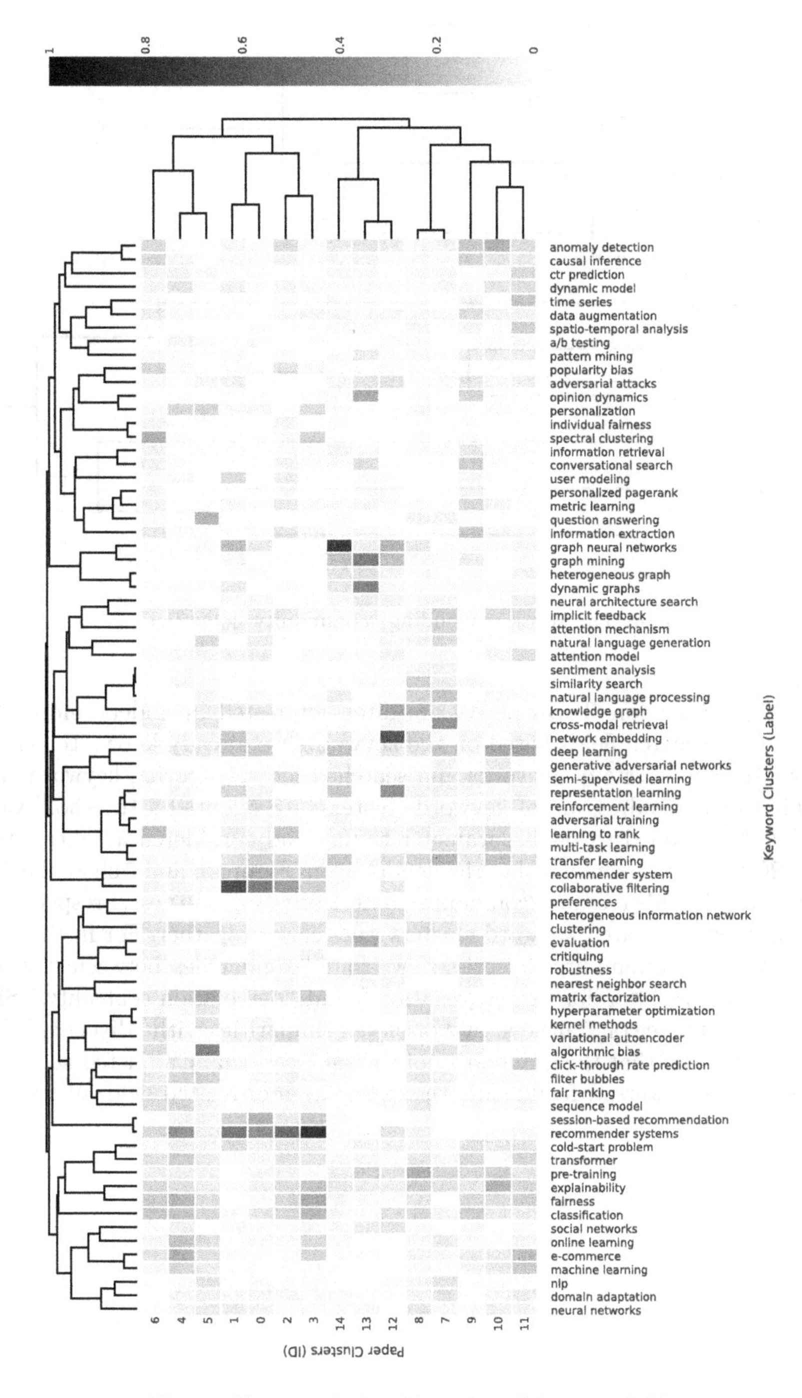

Fig. 5. Cluster relationships when θ is set to 0.20

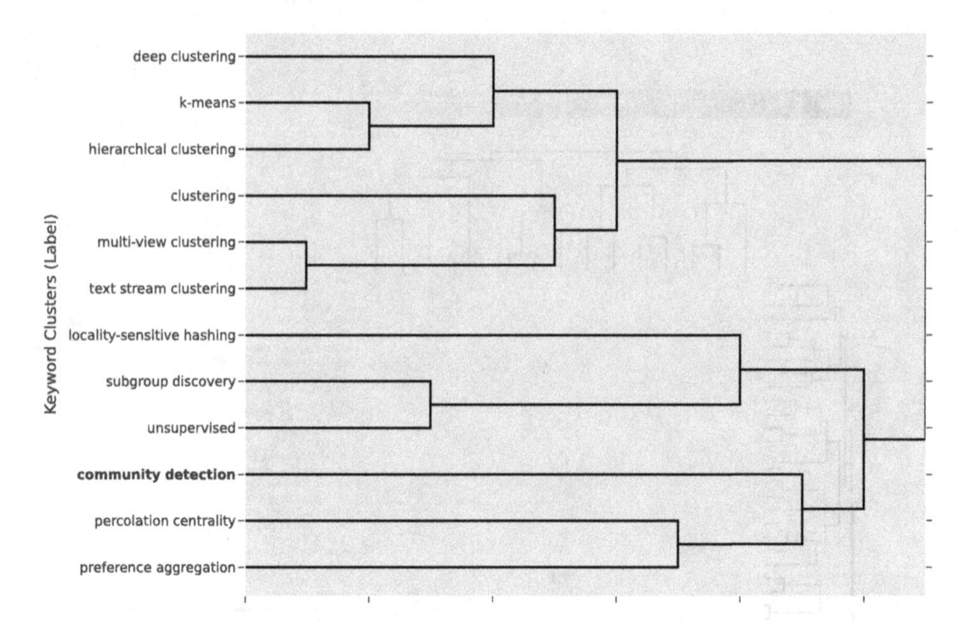

Fig. 6. The subtree including "community detection"

Similarity Between Topics. We are often interested in understanding the surrounding research fields based on the keywords we already know. To achieve this, we specified certain keywords and searched for neighboring keyword clusters within the dendrogram of keyword clusters. We adopted a threshold value of $\theta = 0.72$ obtained through parameter tuning and considered a subtree that traced 200 divisions from a leaf in the dendrogram. Figures 6 and 7 show the dendrograms of the subtrees for "community detection" and "fairness," respectively. Overall, strongly related keyword clusters were closer together. For instance, in the subtree of "community detection," we found similarities between "locally-sensitive hashing," "preference aggregation," and "percolation centrality." Similarly, in the subtree of "fairness," we observed similarities with "diversity." It is worth mentioning that the granularity of the subtrees can be adjusted according to the analyst's interest, facilitating the exploration of related fields of interest.

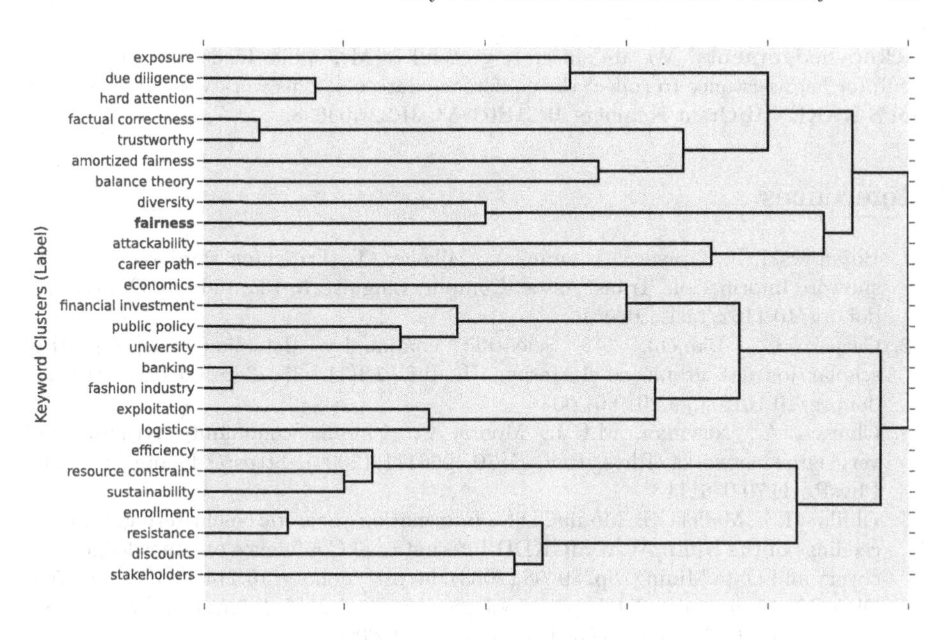

Fig. 7. The subtree including "fairness"

5 Conclusion

In this study, we propose a robust framework for discovering research fields by leveraging auxiliary knowledge from papers and words while addressing the challenges posed by lexical variations. By generating embedding vectors from the three external domains, we effectively incorporated external classification knowledge and applied HICCAM to cluster papers and keywords. A quantitative comparison with the base clusters generated by the citation-based method showed that the embedding vectors generated by this method, which can understand the context of sentences, sufficiently contribute to field detection. Furthermore, a qualitative analysis of the clustering results aligned with our domain knowledge, highlighting the scalability of cluster granularity and offering a highly interpretable method of discovery.

However, there are a few drawbacks. Firstly, a more reliable quantitative comparison with the "true" paper communities is desirable. As our study collected papers independently, assigning true labels to the papers was not feasible. In the future, a more accurate benchmark can be established using annotation through crowdsourcing or other means. Second, a comparison with existing clustering frameworks specifically tailored for research field discovery is required. Our study focused on applying HICCAM, a successful document-clustering approach, to article field surveys. It is important to address this issue in future research.

Acknowledgements. We are sincerely grateful to Ms. Akiko Ikeda, our laboratory staff, for her assistance to collect the conference datasets. This work was supported by JSPS KAKENHI Grant Numbers JP21H03553, JP22H03698.

References

1. Bojanowski, P., Grave, E., Joulin, A., Mikolov, T.: Enriching word vectors with subword information. Trans. Assoc. Comput. Linguist. **5**, 135–146 (2017). https://doi.org/10.1162/tacl_a_00051
2. Carusi, C., Bianchi, G.: Scientific community detection via bipartite scholar/journal graph co-clustering. J. Inf. **13**(1), 354–386 (2019). https://doi.org/10.1016/j.joi.2019.01.004
3. Clauset, A., Newman, M.E.J., Moore, C.: Finding community structure in very large networks. Phys. Rev. E **70**, 066111 (2004). https://doi.org/10.1103/PhysRevE.70.066111
4. Dhillon, I.S., Mallela, S., Modha, D.S.: Information-theoretic co-clustering. In: Proceedings of the Ninth ACM SIGKDD International Conference on Knowledge Discovery and Data Mining, pp. 89–98 (2003). https://doi.org/10.1145/956750.956764
5. Ding, Y.: Community detection: topological vs. topical. J. Informetrics **5**(4), 498–514 (2011). https://doi.org/10.1016/j.joi.2011.02.006
6. Haji, S.H., Jacksi, K., Salah, R.M.: A semantics-based clustering approach for online laboratories using k-means and HAC algorithms. Mathematics **11**(3) (2023). https://doi.org/10.3390/math11030548
7. Mehrabi, N., Morstatter, F., Peng, N., Galstyan, A.: Debiasing community detection: the importance of lowly connected nodes. In: Proceedings of the 2019 IEEE/ACM International Conference on Advances in Social Networks Analysis and Mining, pp. 509–512 (2020). https://doi.org/10.1145/3341161.3342915
8. Nair, S.R., Gokul, G., Vadakkan, A.A., Pillai, A.G., Thushara, M.: Clustering of research documents - a survey on semantic analysis and keyword extraction. In: 2021 6th International Conference for Convergence in Technology (I2CT), pp. 1–6 (2021). https://doi.org/10.1109/I2CT51068.2021.9418197
9. Neelakantan, A., et al.: Text and code embeddings by contrastive pre-training (2022)
10. Sugahara, K., Okamoto, K.: Hierarchical co-clustering with augmented matrices from external domains. Pattern Recogn. **142**, 109657 (2023). https://doi.org/10.1016/j.patcog.2023.109657
11. Vahidnia, S., Abbasi, A., Abbass, H.A.: Embedding-based detection and extraction of research topics from academic documents using deep clustering. J. Data Inf. Sci. **6**(3), 99–122 (2021). https://doi.org/10.2478/jdis-2021-0024
12. Wang, Y., Xu, W., Jiang, H.: Using text mining and clustering to group research proposals for research project selection. In: 2015 48th Hawaii International Conference on System Sciences, pp. 1256–1263 (2015). https://doi.org/10.1109/HICSS.2015.153
13. Weißer, T., Saßmannshausen, T., Ohrndorf, D., Burggräf, P., Wagner, J.: A clustering approach for topic filtering within systematic literature reviews. MethodsX **7**, 100831 (2020). https://doi.org/10.1016/j.mex.2020.100831
14. White, S., Smyth, P.: A spectral clustering approach to finding communities in graphs. In: Proceedings of the 2005 SIAM International Conference on Data Mining (SDM), pp. 274–285 (2005). https://doi.org/10.1137/1.9781611972757.25

An End-to-End Intent Recognition Method for Combat Drone Swarm

Hui He[1], Zhihong Peng[1(✉)], Peiqiao Shang[1], Wenjie Wang[1], and Xiaoshuai Pei[2]

[1] Beijing Institute of Technology, Beijing 100081, China
hh41723074@163.com
[2] China Electronics Technology Group Corporation, Beijing 100043, China

Abstract. In the field of intent recognition of combat drone swarm, traditional methods are based on the data characteristics which are only from a single target and at a single moment. It is difficult to capture the feature information of the entire swarm on time series. This paper proposes an end-to-end UAV swarm intent recognition method. Firstly, the distance threat coefficient and angle threat coefficient between UAVs are used to model the graph structure data of UAV swarm. Secondly, a novel deep learning method based on graph attention network, graph pooling method and gated recurrent unit (GAT-AP-GRU) is designed. This network can process the graph structure data obtained by modeling and identify the intention of the swarm. Experiments comparing with other methods and ablation experiments demonstrate that GAT-AP-GRU outperforms state-of-the-art methods in terms of accuracy of intent recognition.

Keywords: Intent Recognition · Graph Network · Mapping Method

1 Introduction

With the rapid development of intelligent technology and big data technology, the battlefield environment is becoming increasingly complex. Furthermore, there are more factors influencing the depth of war, leading to a constant evolution in military combat requirements. In response, unmanned aerial vehicles (UAVs) have been widely employed in modern local battlefields due to their simple structure and low hardware costs [1]. Currently, UAV confrontation systems have transitioned from initial single-aircraft confrontation modes to a system combat mode composed of multiple aircraft types. These UAV swarms predominantly consist of micro-small UAVs with weak infrared characteristics, small radar reflective surfaces, low flight altitudes, and other attributes that enhance stealth capabilities [2]. Moreover, most of the swarm systems have

Supported by National Natural Science Foundation (NNSF) of China under Grant U22B2058 and U2013602.

B. Xin et al. (Eds.): IWACIII 2023, CCIS 1931, pp. 167–177, 2024.
https://doi.org/10.1007/978-981-99-7590-7_14

self-adaptive capabilities such as autonomous formation and intelligent decision-making, which makes it more difficult to counteract UAV swarms. In the face of all these challenges, intelligent hostile UAVs and their swarm intention recognition technology has become particularly critical. Accurately predicting the enemy's combat intentions in a timely manner is essential for air warfare and defense decision-making systems to make appropriate decisions.

In the field of UAV intent recognition, current approaches primarily focus on recognizing individual UAV. There are two main mainstream approaches [3]: rule-based methods and data-driven deep learning methods. Rule-based methods rely mainly on expert experience and rules, rather than learning the UAV features associated with each intent from data [4]. This approach is useless when there is a limited or incomplete expert experience. Moreover, the generalization performance of this methods is pool [5].

In recent years, data-driven deep learning methods based on data have shown superior performance compared to rule-based methods in terms of robustness and generalization. Some of these approaches use attention mechanism [6], while others employ long short-term memory (LSTM) networks [7]. Attention-based methods mimic the visual attention mechanism of living creatures, capturing correlations between different features of the UAV and highlighting key features that determine intent. LSTM-based methods focus on temporal features of intent and can capture hidden feature information from previous time series.

However, in modern warfare, where operational efficiency and risk reduction are critical considerations, the use of multiple UAVs for cooperative missions has become increasingly common. As a result, the focus of intent recognition research should shift from individual unit to swarms. Currently, there are two limitations in solving the intention recognition problem for UAV swarms. Firstly, current methods struggle to effectively construct the interaction relationship between UAVs, which makes it difficult to consider the impact of swarm interaction structure information on feature extraction. Secondly, current methods lack effective pooling methods for obtaining group-level feature information from individual-level feature information.

This paper presents an end-to-end intent recognition method for combat UAV swarm. Firstly, a threat factor-based graph construction method is proposed to reflect the interaction of UAV swarm in the form of graph structure. Additionally, the method proposes an innovative swarm intent recognition framework based on Graph Attention Networks (GAT) [8] and Gated Recurrent Units (GRU) [9]. This framework, named Attention-Pool based GAT and GRU Framework (GAT-AP-GRU), incorporates a pooling mechanism to effectively extract group-level feature information from individual-level feature information. The results of ablation experiments demonstrate the effectiveness of the proposed approach in improving UAV swarm intent recognition performance.

This paper will be introduced in the following parts. Section 2 presents an overview of UAV intent recognition. Section 3 presents the proposed end-to-end intent recognition approach. Experimental results and result analysis are given in Sect. 4, and conclusions are given in Sect. 5.

2 Related Work

Current intention recognition models can be divided into two main categories: one is rule-based approaches and the other is data-driven deep learning approaches.

Lei [10] builds Bayesian networks based on expert knowledge and uses directed graphs to characterize the relationship between features and intentions. Chang [11] used intuitionistic fuzzy information processing rules as the basis for intention recognition of targets based on gray correlation group decision making. Cao [12] combines the obtained high-dimensional spatial data similarity and attribute similarity to perform intent recognition of targets by D-S evidence theory. The rule-based methods have been proven effective in their respective tasks. However, their dependence on rules limits their ability to generalize well and is not suitable for increasingly complex battlefield environments.

In recent years, data-driven deep learning methods have been widely applied in the field of intent recognition. For example, Liu [13] introduced the Long Short Term Memory (LSTM) to extract temporal features of drones and used decision trees to identify target intent. Teng [14] used Bi-directional gated recurrent unit (BiGRU) to synthesize information from the entire time series for intent recognition of UAVs at historical moments. Later, Teng [15] used Temporal Convolutional Network (TCN) network to extract high frequency features from the data and speed up the training process. Liu [16] introduced the attention mechanism to assign different weights to different features, representing the importance of each feature for intent classification. However, none of these methods can effectively reflect the influence of swarm interactions on feature extraction.

3 General Framework of the End-to-End Intent Recognition Method

3.1 Problem Definition

The background is assumed to be that there are N aircraft numbered $uav_i(i = 1, 2, \cdots, N)$ coordinating to attack a ground target. The physical properties of uav_i at time-step t are denoted as x_i^t. The research problem of this paper can be defined as solving the intent of the UAVs swarm at time-step m using the given x_i^t at time-steps $t = 1, 2, \cdots, m$.

3.2 Model Architecture

In this section, a general framework of the end-to-end intent recognition approach is presented. There are two components of the method: a graph building module and GAT-AP-GRU method. The graph building module transforms the data structure of a swarm into a graph structure based on the interaction of UAVs within the swarm. The GAT-AP-GRU method consists of two components: the Feature Extraction Module in extraction phase and the Intention

Prediction Module in recognition phase. The components are shown in Fig. 1. The GAT module extracts the features of each UAV based on the graph structure of the swarm by combining the information of its neighbors through the graph attention mechanism. The attention-pooling module pays attention to the association relationships which are significant to pool and obtain the embedding of the swarm. The recognition module receives pooling features to predict the intent of the swarm. This end-to-end intent recognition method will be described in detail in the following sections.

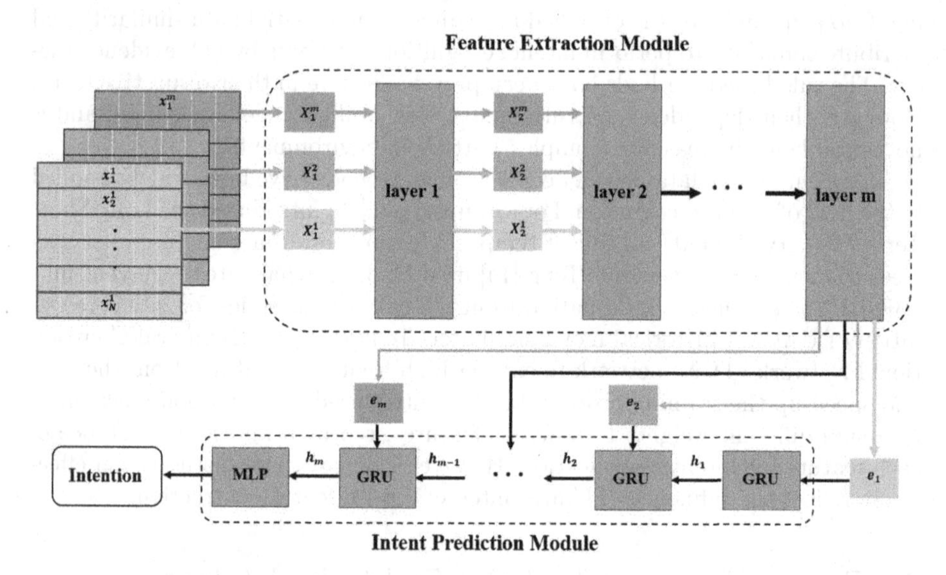

Fig. 1. Pipeline of the GAT-AP-GRU network, which contains 2 parts: Feature Extraction Module and Intent Prediction Module. The 'layer' in Feature Extraction Module is composed of GAT module and attention-pooling module. The architecture of layers will be shown in Fig. 2.

3.3 Mapping Method

The mapping method is designed to find the graph corresponding to the interaction relationships within the swarm. The nodes in the graph represent individual UAVs and the edges indicate the existence of interaction relations between connected UAVs. The adjacency matrix $A \in R_{n*n}$ is used to denote the structure of the constructed graph, where the relation value A_{ij} indicates the existence of an edge between uav_i and uav_j. Considering that the threat coefficient represents the degree of threat posed by a drone, it reflects to some extent the collaborative relationship within a swarm [17]. Therefore, the similarity of threat coefficients is used to construct a graph.

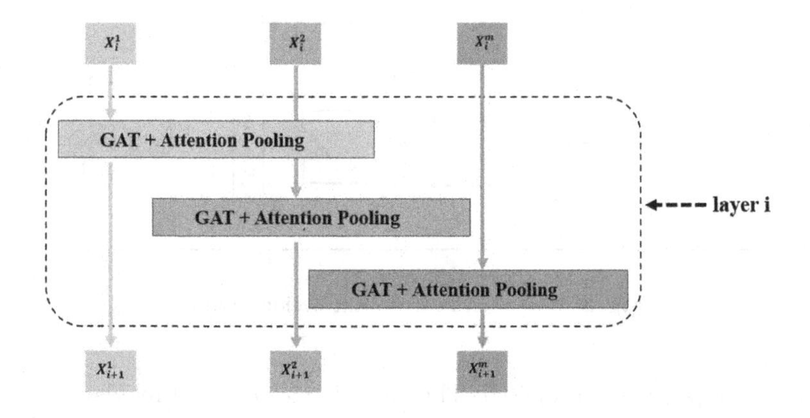

Fig. 2. Layer i in Feature Extraction Module

The distance threat coefficient t_d and the angular threat coefficient t_g are used for map building. For uav_i, the equations are expressed as follows:

$$t_{di} = \begin{cases} 0.05 & , \quad d_m < d \\ 0.05 + 0.95 * \dfrac{d - d_e}{d_m - d_e}, & d_m \geq d \end{cases} \tag{1}$$

$$t_{gi} = \frac{\sqrt{\alpha}}{2} \tag{2}$$

$$t_{si} = t_{di} * exp\left(t_{gi}\right) \tag{3}$$

where d denotes the distance between the uav_i and the ground target, d_e denotes the effective attack range of the uav_i, d_m denotes the maximum attack range of the uav_i, and α denotes the directional angle between the uav_i and the ground target, t_{si} denotes the threat coefficient of uav_i.

The threat factor similarity between uav_i and other UAVs (e.g. uav_i) is defined as follows:

$$f\left(uav_i,\ uav_j\right) = \left|\frac{t_{si} - t_{sj}}{t_{si} + t_{sj}}\right| \tag{4}$$

In experiments, the following function is used to calculate A:

$$A_{ij} = ceil\left(\frac{\exp\left(f\left(uav_i,\ uav_j\right)\right)}{\sum_{j=1}^{N} \exp\left(f\left(uav_i,\ uav_j\right)\right)} - \lambda\right) \tag{5}$$

where $ceil()$ denotes the upward rounding function and λ is the empirical parameter, which is set to 0.25 here.

3.4 Feature Extraction Module

The intention of a drone is influenced not only by its own state but also by the actions of other drones in the same swarm. To obtain accurate drone embeddings and swarm embeddings, a feature extraction module based on GAT and

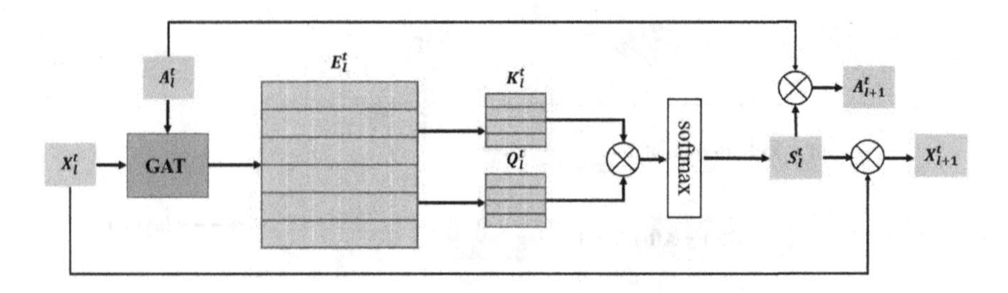

Fig. 3. Layer i in Feature Extraction Module

self-attentive pooling is proposed. This module includes multiple layers, whose architecture is shown in Fig. 3. When the final pooling output is a single vector, indicating the completion of the extraction module's work. It outputs the embedding e_t of the swarm corresponding to the current time step.

The workflow of the module is described as follows, using layer l as an example.

First, the embeddings of the input nodes at each layer need to be extracted. The adjacency matrix corresponding to the input graph at layer l is assumed as A_l^t, and the feature matrix composed of all nodes is X_l^t. Then, the extraction operation can be performed to obtain the embeddings matrix E_l^t for all input nodes. The extraction operation is as follows:

$$E_l^t = GAT\left(X_l^t,\ A_l^t\right) \tag{6}$$

In order to identify the intent of the swarm, it is necessary to obtain the feature information of the swarm as well. To address this requirement, a novel self-attentive pooling module is proposed. This module takes as input the adjacency matrix A_l^t of the current layer's graph and the embeddings matrix E_l^t from the GAT output. The output of the module are the adjacency matrix $A_{(l+1)}^t$ and the input feature matrix $X_{(l+1)}^t$ for the next layer's pooled graph.

This module has three key training parameters, denoted as W_k^l, and W_q^l. The dimensions of these parameters control the number of nodes at the next layer. K_l^t and Q_l^t represent the query matrix and key matrix corresponding to E_l^t. Additionally, S_l^t represents the pooling matrix. The pooling operation flow can be mathematically represented by the following equation:

$$K_l^t,\ Q_l^t = \left[W_k^l,\ W_q^l\right] E_l^t \tag{7}$$

$$S_l^t = softmax\left(\frac{Q_l^t * (K_l^t)^T}{\sqrt{d_k}}\right) \tag{8}$$

$$X_{l+1}^t = \left(S_l^t\right)^T * E_l^t \tag{9}$$

$$A_{l+1}^t = \left(S_l^t\right)^T * A_l^t * S_l^t \tag{10}$$

where d_k is the length of the key matrix that guarantees the stability of the computed attention values.

Similar to [18], in order to reduce the training difficulty, the loss function to be minimized in the training is set as:

$$L = \left\| A_l^t, S_l^t \left(S_l^t\right)^T \right\|_F \tag{11}$$

where $||\cdot||_F$ denotes the Frobenius norm.

3.5 Intent Prediction Module

Intentions are frequently manifested through a sequence of actions, and temporal features play a significant role in capturing the correlation between these actions and intentions. The intent recognition module utilizes a GRU network to process the temporal information of the data and capture the temporal correlation between swarm intent and temporal features. The structure of the intent recognition module is illustrated in Fig. It receives the pooled output of embeddings $e_t(t = 1, 2, \cdots, m)$ for swarm at each time steps and generates the GRU hidden state $h_t(t = 2, 3, \cdots, m)$ at each time step. The formula is as follows:

$$h_t = GRU\left(h_{t-1}, e_t\right) \tag{12}$$

Afterwards, h_m is fed into a multilayer perceptron $MLP(\cdot)$ to obtain the predictive intent labels matrix Y_i of the swarm at time step m. In experiments, m is set to 20.

$$Y_i = MLP\left(h_m\right) \tag{13}$$

The cross-entropy loss function is employed to accurately identify the intent of the target swarm.

4 Experiments

4.1 Data and Environment

The experimental data were gathered from a near-shore air defense simulation system and labeled based on expert experience. The dataset consists of 5000 samples, with 4000 samples designated for training and 1000 samples for testing. The data is structured with a time step of 20, and the initial feature dimension is 9.

The experimental environment is set up as follows: an Intel Xeon Platinum 8375C CPU, 96 GB of memory, GeForce RTX 4090 graphics card, operating system Ubuntu 20.04.6, and CUDA 12.0.

4.2 Evaluation Metric

The Receiver Operating Characteristic (ROC) curve is a graphical representation that illustrates the relationship between the true positive rate (TPR) and the false positive rate (FPR) for different classification thresholds. The area under the ROC curve (AUC) is a metric that quantifies the overall classification performance of a model. In this experiment, AUC is utilized to evaluate the classification ability of various models for different intention classes. Since the experiment involves a four-category problem, we calculate the AUC for each intention category by treating all other intentions as a single category. Additionally, the accuracy rate is employed as a general evaluation metric to assess the effectiveness of the model.

$$TPR = \frac{TP}{TP + FN} \tag{14}$$

$$FPR = \frac{FP}{FP + TN} \tag{15}$$

$$acc = \frac{TP + TN}{TP + TN + FP + FN} \tag{16}$$

4.3 Baseline

To showcase the superiority of the proposed model, it is compared with the following model:

MLP [19]: This method utilizes a multilayer perceptron (MLP) for intent recognition and employs the Adam optimization algorithm.

SAE [20]: This method employs the Auto-Encoder Neural Network (AENN) as the fundamental unit to construct a Stacked Auto-Encoder (SAE) which can temporally encode the original features. The method applies a logistic regression classifier for intent recognition.

LSTM [13]: This method is based on a vanilla LSTM model, which takes into account the impact of incomplete information.

BiGRU-Attention [16]: In this method, BiGRU is employed to capture and combine information from all preceding and succeeding time series in order to extract the embedding of the current moment. Additionally, instead of utilizing a fully connected layer, the attention mechanism is employed for intention recognition.

Attention-TCN-BiGRU [15]: In contrast to BiGRU-Attention, this approach incorporates TCN for further feature extraction of the data.

4.4 Result

The data are labeled in four categories: takeoff, penetration, attack and retreat. The AUC metric is used to evaluate the effectiveness of the proposed method and baseline methods for recognition on different intentions. The results are

Table 1. The results of the comparative experiments.

	MLP	SAE	LSTM	BiGRU-Attention	Attention-TCN-BiGRU	**Ours**
takeoff	78.2	85.2	86.1	89	89	**90.6**
penetration	79.3	87.9	91.9	93.7	94.5	**97.3**
attack	78.3	85.5	85.5	87.3	86.2	**89.7**
retreat	94.3	92.1	96.5	96.8	**98.3**	97.2
avg	82.5	87.7	90.0	91.7	92.0	**93.7**

shown in Table 1. The bold font in the table indicates the model with the best performance in each intention.

As anticipated, the methods that incorporate temporal information exhibit higher effectiveness compared to those that do not. Furthermore, the methods utilizing attention mechanisms demonstrate superior performance compared to those without attention mechanisms. It becomes apparent that all the methods exhibit lower accuracy in identifying the attack and surprise intentions. This discrepancy could be attributed to the fact that the presentation of these two intentions is more similar, making it challenging for the models to distinguish their respective features and learn them accurately.

Among all the baseline methods, the Attention-TCN-BiGRU achieves the highest average AUC. By comparison, the proposed method surpasses the accuracy of the Attention-TCN-BiGRU. The experimental results clearly demonstrate that the proposed model exhibits a significantly improved prediction accuracy compared to the other baseline methods.

In order to further highlight the effectiveness of the proposed method, it is compared with GAT-GRU, Attention-Pool-GRU, and GRU in terms of accuracy. Among these methods, GAT-GRU and GRU utilize the average pooling strategy instead of the proposed pooling method. The results of the ablation experiments are shown in Table 2. It becomes evident that the proposed method enable the network to learn more accurate intention features, leading to a significant improvement in intention recognition accuracy.

Table 2. The results of the ablation experiments.

Model Composition Structure			acc	loss
GAT	Attention Pool	GRU		
✔	✔	✔	95.6	0.126
✔		✔	92.1	0.196
	✔	✔	89.7	0.211
		✔	84.5	0.439

5 Conclusion

A novel end-to-end intent recognition method has been proposed, that outperforms state-of-the-art methods in the field of UAV swarm intent recognition. The proposed model effectively captures intra-swarm interactions by leveraging threat coefficient similarity, which enable feature aggregation among neighboring UAVs. Furthermore, an attention mechanism-based pooling method has been proposed, which takes into account both the structural information of swarm and the criticality of individual UAVs during the pooling operation. Through comparisons with other methods and ablation experiments, the proposed model has been proven to enhances the accuracy of UAV swarm intent recognition.

References

1. Lehto, M., Hutchinson, W.: Mini-drone swarms. J. Inf. Warfare **20**(1), 33–49 (2021)
2. Whelan, J., Almehmadi, A., El-Khatib, K.: Artificial intelligence for intrusion detection systems in unmanned aerial vehicles. Comput. Electr. Eng. **99**, 107784 (2022)
3. Xu, H., Zhao, J., Chen, L., Tan, W., Zhang, H.: A review of methods of battlefield target combat intention recognition. In: International Conference on Autonomous Unmanned Systems, pp. 3686–3696. Springer, Cham (2022). https://doi.org/10.1007/978-981-99-0479-2_340
4. Liu, J.: Air target intention recognition based on incomplete multi-granulation rough set. In: 2022 IEEE Asia-Pacific Conference on Image Processing, Electronics and Computers (IPEC), pp. 944–947. IEEE (2022)
5. Wang, Y., Wang, J., Fan, S., Wang, Y.: Quick intention identification of an enemy aerial target through information classification processing. Aerosp. Sci. Technol. **132**, 108005 (2023)
6. Tan, B., Li, Q., Zhang, T., Zhao, H.: The research of air combat intention identification method based on bilstm+ attention. Electronics **12**(12), 2633 (2023)
7. Wang, P., Wang, Y., Gong, X.: An air target tactical intention recognition method based on the fusion deep learning network model. In: Second International Conference on Electronic Information Engineering, Big Data, and Computer Technology (EIBDCT 2023), vol. 12642, pp. 605–613. SPIE (2023)
8. Veličković, P., Cucurull, G., Casanova, A., Romero, A., Lio, P., Bengio, Y.: Graph attention networks. arXiv preprint arXiv:1710.10903 (2017)
9. Chung, J., Gulcehre, C., Cho, K., Bengio, Y.: Empirical evaluation of gated recurrent neural networks on sequence modeling. arXiv preprint arXiv:1412.3555 (2014)
10. Lei, Z., Dong, Z.m., Wu, D.y.: Target tactical intention recognition based on fuzzy dynamic bayesian network. In: 2019 International Conference on Modeling, Analysis, Simulation Technologies and Applications (MASTA 2019), pp. 241–244. Atlantis Press (2019)
11. Chang, T., Kong, W., Dai, W.: A threat assessment method based on target combat intent information fusion. Control Decision Making **34**(3), 591–601 (2019)
12. Cao, S., Liu, Y., Xue, S.: Improve the target intent recognition method for high-dimensional data similarity. Sensors Microsystems **36**(5), 25–28 (2017)
13. Liu, Z., Chen, M., Wu, Q., Chen, S.: Prediction of unmanned aerial vehicle target intention under incomplete information. Scientia Sinica Inform. **50**(5), 704–717 (2020)

14. Teng, F., Song, Y., Wang, G., Zhang, P., Wang, L., Zhang, Z., et al.: A gru-based method for predicting intention of aerial targets. Computational Intelligence and Neuroscience 2021 (2021)
15. Teng, F., Song, Y., Guo, X.: Attention-tcn-bigru: an air target combat intention recognition model. Mathematics **9**(19), 2412 (2021)
16. Teng, F., Guo, X., Song, Y., Wang, G.: An air target tactical intention recognition model based on bidirectional gru with attention mechanism. IEEE Access **9**, 169122–169134 (2021)
17. Liu, H., Ma, Z., Deng, X., Jiang, W.: A new method to air target threat evaluation based on dempster-shafer evidence theory. In: 2018 Chinese Control And Decision Conference (CCDC), pp. 2504–2508. IEEE (2018)
18. Ying, Z., You, J., Morris, C., Ren, X., Hamilton, W., Leskovec, J.: Hierarchical graph representation learning with differentiable pooling. Advances in neural information processing systems 31 (2018)
19. Qu, C., Guo, Z., Xia, S., Zhu, L.: Intention recognition of aerial target based on deep learning. Evolutionary Intelligence, pp. 1–9 (2022)
20. Ou, W., Liu, S., He, X., Guo, S.: Tactical intention recognition algorithm based on encoded temporal features. Command Control Simul. **38**(6), 36–41 (2016)

An Attention Detection System Based on Gaze Estimation Using Self-supervised Learning

Xiang-Yu Zeng[1,3,4], Bo-Yang Zhang[1,3,4], and Zhen-Tao Liu[2,3,4](✉)

[1] School of Future Technology, China University of Geosciences, Wuhan 430074, China
[2] School of Automation, China University of Geosciences, Wuhan 430074, China
liuzhentao@cug.edu.cn
[3] Hubei Key Laboratory of Advanced Control and Intelligent Automation for Complex Systems, Wuhan 430074, China
[4] Engineering Research Center of Intelligent Technology for Geo-Exploration, Ministry of Education, Wuhan 430074, China

Abstract. With the rapid development of online education around world, it becomes difficult for teachers to evaluate the effect of learning during the online class. To solve this problem, an attention detection system based on gaze estimation using self-supervised learning is proposed. Self-supervised learning is used to train the gaze estimation network and it achieved an average gaze angle error of 6.5° on the MPIIFaceGaze dataset. Cross dataset testing was performed by 5 test subjects in the RT-GENE dataset and all samples in the Columbia dataset, in which achieved average gaze angle error 13.8° and 5.5°. It is found that the network has good cross dataset generalization performance, but it is susceptible to factors such as feature occlusion and low image clarity. Then, we developed an attention detection system which has functions of unmanned detection, eye state detection, and gaze detection. Online classroom experiments were conducted to verify the effectiveness of the proposed method, in which the average accuracy of the detection of distraction is around 60%.

Keywords: Online Education · Attention Detection · Gaze Estimation · Self-supervised Learning

1 Introduction

Nowadays, online education has become popular with a fast-paced trend. However, teachers not only need to pay attention to the lecture content, but also take into account the learning situation of students. Teachers and students cannot directly communicate face-to-face, making it difficult for teachers to communicate timely. It is difficult to determine whether students are actively participating in teaching, even if they are receiving feedback from teaching [1–3]. To solve this

B. Xin et al. (Eds.): IWACIII 2023, CCIS 1931, pp. 178–188, 2024.
https://doi.org/10.1007/978-981-99-7590-7_15

problem, there are many solutions for online education, such as student's attention detection and so on. With external assistance, teachers can focus more on teaching the content well, rather than focusing too much on observing students' state or understanding students' listening status through language feedback and other information.

Attention analysis is conducted through the visual features of human external expression, including body posture, head posture, gaze direction, and degree of eye opening and closing. In teaching, teachers usually use these external visual features to analyze students' attention concentration level. However, gaze is a component of the human eye state, which can quickly and efficiently reflect a person's behavioral intention. It is a key component of future intelligent and efficient human-machine interaction systems. The interaction between people can be facilitated by accurate gaze data, which can help people interact with machines and provide a non-contact interactive experience [4–8]. In the field of automatic driving and auxiliary driving, the driver's gaze can be used to judge whether the driver is focused and whether the driver is sleep-deprived driving, and the driving safety can be ensured by timely reminding or automatic driving succession [9]. Gaze estimation can also be applied to intelligent recommendation systems to locate customers' interests by analyzing their gaze, make corresponding recommendations [10].

In recent years, with the application of deep learning methods in the field of gaze estimation, the estimation accuracy of this method has been greatly improved. Zhang et al. [11] tried to use Convolutional neural network for the gaze estimation task. The network takes the eye image as the input, splices the head posture information with the extracted eye features, and finally regresses to obtain the gaze, and they [13] replaced the previous shallow convolutional network with VGG16, greatly improving the model accuracy. Chen et al. [12] were inspired by the asymmetry of images cropped from both eyes, and used this asymmetry to optimize the results of line of sight estimation, then used a shallower neural network to estimate head posture.

However, the success of gaze estimation in deep learning depends on cumbersome and accurate labeling. When there is a lack of precise label annotations, the performance of the training model is hindered. Researchers are beginning to use self-supervised learning methods to solve problems. Wang et al. [14] propose a novel gaze adaptation approach, namely Contrastive Regression Gaze Adaptation, for generalizing gaze estimation on the target domain in an unsupervised manner. Farkhonde et al. [15] proposed an equivariant version of online clustering based self supervised method SwAV, to learn more information representation for gaze estimation. Jinda et al. [16] proposed a simple contrastive representation learning framework for lgaze estimation called GazeCLR. It utilizes multi-view data to improve invariance, relying on selected data that does not change the gaze direction, and utilizes data augmentation technology for learning.

In this paper, we propose a novel method that contributes to gaze estimation algorithms. Figure 1 shows the overall framework of our method. We simplify the gaze as the ray from the midpoint of the line connecting the human eyes to

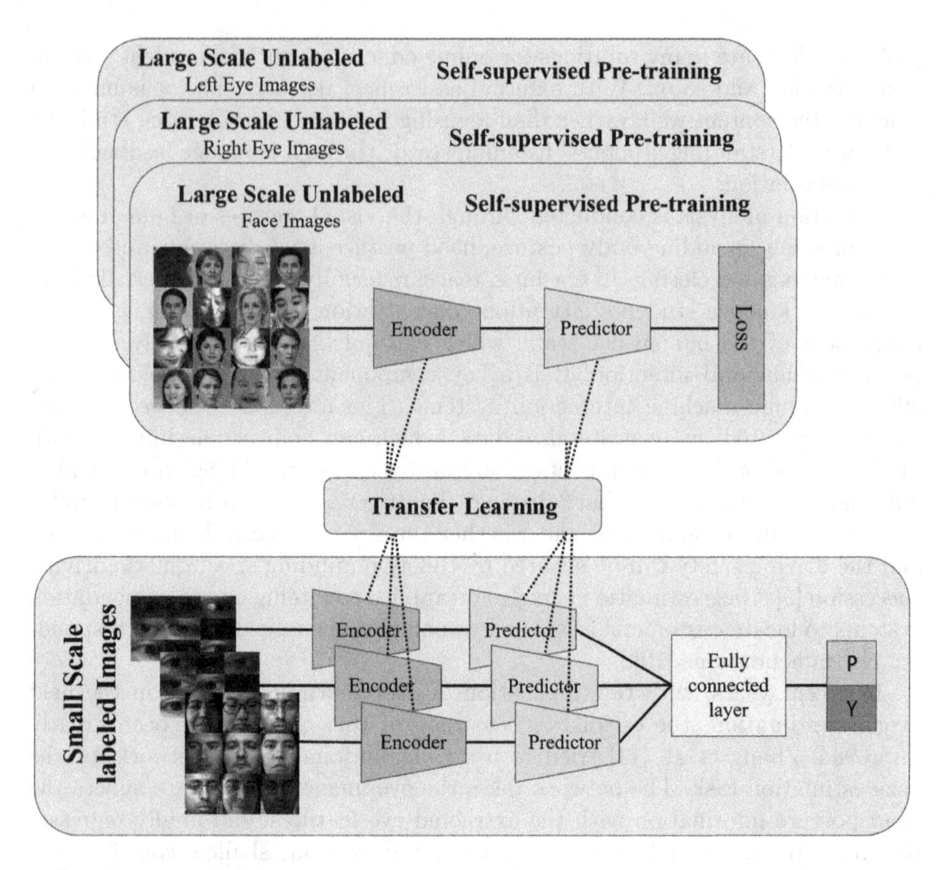

Fig. 1. Framework overview of our method. (a) The self-supervised pre-training is performed through self-supervised contrastive learning and train face and two eyes pre-training models. (b) Freeze the weights of these models from the self-supervised pre-training. (c) Fusion of face feature and two eyes features from pre-training models and train the gaze estimation algorithm using these models.

the target point of gaze, which can be decomposed into angles in the vertical and horizontal directions in the spatial coordinate system. Therefore, the gaze estimation algorithm estimates a total of two values (i.e., yaw and pitch).

On the basis of previous gaze estimation algorithms, an attention detection system has been built for students in online education, which includes unmanned detection, eye state detection, and gaze estimation. The system performs detection based on face images and then analyses the student's attention state for each frame in real time based on the detection results of each module.

The rest of this paper is organized as follows. Section 2 briefly introduces the specific architecture of self-Supervised gaze estimation algorithm. Section 3 gives a detailed introduction to the attention detection system. The experimental results of attention detection system are given in Sect. 4.

2 Framework of Gaze Estimation

The overall framework of self-Supervised gaze estimation mainly consists of two processes. The first part is self-supervised contrastive learning pre-training. The second part is gaze regression using transfer learning.

2.1 Contrastive Learning Pre-training

Contrastive learning pre-training using SimSiam structure. SimSiam networks have become a common structure among various unsupervised visual representation learning models recently [17], which maximize the similarity between two enhancements of an image, avoid collapse solutions under certain conditions, and use simple networks to learn more meaningful feature representations. Structure is shown in the Fig. 2.

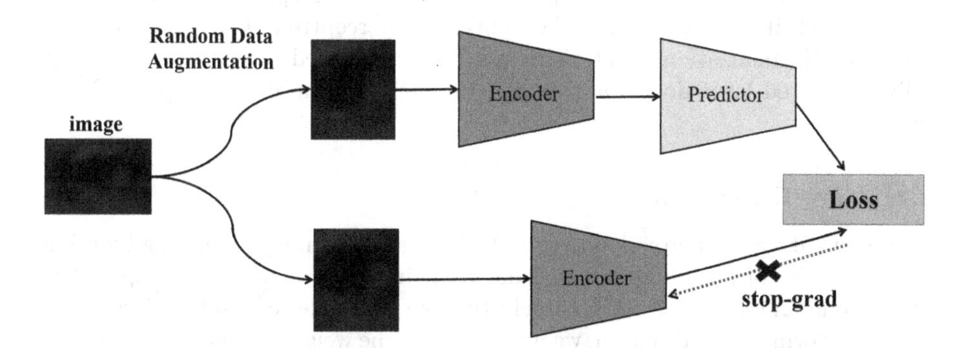

Fig. 2. Framework of SimSaim method. There are also two identical Predictors, while the down side is not drawn. The two encoders are identical. Loss is the comparison of the feature vectors output by the predictor, and it is the negative cosine similarity between the encoder output z and the predictor output p.

The SimSiam network input is an unlabeled image, and then the input image is randomly enhanced with data to generate two images. Then they are put into the up and down encoders separately. Each encoder receives no gradient from the first term, but it receives gradients from the second term. The specific formula is

$$\mathcal{D}(p_1, z_2) = -\frac{p_1}{\|p_1\|_2} \cdot \frac{z_2}{\|z_2\|_2} \tag{1}$$

$$\mathcal{L} = \frac{1}{2}\mathcal{D}(p_1, \text{stopgrad}(z_2)) + \frac{1}{2}\mathcal{D}(p_2, \text{stopgrad}(z_1)) \tag{2}$$

where p_1 and z_1 is output of encoder and predictor on the up side of the network respectively, p_2 and z_2 is the output of the encoder and predictor on the down side of the network respectively.

During the training process, input is unlabeled images, utilizing a large number of unlabeled face and eye images publicly available on the internet, which approximately 50000 face and eye images each. Finally, three pre-trained networks, namely face, left eye, and right eye, were obtained through training.

Random Data Augmentation. After inputting an unlabeled image, random horizontal transformation, brightness contrast and saturation adjustment, and grayscale enhancement will be randomly performed.

Encoder. The Encoder is composed of ResNet-20 as the backbone network and a projection multi-layer perceptron MLP head. The MLP of the projection head has three fully connected layers, which apply the BatchNormalization function to each fully connected layer, including its output fully connected layer, which does not use the ReLU function.

Predictor. The predictor is a separate multi-layer perceptron, Unlike the projection head, it has two layers. The multi-layer perceptron of the predictor has the BatchNormalization function to the fully connected layer, but its output fully connected layer does not have the BatchNormalization function or ReLU function.

2.2 Gaze Estimation

Gaze estimation is a regression task with the goal of learning a mapping function $\mathcal{H} : x \rightarrow g$, it converts high-dimensional RGB images $x \in \mathbb{R}^{H \times W \times 3}$ map to low dimensional 2D angles [18–20], namely the two 2D angles of pitch and yaw.

To perform gaze estimation, we first initialize the weights of the backbone network using the pre-training weights learned through self-supervised contrastive learning pre-training in Sect. 2.1. The features extracted from the three pre-training models are directly concatenated, then fed into the fully connected layer, and finally regressed to obtain the gaze value.

Regression. Specifically, in the regression process, the loss between the estimated angle and the actual angle is utilized to refine the entire network and ultimately obtain the predicted gaze value. The Loss function of gaze estimation model uses L2 Loss function for network training, as

$$L_2(I) = \frac{1}{N} \sum_{n=1}^{N} \|g_g - g_p\|^2 \quad (3)$$

where g_p is the predicted gaze value from the image and g_p is the actual gaze value.

In the training, we use MPIIFaceGaze dataset to train our model. For dataset, approximately 35000 eye and face images were taken as the training set, the first 5000 eye and face images were used as the validation set, and the first 5000–10000 eye and face images were used as the test set. Finally, the test was conducted on the test set, and the average angle error of the network was 6.5°.

MPIIFaceGaze Dataset. This processed dataset contains face and eye images. It also provides gaze directions of each face. This dataset is an extensive version of MPIIGaze. So we follow [21] to process dataset. We also use the evaluation dataset of MPIIGaze for evaluation. Finally, our dataset consists of 3,000 eye images each (i.e., 1500 left and 1500 right) of 15 people, and for each eye image, we get the corresponding face image. So we obtain approximately 45000 eye and face images each.

Cross Dataset Testing. In order to comprehensively evaluate the performance of the gaze estimation network, the network trained on the MPIIFaceGaze dataset was validated on other gaze estimation datasets. The validation data were randomly selected from 3 test subject images in 5 test subjects in the RT-GENE dataset and all samples in the Columbia dataset.

The test results are shown in Table 1.

Table 1. Cross Datasets Average Angle Error.

Datasets	Average Angle Error
RT-Gene	13.8°
Columbia	5.5°
MPIIFaceGaze	6.5°

From the results, it can be found that the network performance on the RT-GENE dataset has a significant difference compared to other datasets, mainly due to the low face clarity of the RT-GENE dataset and the significant impact of the signal-to-noise ratio of the image.

Due to the equipment worn by the test personnel during the collection process of the RT-GENE dataset blocking some faces, the impact on faces with lower clarity during network operation is increased. Tests on the Columbia dataset showed that the gaze estimation for faces with high definition and small head pose changes is more accurate, with an average test error smaller than the errors on the other two datasets.

Overall, our method has good generalization performance across datasets, but is susceptible to factors such as feature occlusion and low image clarity.

3 Attention Detection System

The attention detection system is designed for online classrooms, it collects learners' learning videos through the built-in webcam on their computer, and then the system analyzes the attention state of each frame in real-time.

The system is developed by using Qt5 in Python, and it can complete real-time detection and operation only on computers equipped with cameras and

Windows systems, with strong applicability, as shown in Fig. 3. During the detection, only the system software needs to be started and run before the detection can be enabled. The background detection of the system will not affect the normal operation of the teaching platform (such as Tencent Meeting, Zoom, DingTalk).

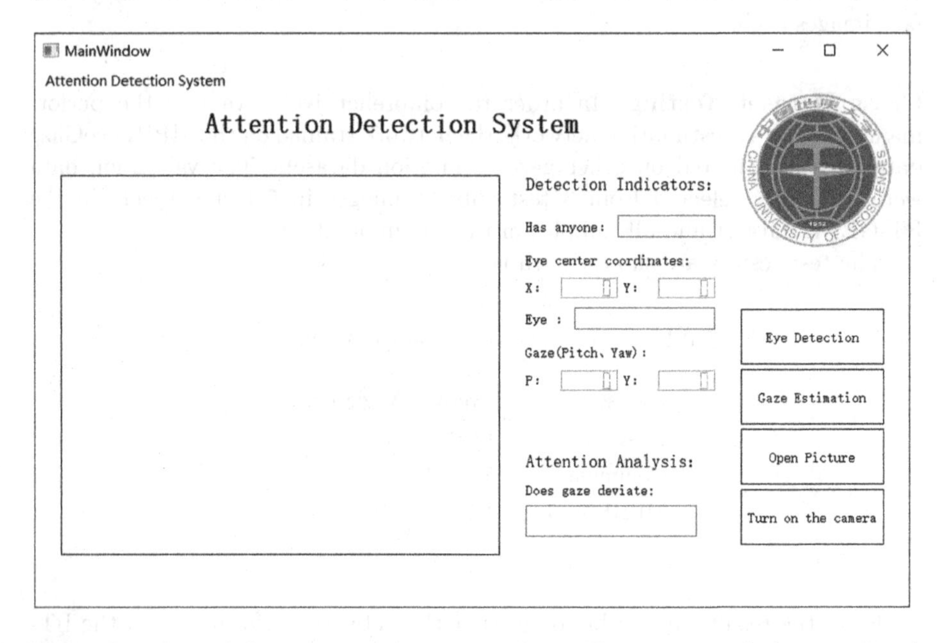

Fig. 3. Interface of the attention detection system. The system includes unmanned detection, eye state detection, and gaze detection. Each detection module detects based on face image and eye images, and obtains the detection results of each module.

By drawing inspiration from the principle of the PERCLOS (Percentage of Eye Closure) algorithm, this paper sets up a timer to determine whether the subject is focused or not based on the proportion of unmanned frames, the proportion of frame deviation from the gaze, and the proportion of frame closure of the human eye in one cycle of the timer. Using the proportion of frames detected by specific events in one round of detection as the basis for determining attention.

The unmanned detection module is used to determine whether students are absent during class. The gaze estimation module is used to determine students' gaze during class. When one of the Pitch and Yaw angles of the gaze in a single frame during the detection process is greater than 35°, it is judged that the gaze deviates from this frame. The eye state detection module is used to determine the opening and closing status of students during class. And one of the criteria for judging attention dispersion is separation, deviation of sight, and closure of the human eye.

Firstly, when no one is detected, the indicator of whether there is a student will be updated and displayed as no one. When someone is detected, the indicator is present will be updated and displayed. Then the system performs gaze estimation. It predicts the two angles of the gaze, and updates the gaze angles in the indicators, then visualizes the person's gaze in the image. Finally, the system performs eye opening and closing detection. It provides the coordinate points of the center points of the left and right eyes in the image, and updates the detection results in the upper left corner of the image. It also updates the opening and closing status of the human eye in real-time in the indicators.

4 Experiments

The experimental environment is a real online classroom. During the experiment, videos of online teaching scenes were first collected and recorded through a camera, followed by an experiment after class. Finally, the videos used for the experiment were recorded for 10 h, with each class lasting half an hour. There are 20 courses in total. And there is a screen shot of the experimental video as shown in Fig. 4.

Fig. 4. A screen shot of the online education video.

The experimental equipment is a desktop computer or laptop with a regular camera, and there is only one student in front of each device. The number of volunteers participating is 10, with a gender ratio of 1:1. The tester can freely adjust the head position within a range of 40–60 cm from the device screen. Experiments were performed on a computer with Microsoft Windows10 64-bit system, 16 GB RAM, AMD Ryzen 7 4800H CPU, NVIDIA GTX1650 GPU, and about 300000 pixel camera. The experimental platform is based on TensorFlow deep learning framework.

To verify the accuracy of attention detection in this paper, the recorded video is first divided into five segments. Each segment randomly captures the recorded video with the required time for one timer cycle for one experiment, and each segment undergoes 20 repeated experiments. If there are ten testers in each video, a total of 200 experiments will be conducted for each segment. The experimental process is shown in Fig. 5. The experimental results are shown in Table 2, and only the results of attention distraction are counted.

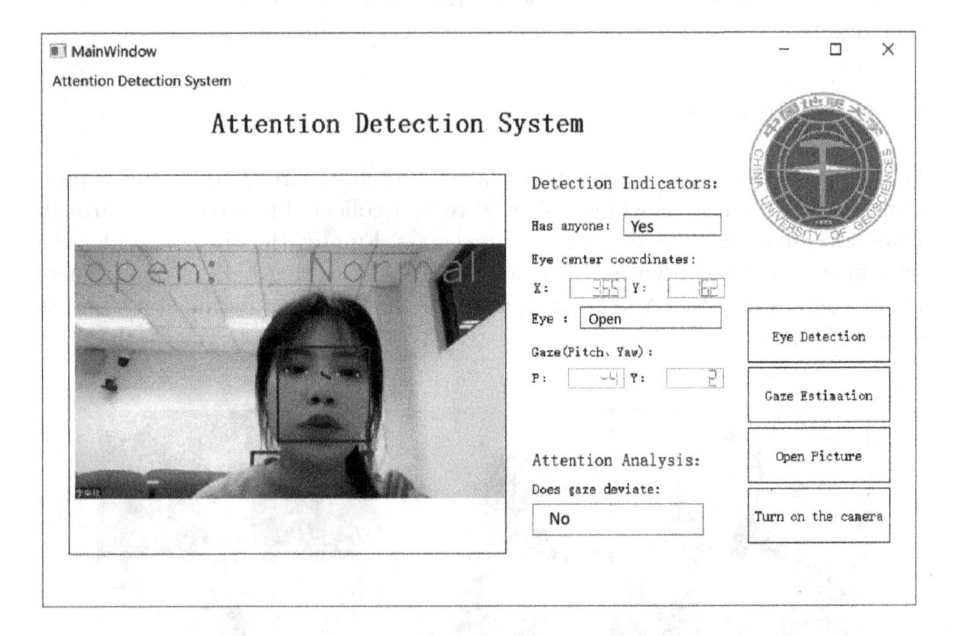

Fig. 5. A screen shot of the experiment on one student.

Table 2. Experiments Results of Student Attention Detection.

	Attention Dispersion Times	Detection Number of Errors	Missed Detection Times	Accuracy
The first video	55	17	3	63.6%
The second video	30	5	5	66.7%
The third video	73	24	9	54.8%
The fourth video	66	22	5	59.1%
The fifth video	56	15	4	66.1%
The sixth video	280	83	26	61.1%

From Table 2, it can be observed that the average accuracy of the attention detection algorithm in this paper is around 60%. Our method has fewer missed detections. The algorithm integrates multiple detection modules and focuses on multiple visual features. Because the manifestation of attention distraction is diverse, single feature detection only focuses on changes in a certain visual feature, ignoring other key features that can represent attention distraction. Therefore, this algorithm greatly improves the above problem.

5 Conclusion

This paper focuses on the current needs of online education teaching work, using gaze estimation method based on comparative learning to obtain students' gaze value and other detection module to get student's states. Then, based on the correlation between students' states during class and the classroom, a quantitative analysis of students' attention is carried out. A complete and feasible recognition method is proposed, and an online classroom human-computer interaction experiment is conducted to verify it. While the attention detection system in this paper can accomplish attention detection, the experimental results are not ideal. Even though the experimental results are not very satisfactory, the detection task can still be accomplished within an acceptable range. Consequently, the work done in this paper still has a lot of room for further improvement.

In the teaching scenario of online classrooms, the installation position of student cameras cannot be the same as the settings in the dataset in real scenarios, resulting in more complex changes in students' gaze during class. So we can use more gaze estimation data to generalize estimation models. And the attention detection system only considers three simple modules. Therefore, other related detection modules such as gaze area analysis, fixation point, and behavior analysis can be considered to improve the detection ability of the attention detection system.

References

1. Paudel, P.: Online education: benefits, challenges and strategies during and after COVID-19 in higher education. Int. J. Stud. Educ. **3**(2), 70–85 (2021)
2. Chen, T., Peng, L., Yin, X., et al.: Analysis of user satisfaction with online education platforms in China during the COVID-19 pandemic. Healthcare **8**(3), 200 (2020). Multidisciplinary Digital Publishing Institute
3. Cazzato, D., Leo, M., Distante, C., et al.: When i look into your eyes: a survey on computer vision contributions for human gaze estimation and tracking. Sensors **20**(13), 3739 (2020)
4. Smith, B.A., Yin, Q., Feiner, S.K., et al.: Gaze locking: passive eye contact detection for human-object interaction. In: Proceedings of the 26th Annual ACM Symposium on User Interface Software and Technology, pp. 271–280 (2013)
5. Ahmed, F.R.A., et al.: Analysis and challenges of robust E-exams performance under COVID-19. Results Phys. **23**, 103987 (2021)
6. Vidal, M., Turner, J., Bulling, A., et al.: Wearable eye tracking for mental health monitoring. Comput. Commun. **35**(11), 1306–1311 (2012)

7. Chirumamilla, A., Sindre, G., Nguyen-Duc, A.: Cheating in e-exams and paper exams: the perceptions of engineering students and teachers in Norway. Assess. Eval. High. Educ. **45**(7), 940–957 (2020)
8. Bawarith, R., Abdullah, A., et al.: E-exam cheating detection system. Int. J. Adv. Comput. Sci. Appl. **8**(4), 176–181 (2017)
9. Vicente, F., Huang, Z., Xiong, X., et al.: Driver gaze tracking and eyes off the road detection system. IEEE Trans. Intell. Transp. Syst. **16**(4), 2014–2027 (2015)
10. Ahn, H.: Non-contact real time eye gaze mapping system based on deep convolutional neural network. arXiv preprint arXiv:2009.04645 (2020)
11. Zhang, X., Sugano, Y., Fritz, M., et al.: Appearance-based gaze estimation in the wild. In: Proceedings of the IEEE Conference on Computer Vision and Pattern Recognition, pp. 4511–4520 (2015)
12. Cheng, Y., Zhang, X., Lu, F., et al.: Gaze estimation by exploring two-eye asymmetry. IEEE Trans. Image Process. **29**, 5259–5272 (2020)
13. Zhang, X., Sugano, Y., Fritz, M., et al.: MPIIGaze: real-world dataset and deep appearance-based gaze estimation. IEEE Trans. Pattern Anal. Mach. Intell. **41**(1), 162–175 (2019)
14. Wang, Y., Jiang, Y., Li, J., et al.: Contrastive regression for domain adaptation on gaze estimation. Proceedings of the IEEE Conference on Computer Vision and Pattern Recognition, pp. 19354–19363 (2022)
15. Farkhondeh, A., Palmero, C., Escalera, S., et al.: Towards self-supervised gaze estimation. In: Proceedings of the IEEE Conference on Computer Vision and Pattern Recognition, pp. 10974–19387 (2022)
16. Jindal, S., Manduchi, R.: Contrastive representation learning for gaze estimation. In: Proceedings of the IEEE Conference on Computer Vision and Pattern Recognition, pp. 13404–13421 (2022)
17. Chen, X., He, K.: Exploring simple Siamese representation learning. In: Proceedings of the IEEE Conference on Computer Vision and Pattern Recognition, pp. 15750–15758 (2021)
18. Wang, K., Ji, Q.: Real time eye gaze tracking with 3d deformable eye-face model. In: Proceedings of the IEEE International Conference on Computer Vision, pp. 2003–1011 (2017)
19. Sun, L., Liu, Z., Sun, M.T.: Real time gaze estimation with a consumer depth camera. Inf. Sci. **320**, 346–360 (2015)
20. Yang, H.D., Lee, S.W.: Reconstruction of 3D human body pose from stereo image sequences based on top-down learning. Pattern Recogn. **40**(11), 3120–3131 (2007)
21. Zhang, X., Sugano, Y., Fritz, M., et al.: It's written all over your face: full-face appearance-based gaze estimation. arXiv preprint arXiv:1611.08860 (2016)

Effects of Pseudo Labels in Pose Estimation Models Using Semi-supervised Learning

Harunobu Ariga(✉) and Yuki Shinomiya

Shizuoka Institute of Science and Technology, Shizuoka 437-0032, Japan
{2018004.ah,shinomiya.yuki}@sist.ac.jp

Abstract. This paper aims to analyze the effects of pseudo labels for training pose estimation models using semi-supervised learning. One of the problems in training pose estimation models is the high cost of labeling, which requires keypoint annotation of training data. Pseudo-labels estimated by other pose estimator models are able to improve the performance of another estimator. This paper investigates the trade-off between the number of annotations by humans and pseudo labels given by the model.

Keywords: Pose estimation · Semi-supervised leaning · Keypoint detection

1 Introduction

Recently, surveillance cameras and drive recorders have been widely used for crime prevention, biometric identification, and automated driving, and the demand for analysis and automated processing of these camera images is increasing. In such cases, estimating the location of a person, or estimating or detecting the person itself is used as a useful method. These methods usually include pose estimation, person estimation, and detection.

One of the problems in training pose estimation models is the high cost of labeling, which requires keypoint annotation of training data. This paper investigates the trade-off between the number of annotations and estimation accuracy using semi-supervised learning. The estimation accuracy is evaluated using the Common Objects in COntext (COCO) dataset [5].

In this paper, we investigated the effect of using pseudo labels on keypoint detection in pose estimation, and this method is effective for pose estimation.

This paper is organized as follows: Sect. 2 describes related research, Sect. 3 describes the proposed method, Sect. 4 describes the experimental setup, and Sects. 5 and 6 describe the experimental results and discussion.

2 Related Works

In pose estimation, keypoint detection methods are mainly used to estimate the posture of a human body by extracting skeletal information from the shoulders,

B. Xin et al. (Eds.): IWACIII 2023, CCIS 1931, pp. 189–199, 2024.
https://doi.org/10.1007/978-981-99-7590-7_16

Fig. 1. Example of keypoint inference of images in the COCO dataset by YOLO.

elbows, wrists, and other parts of the human body. The most popular models for skeleton estimation are Openpose [1] and YOLO [6], which is a typical detection model used mainly for object detection tasks. YOLO infers a total of 17 points of skeletal information (Fig. 1) from "nose, shoulders, wrists," etc., and returns a set of parameters (x-coordinate, y-coordinate, and confidence) for each key point.

The MS COCO dataset [5] is used as the dataset for training. In this dataset, keypoint information is labeled as training data, and each keypoint information holds three parameters such as (x-coordinate, y-coordinate, and visibility). Therefore, visibility is a manually labeled parameter. There have been many efforts to study the human skeleton using keypoint detection. Table 1 shows the label name of the keypoints and the corresponding indices.

In these efforts, the main problems with keypoint detection methods are the inability to make correct inferences due to missing keypoint information caused by occlusion or low accuracy, and the occurrence of false positives. Yamakawa et al. [8], for example, use time series information to supplement keypoint information from frames before and after the missing information to determine whether the keypoint information is correct or incorrect. On the other hand, Kushizaki et al. [3] showed that the identification rate increased by performing shaping processing under certain conditions for false positives.

Table 1. Keypoint and number assignment by YOLO.

label	index	label	index
nose	1	left hand	10
left eye	2	right hand	11
right eye	3	left hip	12
left ear	4	right hip	13
right ear	5	left knee	14
left shoulder	6	right knee	15
right shoulder	7	left heel	16
left elbow	8	right heel	17
right elbow	9		

2.1 Semi-supervised Learning

In this paper, we experiment with semi-supervised learning, a method for efficiently learning large amounts of unsupervised data from small amounts of supervised data. In particular, we use a method called self-training [7,9]. Self-training is a method to recursively train on the inference results as pseudo-labels [4,7] by using a learner that has trained on the training data. In this paper, we use test data from the MS COCO dataset described above for training.

3 Proposal Learning Procedure

In this paper, we investigate the ratio between the number of annotations and the number of pseudo-labels by semi-supervised learning, and whether there is a trade-off relationship. As mentioned above, adding annotations to data is a very costly human task, given the number of supervised data required.

We use mechanically determined labels which are used as pseudo-label data. By using both pseudo-label data and training data for m1, we thought that the cost of annotation could be reduced. The ratio of these data to be trained is varied, and the one with the best learning outcome is investigated (Fig. 2). we denote m1 as the result of learning only the training data and m2 as the result of retraining with pseudo-label data.

4 Experiments

4.1 Dataset

For the experiments, data with and without annotations were separated. The data were randomly selected from the MS COCO dataset, and the number of images was increased in steps of 100, 200, 300, 400, and 500 for each image. From semi-supervised learning, the weights learned from 100 to 500 pieces of annotated data are each inferred from unannotated data, and the 100 to 500 pieces are relearned as pseudo-label data. To show the results.

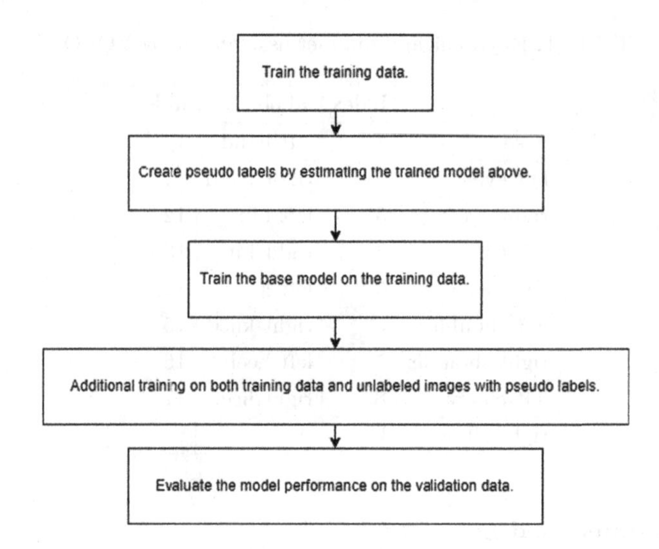

Fig. 2. The learning procedure in this paper.

4.2 Parameter Setup

In this paper, YOLOv8 [2] is used for keypoint detection and inference. the parameters retained by each point in the pseudo-label data are (x-coordinate, y-coordinate, and confidence), while the parameters retained by the training data are (x-coordinate, y-coordinate, and visibility), the data format differs for the confidence and visibility portions.

Confidence is a continuous value of 0 and 1, and visibility is stored as a discrete value between 0, 1, and 2. Assuming the threshold value is 0.5 if the confidence level is less than the threshold, visibility is treated as 0, and if it is greater than the threshold, visibility is treated as 1.

Visibility is divided into 0, 1, and 2 within YOLO. 0, 1, and, 2 mean "neither visible nor predictable", "not visible but predictable", and "visible" respectively. Keypoints with visibility 0 are not used for training. Therefore, since the training does not distinguish between 1 and 2, the pseudo-label processing is aimed at converting to 0 and 1 with 0.5 as the threshold value.

For m1, the initial weights were set randomly. For m2, the trained weights m1 were used as the initial weights. For both m1 and m2, we follow a training strategy of a cosine weight decay with an initial lr 0.01 and a final lr 10^{-5}, and the batch size was set to 16.

yolov8n-pose and yolov8s-pose without pre-trained weights were used for training. For convenience, we refer to yolov8n-pose as model-n and yolov8s-pose as model-s.

4.3 Epochs Normalization

In this experiment, the number of images of training data and pseudo-label data used in each experiment is different, which may cause differences in the number of batches for the same number of epochs and affect the experimental results. Therefore, a reference number of epochs was set for each of m1 and m2, and the number of epochs for each of them. For m1, we consider 5 cases with 100 to 500 images as "n". The reference is a training with 500 images, with epochs set to 100 and batch set to 16.

This results in 3,100 steps, and the following equation is formulated to approximate the number of steps even in the other conditions.

$$\#\text{steps} \approx \frac{n}{\#\text{batch}} \tag{1}$$

$$\#\text{epochs} \approx \frac{3,100}{\#\text{steps}} \tag{2}$$

For m2, in addition to 100 to 500 pieces of training data, 100 to 500 pieces of pseudo-label data as "n'" are added to the training, respectively. The standard is 1,000 images of 500 pieces of training data and 500 pieces of pseudo-label data, and epochs and batch size is set to 20 and 16, respectively. This results in 20,000 steps, and the following equation is formulated to approximate the number of steps even in the other conditions.

$$\#\text{epochs} \approx \frac{1,250 \times \#\text{batch}}{(n+n')} \tag{3}$$

4.4 Evaluation

For each evaluation, we evaluate the performance by precision, recall, F-measure, and mAP50. The F-measure is the harmonic mean of the precision and recall, the mAP50 is a measure of average precision with a threshold of 50% accuracy for IoU from semi-supervised learning. IoU is a measure of the degree of overlap between correct and guess labels.

5 Experimental Results and Discussions

The results of the evaluation of F-measure by model-n are shown in Table 2 and by model-s in Table 3. model-n shows a trend of increasing accuracy for m1 basically in proportion to the number of images, When retrained, model-n shows an increasing trend in the F-measure, and with respect to the performance ratio of each training data to the number of pseudo-label data, the ratio of data outputting the best results for the pseudo-label data tends to decrease as the number of training data increases. For model-s, m1, on the other hand, the accuracy varied with the number of images, and the retraining results showed a proportional relationship between the number of training data and the number

Fig. 3. Performance comparison of pseudo-label data with different number of data for training in model-n. The left figure shows the case with 200 training data, and the right figure shows the case with 300 training data.

of pseudo-label data. For model-n and model-s, the peak number of pseudo-label data relative to the number of training data was 200 and 500, respectively.

In addition, the comparison of pseudo-label data. Figure 3 shows an example plot of pose estimation for the model-n case, when the same number of pseudo-label data were used for training data trained under different conditions. In the plots, the same images are extracted for comparison. From these plots, we can confirm that "points that were not estimated are estimated" and "people who were not estimated are estimated" in the case of 300 pieces of training data compared to the case of 200 pieces of training data.

Similarly, in Fig. 4, we plot an example of a plot of pose estimation when the same number of pseudo-label data are used for the training data with different conditions in the mode-s case. The plots show images for the same case as model-

Fig. 4. Performance comparison of pseudo-label data with different number of data for training in model-s. The left figure shows the case with 200 training data, and the right figure shows the case with 300 training data.

n. The results show that under the same conditions, the difference in models did not lead to an accuracy advantage.

Although no model-specific differences in learning were observed during m1 training, there is a clear difference between model-n and model-s in the relationship between the number of training data and the number of pseudo-label data. Since a proportional relationship exists between the number of training data and the number of pseudo-label data in model-s and the ratio is close to 1:1, it is possible that the allowable number of pseudo-label data is higher due to the larger capacity compared to model-n. The reasons for this tendency are that when pseudo-label data become large for a small amount of training data, the pseudo-label data may become noise due to low accuracy in estimating pseudo-labels, or overlearning may occur due to the size of the epoch in relation to the amount of data.

On the other hand, in model-n, the tendency of the smaller the number of training data, the larger the number of pseudo-label data, which may indicate that there are not enough learning opportunities for the model (Figs. 5, 6, 7 and 8 and Tables 4 and 5).

Table 2. Comparision of F-measure in model-n

# of annotations	# of training samples for the base model				
	100	200	300	400	500
0 (Baseline)	0.1419	0.1513	0.1885	0.2361	0.1767
100	0.1663(+0.0244)	0.1472(−0.0041)	0.2416(+0.0531)	0.2278(−0.0083)	0.2492(+0.0725)
200	0.1840(+0.0421)	0.1768(+0.0255)	0.2287(+0.0402)	0.2390(+0.0029)	0.2780(+0.1013)
300	0.1969(+0.0550)	0.1659(+0.0146)	0.2767(+0.0882)	0.2638(+0.0277)	0.2657(+0.0890)
400	0.1760(+0.0341)	0.1718(+0.0205)	0.2373(+0.0488)	0.2603(+0.0242)	0.2554(+0.0787)
500	0.1787(+0.0368)	0.1984(+0.0471)	0.2559(+0.0674)	0.2513(+0.0152)	0.2349(+0.0582)

Table 3. Comparision of F-measure in model-s

# of annotations	# of training samples for the base model				
	0	1	2	3	4
0 (baseline)	0.1588	0.1128	0.2128	0.1768	0.2194
100	0.1510(−0.0078)	0.1986(+0.0858)	0.2514(+0.0386)	0.2476(+0.0708)	0.2311(+0.0117)
200	0.1561(−0.0027)	0.1779(+0.0651)	0.2108(−0.0020)	0.2805(+0.1037)	0.2326(+0.0132)
300	0.1271(−0.0317)	0.1753(+0.0625)	0.2271(+0.0143)	0.2842(+0.1074)	0.2723(+0.0529)
400	0.1279(−0.0309)	0.1601(+0.0473)	0.2330(+0.0202)	0.2948(+0.1180)	0.2598(+0.0404)
500	0.0983(−0.0605)	0.1842(+0.0714)	0.1454(−0.0674)	0.2543(+0.0775)	0.2750(+0.0556)

Table 4. Comparision of mAP50 in model-n

# of annotations	# of training samples for the base model				
	100	200	300	400	500
0 (baseline)	0.0507	0.0537	0.0798	0.1406	0.0964
100	0.0652(+0.0144)	0.0577(+0.0040)	0.1306(+0.0509)	0.1475(+0.0069)	0.1362(+0.0398)
200	0.0743(+0.0236)	0.0882(+0.0345)	0.1395(+0.0597)	0.1653(+0.0247)	0.1565(+0.0601)
300	0.1048(+0.0541)	0.0761(+0.0224)	0.1602(+0.0805)	0.1587(+0.0182)	0.1461(+0.0497)
400	0.0898(+0.0391)	0.0930(+0.0394)	0.1150(+0.0353)	0.1595(+0.0189)	0.1391(+0.0427)
500	0.0989(+0.0481)	0.0863(+0.0326)	0.1323(+0.0525)	0.1578(+0.0172)	0.1402(+0.0438)

Table 5. Comparision of mAP50 in model-s

# of annotations	# of training samples for the base model				
	0	1	2	3	4
0 (baseline)	0.0820	0.0503	0.1033	0.0990	0.1242
100	0.0729(−0.0091)	0.1130(+0.0627)	0.1311(+0.0278)	0.1626(+0.0636)	0.1276(+0.0034)
200	0.0647(−0.0173)	0.0896(+0.0393)	0.1239(+0.0206)	0.1790(+0.0800)	0.1470(+0.0228)
300	0.0825(+0.0005)	0.0741(+0.0238)	0.1207(+0.0174)	0.1694(+0.0704)	0.1770(+0.0528)
400	0.1124(+0.0304)	0.0888(+0.0385)	0.1233(+0.0200)	0.1873(+0.0883)	0.1830(+0.0588)
500	0.0628(−0.0192)	0.0715(+0.0212)	0.0802(−0.0231)	0.1467(+0.0477)	0.1625(+0.0383)

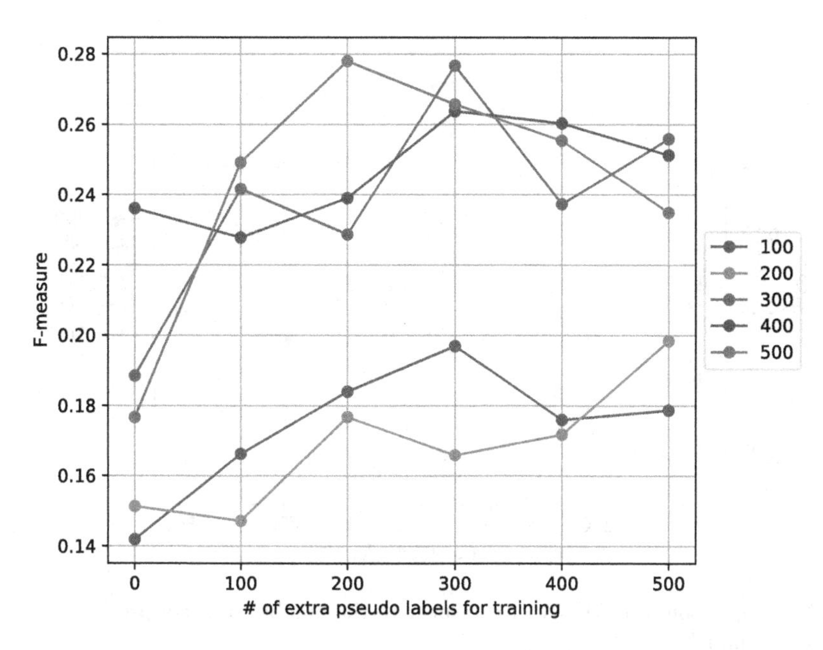

Fig. 5. Comparison of F-measures with respect to the number of extra pseudo labels for training in model-n.

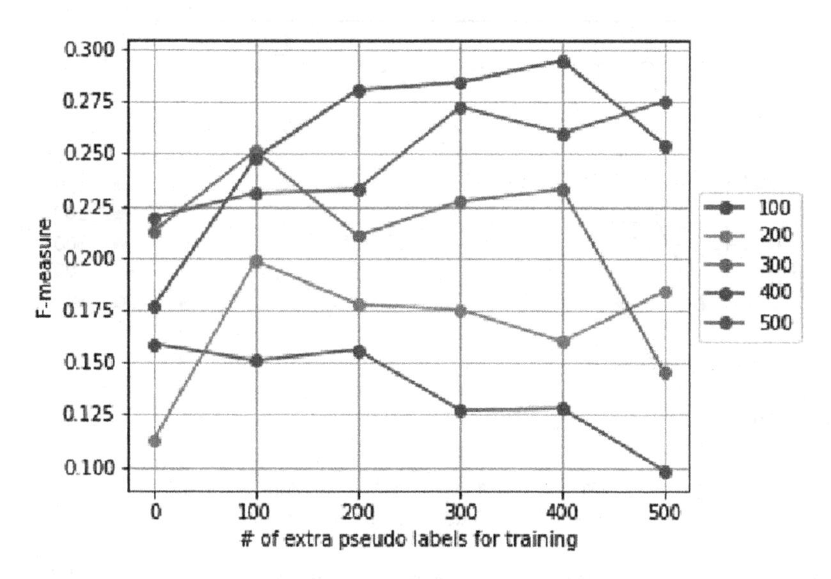

Fig. 6. Comparison of F-measures with respect to the number of extra pseudo labels for training in model-s.

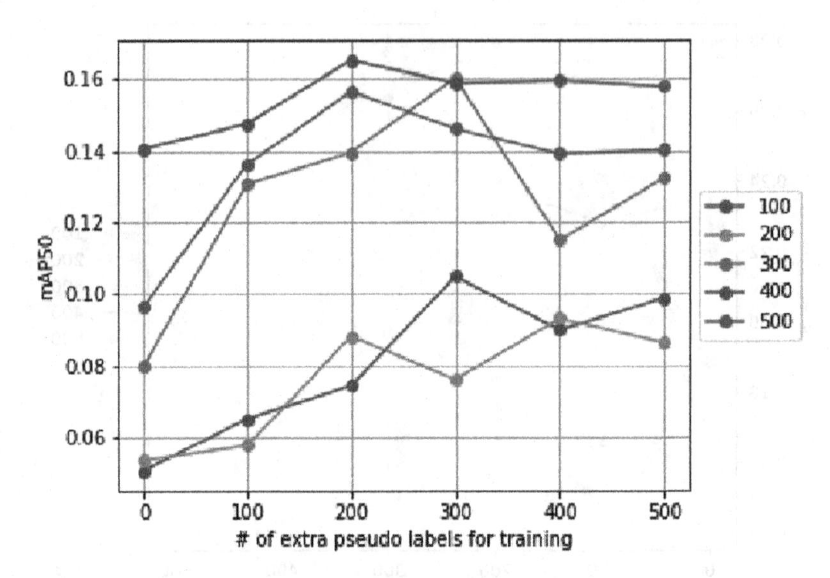

Fig. 7. Comparison of mAP50s with respect to the number of extra pseudo labels for training in model-n.

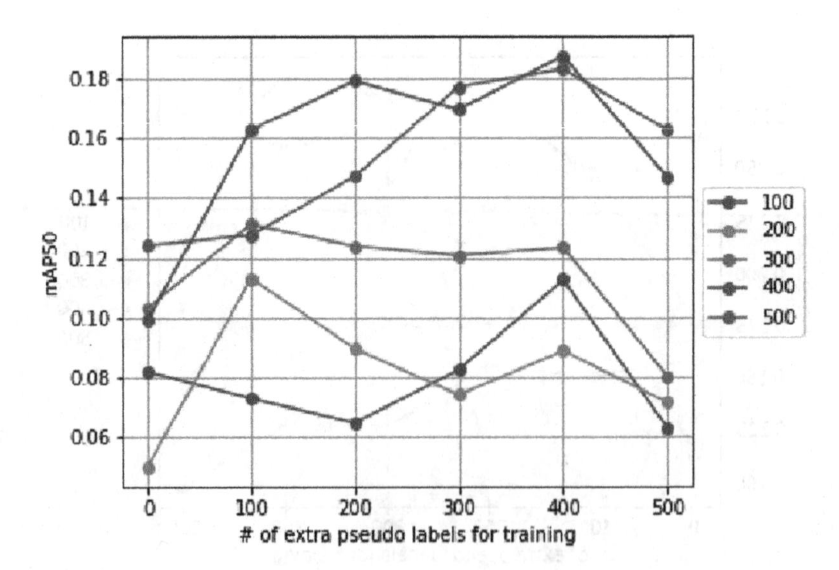

Fig. 8. Comparison of mAP50s with respect to the number of extra pseudo labels for training in model-s.

6 Conclusions and Future Works

This paper, we have investigated the trade-off between the number of training data and the number of pseudo-label data. The results indicate that model-s may be a more suitable model compared to model-n, in which case there may be a proportional relationship between the number of training data and the number of pseudo-label data. We inferred from both models that providing a larger number of pseudo-label data relative to the number of training data may result in noise to the learning process. In our experiments, we have confirmed that the m1 does not detect the person at the stage of assigning pseudo-labels, and that it outputs results with poor accuracy in guessing the keypoints. From there, semi-supervised learning requires a more suitable model and the same number of pseudo-labels as the number of training data in order to achieve high accuracy. In this study, the conversion of visibility and confidence was performed using the threshold value of 0.5. Since these data have not been handled strictly and clear standards have not been confirmed, more efficient and useful conversions are needed.

In addition, in this paper, we divided the keypoint information by person and had them perform calculations, but we believe that by dividing the keypoint information by categories such as head, torso, arms, etc., which are generally called "regions", we can obtain even more detailed information and the effect of each region on learning.

Acknowledgement. This work was supported by JSPS KAKENHI Grant Number JP21K17810.

References

1. Cao, Z., Hidalgo, G., Simon, T., Wei, S.E., Sheikh, Y.: Openpose: realtime multi-person 2d pose estimation using part affinity fields (2019)
2. Jocher, G., Chaurasia, A., Qiu, J.: YOLO by Ultralytics (Jan 2023). https://github.com/ultralytics/ultralytics
3. Kushizaki, S., Tauchi, Y., Mizukami, Y.: A study on person identification based on walking motion. Tech. Rep. 31, Graduate School of Science and Engineering, Yamaguchi University (Mar 2018)
4. Lee, D.H.: Pseudo-label: the simple and efficient semi-supervised learning method for deep neural networks (2013)
5. Lin, T., et al.: Microsoft COCO: common objects in context. CoRR abs/arXiv: 1405.0312 (2014)
6. Redmon, J., Divvala, S., Girshick, R., Farhadi, A.: You only look once: unified, real-time object detection. arxiv:1506.02640 (2015)
7. Xie, Q., Luong, M.T., Hovy, E., Le, Q.V.: Self-training with noisy student improves imagenet classification. arXiv preprint arXiv:1911.04252 (2019)
8. Yamakawa, A., Ishikawa, T., Watanabe, H.: Improving pose estimation models using time series correlation. In: The 82nd national Convention of IPSJ, vol. 2020(1), 249–250 (2020). https://cir.nii.ac.jp/crid/1050292572112028032
9. Zhu, X.J.: Semi-supervised learning literature survey. University of Wisconsin-Madison Department of Computer Sciences, Tech. rep. (2005)

Sequential Masking Imitation Learning for Handling Causal Confusion in Autonomous Driving

Huanghui Zhang[1] and Zhi Zheng[1,2](✉)

[1] College of Computer and Cyber Security, Fujian Normal University, Fuzhou 350117, China
zhengz@fjnu.edu.cn

[2] College of Control Science and Engineering, Zhejiang University, Hangzhou 310027, China

Abstract. Training agents for autonomous driving using imitation learning seems like a promising way since its only requirement is the demonstration from expert drivers. However, causal confusion is a problem existing in imitation learning, which is that with more features offered, an agent may perform even worse. Here, we aim to augment agents' imitation ability in driving scenarios under sequential setting, by a novel method we proposed: Sequential Masking Imitation Learning(SEMI). First, we train a Vector Quantised-Variational AutoEncoder(VQ-VAE) to encode a sequence of images into a latent representation with discrete codes. After that we deploy several masks on the encoded images, the masks here will randomly hide some semantic objects in the encoded images. Finally, we design the behavior clone network as a predictor of expert action, using an encoded and masked image sequence as input, encouraging the network to make expert-like predictions when some partition of information about the environment is missing. The masking procedure in SEMI helps the imitator identify the contribution of each encoded feature to the expert's prediction. We demonstrate that this method could alleviate causal confusion in driving simulation by deploying it to the CARLA simulator, and compared it with other methods. Experimental results show that SEMI can effectively reduce confusion in autonomous driving. The agent trained with SEMI method reduces the collision rate by 45% compared to methods without masking procedure, and obtain the highest average survival timesteps among competing methods.

Keywords: Causal confusion · Invariant feature learning · Imitation learning

1 Introduction

Benefiting from the development of perception and computing capabilities, autonomous driving nowadays is able to build a model that can deal with complex situations. Recent research that can transform raw sensor data into a form that helps a model better understand its surrounding, like [1] can translate images captured by a car's front RGB camera, into an overhead map or bird's-eye view images. Combining those progress together, models relying on pre-processed data can perform better than traditional end-to-end models that use raw sensor images as input.

B. Xin et al. (Eds.): IWACIII 2023, CCIS 1931, pp. 200–214, 2024.
https://doi.org/10.1007/978-981-99-7590-7_17

Previous approaches in autonomous driving often took reinforcement learning(RL) methods to train a policy model to get higher rewards. RL methods highly rely on the reward function proposed by researchers, while designing a reward function to guide our model to do what we really want could be difficult [2], especially when the intent of the action is hard to express with mathematical expressions. RL methods also require interacting with the environment while training, which is difficult and unsafe. In addition, RL methods often require a time-consuming reward maximization procedure to have their agents performance guaranteed.

Traditional imitation learning(IL) methods can learn a strategy directly without actual interaction with the environment, using only expert samples organized as state-action sets. Using networks like deep convolutional neural network (CNN) model combine with IL methods like Behavioral Cloning(BC) could produce seemingly well results. However, IL often suffers from a problem for a long time: "causal confusion" [3]. Due to the distribution differences between training and testing states, a imitator may misidentify the real cause of an expert's action and rely on suspicious correlates to make decisions. Effects of the problem showed more obvious and severe when the state information given by the environment is plentiful and the scenario is complex. This confusion led to models learned by IL methods like behavioral cloning, performed poorly when it meet new states different from training samples.

Considering these problems above, we aim to organize the vehicle's surrounding states into semantic bird-eye view images in sequence, design an imitation learning method that encourages imitators to avoid causal confusion and make predictions based on important features.

In this paper, we provide the following contributions.

- We combine semantic bird-eye view images [4], with Object-aware REgularizatiOn (OREO) method from [5] in sequential setting, propose our Sequential Masking Imitation Learning(SEMI) method which show as a robust method for addressing causal confusion in a high-fidelity driving simulation.
- We implement SEMI and train an end-to-end model that can output direct continuous control commands to a vehicle, lead it to follow a pre-set route, unlike many models that rely on much hand-engineered involvement.
- Finally, we deploy our SEMI method on the CARLA simulator [6], test it with several environments that it is unfamiliar with, compare its performance with other methods, analyze its advantages, and demonstrate its consistent ability to perform an expert-like strategy. Experimental results show that the agent trained with SEMI method reduces the collision rate by 45% compare to methods without masking procedure, and obtain the highest average survival timesteps among competing methods.

The remainder of the paper is organized as follows. We review related work in Sect. 2. Our proposed SEMI method and its theory description are presented in Sect. 3. In Sect. 4, we describe the network structure for SEMI in the experiment and give a brief introduction about the simulation setting and the process of collecting data. In Sect. 5, we present the evaluation of our proposed method and compare it with several competing methods. Finally, we conclude the paper in Sect. 6.

2 Related Work

2.1 Pipelines of Autonomous Driving

Many autonomous-driving companies utilize a traditional engineer stack, where the driving problem is divided into perception, prediction, model planning, and control [7]. Solving these sub-tasks requires much hand-engineered involvement, which could be hard and lead to sub-optimal overall performance. Using discrete variables as the operation command is also a prevailing practice in the field of autonomous driving since this can simplify the driving task into a classification problem. However, the prediction of discrete variables that represent the high-level command still needs its downstream module to execute the concrete action. Also, using a finite number of discrete values as the direct control command may be harmful to a driving system's flexibility since operations like setting steering angle and acceleration often require precise control.

End-to-end driving, on the other side, known as mapping raw images to certain control commend, seems like a more promising way to autonomous driving. However, models rely on raw data require tons of samples that could cover a variety of situations like different weathers and different light intensities. Considering the weakness of common end-to-end driving, researchers put much effort to process raw data to improve models' robustness.

Our approach belongs to the end-to-end side of the autonomous driving spectrum, using pre-processed data to simulate the real driving task via imitation learning.

Follow what Chen et al. proposed in [4], we organize our input representation in the form of semantic bird-eye view image by using LiDAR and RGB camera sensors and built-in road maps, as shown in Fig. 1(a).

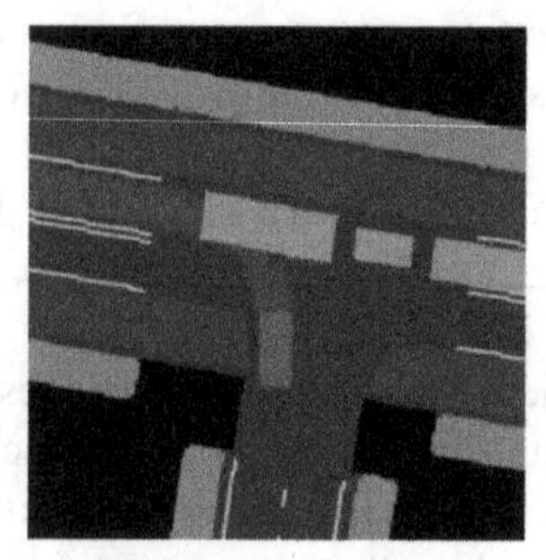

(a) Bird eye view image

(b) RGB camera image

Fig. 1. (a): A sample of bird-eye view images. The red rectangle represents the vehicle controlled by the model, and the green rectangles represents other vehicles. The road painted in blue represents the planned route. (b): The corresponding front camera image of (a). (Color figure online)

2.2 Confusion in Imitation

One branch of IL can be called "Inverse Reinforcement Learning(IRL)". Ever since Ng et al. [8] created the notion of IRL, there have been efforts to guide models in learning the intention behind an expert's action. IRL methods [2,9,10] intend to infer a reward function that can explain an expert's trajectory. However, inferring a reward function could be time-consuming when solving iteratively Markov decision processes(MDP) [2,11], and training a policy network often require interaction with environment, which could be dangerous when applying it to the real world.

Behavioral cloning is the other branch of IL, known as learning only through expert's demonstration [12]. We choose BC because it shows no requirement to interaction with the environment. While learning through only datasets seems stunning fashion, the causality problem in IL can not be ignored due to the distributional shift and the complex state information. When de Haan et al. [3] summarized causal confusion in imitation learning, they proposed to fix it with "Expert query" or "Policy execution", but those seem unpractical when it comes to driving in the real world. Katz et al. [13] use causal reasoning to construct an explanation for an expert's action, and generate a executing plan based on this explanation to carry out the expert's goal. Although this approach can explain an action from an expert properly, it requires the domain authors to enumerate direct causal associations to infer indirect causal relation, which is hard to implement when carrying out a complex job like driving. Hence, we consider methods that could address causal confusion indirectly, like randomly dropping/erasing units from input features to regularize policy [14–16].

Inspired by previous works [3,5,17], we encode raw bird-eye view images into discrete codes through a VQ-VAE [17], organize them in sequence as model input, train the model by BC, and regularize them by randomly masking out semantically similar objects.

Although Park et al. [5] did mention that masking semantic objects from a sequence of observation can be effective in practice, their efforts mainly focused on simple environments with clear goals like Atari games. Here we combine these methods and modify them to reduce the complexity of driving tasks, and address the causal confusion in end-to-end driving simulation.

3 SEMI Methodology

3.1 Semantic Encoder

The semantic encoder is the first part of our SEMI network. By training a VQ-VAE network, we make use of its encoder and vector quantizer to manage a bird-eye view image x which represents a state st [17], into discrete latent representations that can be seen as semantic representations.

VQ-VAE defines a latent embedding space $e \in \mathbb{R}^{K \times D}$, where K is the size of the discrete latent space, and the D is the dimension of each latent embedding vector e_i. The encoder map x into its latent representation $z_e(x)$, where $z_e(x)$ is a set of latent variables. Vector quantizer quantizes $z_e(x)$ to discrete representations $z_q(x)$. Decoder is another part of VQ-VAE, aiming to reconstruct x from $z_q(x)$. Decoder and encoder share the same cookbook $C = \{e_k\}_{k=1}^{K}$ of prototype vectors learned through training. Therefore, the objective of training a VQ-VAE can be formed as minimizing the following term:

$$L_{VQ-VAE} = \log p(x|z_q(x)) + ||sg[z_e(x)] - e||_2^2 + \beta||z_e(x) - sg[e]||_2^2, \quad (1)$$

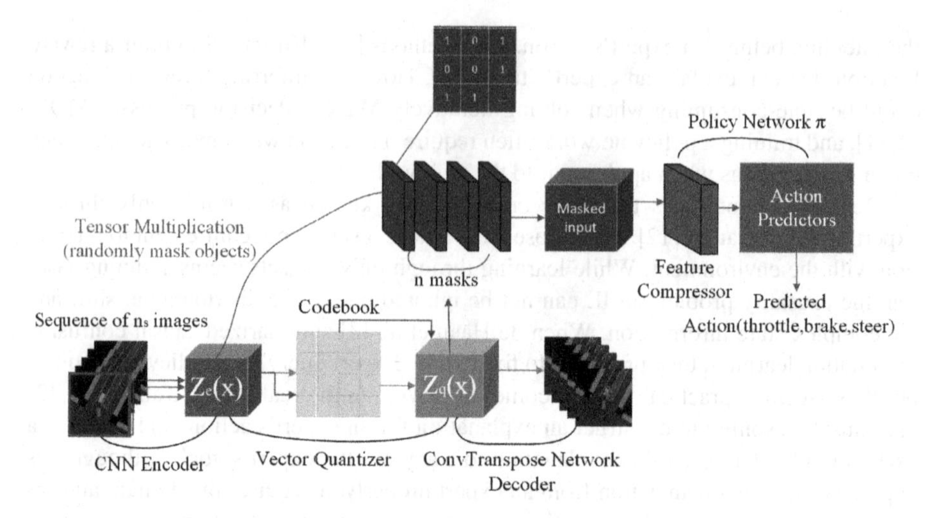

Fig. 2. Overview of our SEMI model. Green arrow represents the process of training VQ-VAE network. Red arrow represents the process of constructing n masks with the first image from sequence. Blue arrow indicates the process of training policy network using masked variable map as input. Notice that once the VQ-VAE trained, we do not need its decoder for imitation learning. (Color figure online)

where sg is the abbreviation of stop gradient operator, and β here is a weight on a commitment loss. We can divide Eq. (1) into three components: reconstruction loss, quantizer loss, and commitment loss. The reconstruction loss optimizes the encoder and decoder. The quantizer loss optimizes the vector quantizer so that it can bring codebook representations closer to the encoder outputs $z_e(x)$. The commitment loss is weight by β to make sure the encoder commits to an embedding and its output does not grow, as described in [17], the resulting algorithm is quite robust to β, so we use $\beta = 0.25$ in our experiments.

3.2 Masking Semantic Objects in Sequential Setting

After training a VQ-VAE network, we exact its encoder and vector quantizer for further use. Here, we first reorganize all bird-eye view images from training dataset with a total length n_t, into n_t sequences that each sequence contains n_s images(c^i represents images from $(i - n_s + 1)$th to ith, and x_j^i represents the jth image of c_i), and denote those sequences as $\{c_i\}_{i=1}^{n_t}$.

From then on, each time when we sample from dataset, we take batch B of sequences as our SEMI model inputs. From our intuition, expert demonstrations organized in sequence can help better reveal an expert's intention in a certain environment and show relations between state and action. However, training a model using samples in the form of sequence, often meet the over-fitting problem. We aim to ease it by randomly masking semantic objects.

Using the mask method from OREO [5], we can produce n masks for every sequence c, where n is a hype parameter pre-defined. By setting the drop probabil-

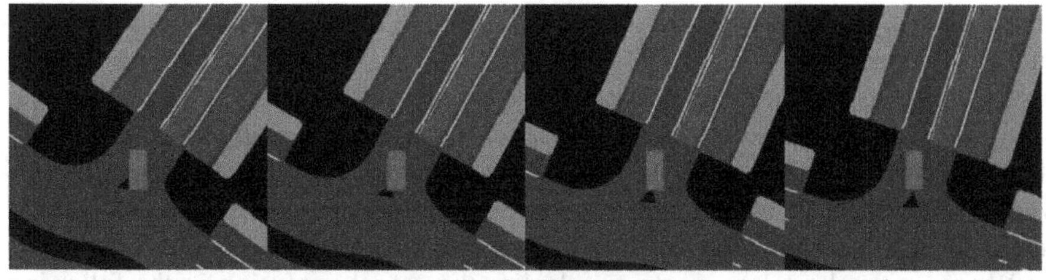

(a) Original images

(b) Masked images

Fig. 3. (a): A sequence of bird-eye view images. (b): A sequence of images reconstructed with the masked latent representation of images from (a). The colored specks in reconstructed images are produced by the decoder during the decode phase, since the masking operation may perturb the latent representation and lead to unstable reconstruction of images. Recall that the red rectangle represents the ego vehicle, and the blue line represents the guidance line. Although the images in (b) are hard to interpret after the randomly masking operation and reconstruction, we can still see that the operation randomly masks the road behind the ego vehicle in the latent representation, while the ego vehicle and the guidance line remain. (Color figure online)

ity p, we sample K binary random variables $b_k \in 0, 1, k = 1, 2, ..., K$ from a Bernoulli distribution with probability $1 - p$. We encode the image sequence c_i into the form of latent variables $z_e(c_i)$ using encoder, then construct n masks $m = (b \times z_{q_k}(x_0^i))_{k=1}^K$ for the sequence c^i, based on the embedding vectors produced by vector quantizer using the sequence's first image x_0^i. After constructing masks, we make tensor multiplication between every $z_e(c_i)$ and masks corresponding to c_i, this will output n new sequences corresponding to c_i, containing $n \times n_s$ masked sets of discrete latent variables.

Conclude the masking process described above: we randomly put several masks to drop some semantic objects from sequential images' discrete latent variables, and produce n new sequences with those variable sets.

Figure 3 is a visualization of masking semantic objects in a sequence of images, we can use it as an illustration of the masking idea. The colored specks in reconstructed images are produced by the decoder during the decode phase, since the masking operation may perturb the latent representation and lead to unstable reconstruction of objects' edges after decoding. As we can see in Fig. 3(b), a random mask is computed and deployed for this particular sequence. Latent variables that correspond to the road texture behind the ego vehicle are masked while the ego vehicle and the guidance line remain. So in the view of human drivers, the masked images can offer nearly the same information as the original images, since the texture of the passed road has little effect on the current action.

By applying several random masks for each sequence, we can examine whether the downstream policy network can still perform well when certain parts of semantic objects as input are masked. With this masking process, the downstream policy network is encouraged to make predictions without relying on features that may lead to confusion, and focus on important features that affect drivers' decisions.

We use several random masks to alleviate the causal confusion in the following theory. If a policy network (the imitator), denote it as $\overline{\pi}$, relies on a set of wrong features(e.g., the road texture behind the ego vehicle, positions of vehicles behind ego vehicle, or the positions of vehicles in the opposite lane) to make its prediction, then when some of these features are masked, the optimizer of $\overline{\pi}$ will update the weight of $\overline{\pi}$ to guide it to make a prediction based on the rest features due to the vast loss computed from the failed prediction.

It is worth noticing that, generally, the number of features that can cause causal confusion and still help $\overline{\pi}$ to achieve well performance in training is less than the number of features that an expert takes into consideration for prediction. In other words, many features in the environment affect the expert's decision, while only a small amount of other features are proxies of the expert's action (features that induce causal confusion). So if these proxy features have been randomly masked in some epoch during training, their effects are diminished and the $\overline{\pi}$ is forced to use other features to predict, therefore these proxy features are no longer consistent shortcuts for $\overline{\pi}$ to make predictions in the training phase. Therefore, the masking process can guide the downstream policy network to have a better understanding of the role of each feature in the prediction tasks.

3.3 Behavior Cloning with Imbalanced Dataset

After masking semantic objects, we connect the n new variable sequences into n maps, each map contains all variables sets of a sequence, and we denote those maps as $\{t_o^i\}_{o=1}^{n_s}$. We take $\{t_o^i\}_{o=1}^{n_s}$ and their corresponding action a_i as our SEMI policy network's input, notice that action a_i is a single control command captured at the time of c_i ended.

With our experience from real-world driving and observation from the simulator, one phenomenon can be learned is that the brake and steer are less triggered compared to the throttle. This phenomenon leads to a situation that is similar to the class imbalance in classification tasks, where the difference between classes in the dataset is imbalanced. Models learned from imbalanced datasets directly, usually show limited performance in generalization.

To improve the performance of our policy network in rare events where brake/steer is required, we first use the undersampling technique in our dataset. By using all of the rare events and reducing the number of abundant events, we keep the ratio of rare events to abundant events at about 1:3 to amplify the importance of braking and steering.

The structure of our policy network can be divided into two parts: feature compressor and action predictors. The compressor receives the map of variables sets and integrates features for predictors. We deploy three action predictors for three parts of action prediction: throttle, brake, and steer, and these predictors will output three scalars as their estimation of expert action, noted as $\hat{throttle}$, $\hat{brake}$, and $\hat{steer}$. The feature compressor and each action predictor are all Multi Layer Perceptrons(MLP).

The input features are first feed into the feature compressor and run through several linear layers and activation layers, then we obtain the compressed features as the output of the feature compressor. After that, the compressed features are passed into each action predictor that outputs a scalar as its prediction. In the deployment stage, the action will be the combination of the three scalars output from these predictors.

For each set of N samples, the policy network π can be optimized by computing and minimizing the mean-squared error(MSE) between action $[\hat{throttle}, \hat{brake}, \hat{steer}]$ predicted by π, and the real expert action$[throttle, brake, steer]$, following the term below:

$$
\begin{aligned}
L_{BC} =& \frac{1}{N}\sum\left(\hat{throttle} - throttle\right)^2 + \frac{1}{N}\sum\left(\hat{brake} - brake\right)^2 \\
&+ \frac{1}{N}\sum\left(\hat{steer} - steer\right)^2 .
\end{aligned} \tag{2}
$$

Algorithm 1: Process of SEMI

Initialize encoder, vector quantizer, decoder, policy π randomly;
Define batch-size B, drop probability p, num-mask n, sequence length n_s;
while not converged **do**
  Sample batch of state images $X \sim$ demonstration;
  foreach x *in* X **do**
    Encode x into $z_e(x)$ using encoder;
    Quantize $z_e(x)$ into $z_q(x)$ using quantizer;
    Decode x from $z_q(x)$ with decoder;
  Optimize parameters by minimizing L_{VQ-VAE} from Eq.1;
end while
Delete decoder;
Organize demonstration images into sequences of images;
while not converged **do**
  Sample batch of sequences of state images and corresponding actions $(C, A) \sim$ demonstration;
  foreach (c_i, a_i) *in* (C, A) **do**
    Encode image sequence c_i into $z_e(c_i)$ using encoder;
    Quantize the encoder's output of first image $z_e(x_0^i)$ into $z_q(x_0^i)$;
    Construct masks $\{m_i\}_{i=1}^{(n_s \times n)}$ based on $z_q(x_0^i)$;
    Mask c_i to new maps $\{t_o^i\}_{o=1}^{n_s}$;
    Take $\{t_o^i\}_{o=1}^{n_s}$ as input of π to predict action $\hat{a_i}$;
  Optimize parameters by minimizing L_{BC} from Eq.2;
end while

Table 1. Hyper parameters of our model.

Hyper parameter	Value
Learning rate:	1e-7
Embedding dimension:	16
Num_embeddings:	128
Num_masks:	4
Sequence_length:	4
Mask_prob:	0.6
Batch_size:	32

4 Experiment

4.1 Network Structure

As shown in Fig. 2, the whole SEMI model can be divided into two parts, the VQ-VAE network and the policy network.

The encoder of VQ-VAE consists of several convolutional layers to amplify the input's channel from 3 to the dimension of embedding and compress information about input. The vector quantizer calculates the discrete latent variable with respect to the output of decoder by finding the nearest embedding vector, then outputs it's index. The decoder contains several transposed convolutional layers corresponding to the encoder. The decoder takes the embedding vector produced by vector quantizer corresponding to the index as input to produce the reconstructed image.

The feature compressor and the action predictors of our SEMI policy network are both made up of MLPs. After flattening data that came from the encoder, we deploy it to pass through linear layers and predict each part of action separately. The hyper parameters we used in the experiment are shown in Table 1.

4.2 Simulation Environment and Data Collection

Based on the code from [4], we collect data and evaluate our proposed SEMI method on the CARLA simulator, along with other methods that we compared.

In order to act as an expert agent driver, we first write code to add manual driving support to the CARLA environment. We then implement an expert agent by having a human driver control the vehicle using an Xbox One controller. The driver will drive following a pre-set trajectory which will guide it to a random destination waypoint. Driver here need to control the vehicle within the correct lane given by trajectory, and avoid collision with other vehicles. We record bird-eye view images and the ego vehicle state (current speed) and control commands for each frame. We run the simulation for about 4 h and generated about 100k images and input commands. To be noticed, we only deploy the agent on the third map of the CARLA simulator for collecting training data, we do this to test the generalization ability of our SEMI method.

For organizing images into sequences, we group n_s successive frames' images and concatenate them together as a wide image to represent a sequence for our SEMI model's input.

4.3 Contrast Experiment

When it comes to comparison, we will not use models that output actions that require lower modules to execute, since those modules are difficult to program for different simulators or real world and our goal is to build an end-to-end model.

Here we train a CNN model to represent traditional BC methods [18]. We train a model using OREO [5] method as well(i.e. compute mask and predict based on a single image). We also train a sequential model without masking semantic objects in the training phase.

5 Results

5.1 Evaluation Procedure

From the maps provided by the CARLA simulator, we select 5 representative maps with different road styles for evaluation of model performance: the first, third, fourth, fifth, and seventh maps were selected. We denote these maps as Town1, Town2, Town3, Town4, and Town5. Each town has 10 different pre-computed routes. We compute each model's average result from these route as their result in a town. The simulation conditions, such as vehicle to driving and weather, remain consistent with the conditions in the recorded samples.

Towns: We will give a brief description of the towns we used for evaluation. Town1 and Town3 are both small simple towns with differences in road design. Town2 is the most complex town in CARLA since it has roundabouts and large junctions. Town4 is a squared-grid town with cross junctions and it has multiple lanes per direction. Town5 is a rural environment with narrow roads, corn, barns, and hardly any traffic lights. However, since the routes in each town are pre-computed randomly, the experimental results are affected by both the ability of the agent and the trait of the given route. Some unique parts in each town are shown in Fig. 4.

Metrics: Similar to Anzalone et al. [19], we evaluate models using four metrics. Collision rate is the crash rate of a single frame. A collision is recorded when the collision detector built into the ego vehicle reports a collision that happens with other vehicles or other still objects in the environment like buildings or roadblocks. Similarity measures the alignment between vehicle and planned road. Average speed is measured by compute the average of every frame's speed. Timesteps are the number of frames without collision.

Process: For every route in a town, each model will run test once. A test will stop when the vehicle runs safely for 2500 frames or collides with the environment or other vehicles. A model's metrics in a town will be computed by the results of 5 routes with the weight of routes: model's surviving frames in one route divided by surviving frames in 5 routes. The total metrics are the sum of weighted metrics from all towns.

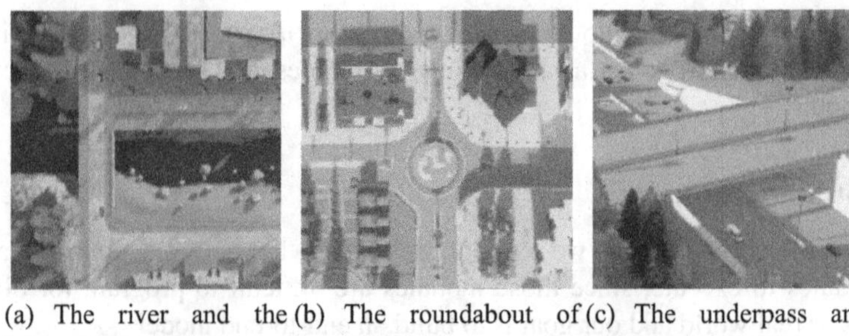

(a) The river and the bridge in Town1. (b) The roundabout of Town2. (c) The underpass and overpass in Town3.

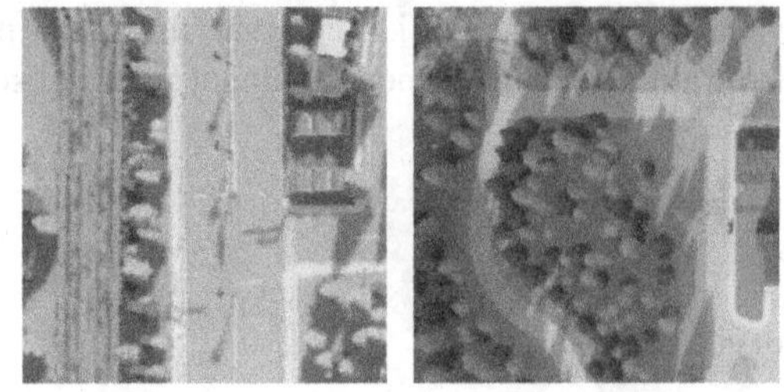

(d) Multi-lane roads and junctions in Town4. (e) Simple junctions and unmarked roads in Town5.

Fig. 4. We pick and show some unique parts in each town we used, such as the bridge in Town1, the roundabout in Town2, the underpass in Town3, the multi-lane roads in Town4, and the unmarked roads in Town5. Notice that these situations except the roundabout are all unfamiliar to the agents since we only collect samples in Town2.

5.2 Discussion

As we can see from Table 2, the SEMI method performs better than its competitors in the metric of collision rate and average survival timesteps, while remains the ability to follow pre-set route. The average speed of our SEMI model is not as fast as its competitors, however, it could be proof that our model tends to brake aggressively to avoid crashing. Our attempt to address causal confusion in imitation learning does improve our model's driving ability. These traits demonstrated that models trained with SEMI method have the ability to perform expert-like strategy.

Models trained by OREO method and the method using sequential samples both show well capabilities in driving following pre-set route. Their results indicate that each part of our SEMI method helps the performance more or less. However, models trained by these methods do not have our model's ability to cope with other vehicles, led to their slightly higher collision rate.

The CNN method without any masking operation does suffer from causal confusion: though it gets low loss in the training phase, it runs terribly in testing scenes. In observation, it's the only model that cannot handle curves and braking well, leading to the shortest survival timestep among the four models.

Of all four metrics we considered in this paper, the collision rate is the most important metric we take into consideration about model performance. Because when it

Table 2. Performance of methods: Object-aware REgularizatiOn(OREO), CNN, Sequential samples Without masking(SW), and our Sequential Masking Imitation Learning(SEMI). Best results are highlighted in bold.

Metric/Method		Town					
		Town1	Town2	Town3	Town4	Town5	**Total**
Collision Rate(‰)	OREO [5]	0.32	0.74	1.19	0.60	0.59	0.64
	CNN [18]	1.28	8.47	1.21	0.86	3.14	1.68
	SW	0.92	0.77	0.54	0.23	0.87	0.59
	SEMI(Ours)	**0.19**	**0.52**	**0.46**	**0.19**	**0.44**	**0.33**
Similarity(%)	OREO [5]	**92.3**	84.6	91.0	94.2	74.4	88.07
	CNN [18]	**92.3**	61.1	91.1	83.1	63.3	83.33
	SW	91.6	**88.7**	89.9	**95.8**	75.6	**89.21**
	SEMI(Ours)	84.8	87.6	**94.5**	93.3	**81.4**	88.41
Average Speed(km/h)	OREO [5]	15.5	**15.2**	**42.9**	22.5	**21.9**	23.34
	CNN [18]	15.5	7.68	26.7	**26.2**	21.7	**23.58**
	SW	**18.0**	13.4	18.8	16.4	15.5	16.12
	SEMI(Ours)	16.5	14.6	19.2	19.9	14	17.11
Average Timesteps	OREO [5]	1216	534	840	1318	1010	984
	CNN [18]	312	118	825	928	318	500
	SW	431	1036	**1324**	1681	918	1078
	SEMI(Ours)	**2047**	**1148**	1303	**2009**	**1352**	**1572**

comes to the deployment and use of autonomous driving, the safety issue is the most cared part. So we can say that the SEMI method we proposed can maintain a safer driving style have more advantages compared to their competitors.

We can also analyze the experiment results based on the traits of each town we evaluated on. As we can see that SEMI model demonstrated its ability to cope with other vehicles and different environments, even when situations of these environments are complicated like Town2 and Town5. While complex traffic does affect the overall performance of models trained from each method, we can still notice that the SEMI model generally maintained a steady driving style, while other competing methods may fail to cope with it, leading to unstable results of these methods in several towns with different difficulties.

Overall, the result demonstrates that SEMI method does handle causal confusion well. An autonomous driving model trained from the SEMI method can manage each part of a direct control command and simulate an expert's behavior well.

5.3 Analysis

Figure 5 is a visualization of the first layer's weight of the trained policy network π from SEMI, use Fig. 1(a) as an example. The elements that do not appear with gray are the features π mainly used for prediction.

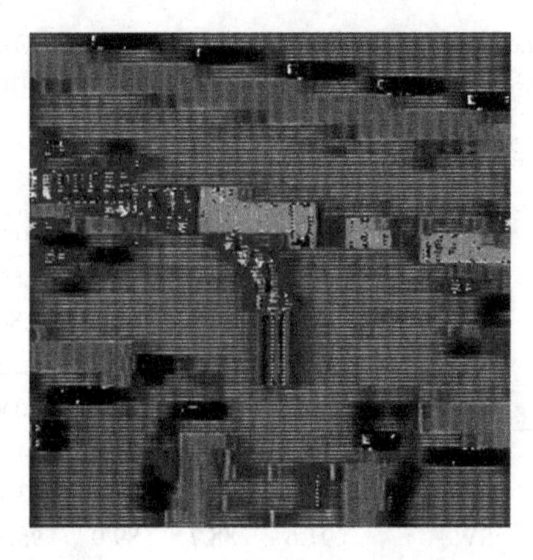

Fig. 5. The image is reconstructed from the combination of the encoded latent representation of Fig. 1(a) and the weight of the first layer in trained policy network π from SEMI, we make several processing to make the decision basis of π more interpretable for readers. Recall that in Fig. 1(a), red rectangle represents the ego vehicle, blue line represents the guidance line, and green rectangles represent other vehicles. The conspicuous elements there are features that π mainly takes into consideration for prediction, while other parts that are colored in gray are features that have little effect on π. (Color figure online)

We can use Fig. 5 for analyzing the decision basis of policy network π that trained with encoding and masking processes as proposed in SEMI. In a situation like Fig. 1(a), π generally takes the important features including the guidance line, the ego vehicle, the vehicles in the guided lane, and the boundary of the road into consideration. These picked features are also often used by human drivers to decide their actions.

The phenomenon that the downstream imitator trained with the SEMI method tends to perform imitation based on real important features can be an explanation for its good scores in several crucial metrics, since it has been proposed in the causal literature that imitators whose decisions rely on the same features that experts use, can eventually obtain expert-like strategies.

6 Conclusion

In this paper, we proposed a method, Sequential Masking Imitation Learning(SEMI) that can train an end-to-end driving model and handle causal confusion by randomly masking semantic objects in a sequence of observation samples. We evaluated our SEMI method on the CARLA simulator and found it outperforms several other behavior clone methods in several metrics we think an expert-like agent should maintain. To be specific, we found that the driving agent trained with our SEMI method showed a preference for taking other vehicles' information into consideration and its willing to brake to avoid collision. These traits, along with the ability to drive following visual guidance, lead our agent equipped with expert-like strategy.

The major limitation of this work is the observation format: the bird-eye view image is hard to obtain in the real world and it contains ground truth information which helps model making decisions. Reducing reliance on such information will be the direction of our future efforts. Another limitation is our model did collide with other vehicles and the environment sometimes, due to the situation surrounding may be unfamiliar to our model, especially when the expert's demonstration had not covered those situations.

Although there are limitations exist, our SEMI method does show a promising way for autonomous driving that requires no interaction with the actual environment and handle the causal confusion in imitation learning.

In future work, we expect to further investigate the causation in autonomous driving, and improve the ability of agent to interpret its surrounding environment. Tools such as causal discovery and counterfactual representation can be used for this purpose.

Acknowledgements. The work is supported by the National Key Research and Development Program of China (No. 2022YFB4201603), the National Natural Science Foundation of China (No. 61873033) and the Natural Science Foundation of Fujian Province (No. 2020H0012).

References

1. Saha, A., Mendez, O., Russell, C., Bowden, R.: Translating images into maps. In: 2022 IEEE International Conference on Robotics and Automation (ICRA). IEEE (2022)
2. Hadfield-Menell, D., Milli, S., Abbeel, P., Russell, S.J., Dragan, A.: Inverse reward design. In: Guyon, I., Luxburg, U.V., Bengio, S., Wallach, H., Fergus, R., Vishwanathan, S., Garnett, R. (eds.) Advances in Neural Information Processing Systems, vol. 30. Curran Associates, Inc. (2017)
3. de Haan, P., Jayaraman, D., Levine, S.: Causal confusion in imitation learning. In: Wallach, H., Larochelle, H., Beygelzimer, A., d'Alché-Buc, F., Fox, E., Garnett, R. (eds.) Advances in Neural Information Processing System, vol. 32. Curran Associates, Inc. (2019)
4. Chen, J., Xu, Z., Tomizuka, M.: End-to-end Autonomous Driving Perception with Sequential Latent Representation Learning. arXiv e-prints arXiv:2003.12464 (Mar 2020)
5. Park, J., et al.: Object-aware regularization for addressing causal confusion in imitation learning. In: Ranzato, M., Beygelzimer, A., Dauphin, Y., Liang, P., Vaughan, J.W. (eds.) Advances in Neural Information Processing Systems, vol. 34, pp. 3029–3042. Curran Associates, Inc. (2021)
6. Dosovitskiy, A., Ros, G., Codevilla, F., Lopez, A., Koltun, V.: CARLA: an open urban driving simulator. In: Proceedings of the 1st Annual Conference on Robot Learning, pp. 1–16 (2017)
7. Zeng, W., Luo, W., Suo, S., Sadat, A., Yang, B., Casas, S., Urtasun, R.: End-to-end interpretable neural motion planner. In: 2019 IEEE/CVF Conference on Computer Vision and Pattern Recognition (CVPR). pp. 8652–8661 (June 2019). https://doi.org/10.1109/CVPR.2019.00886
8. NG, A.: Algorithms for inverse reinforcement learning. In: Proceedings of of 17th International Conference on Machine Learning, vol. 2000, pp. 663–670 (2000)
9. Abbeel, P., Ng, A.: Apprenticeship learning via inverse reinforcement learning, p. 1. ACM Press (2004). https://doi.org/10.1145/1015330.1015430
10. Ratliff, N., Bagnell, J., Zinkevich, M.: Maximum margin planning. In: Proceedings of the 23rd international conference on Machine learning - ICML 2006, pp. 729–736. ACM Press (2006). https://doi.org/10.1145/1143844.1143936

11. Ziebart, B., Maas, A., Bagnell, J., Dey, A.: Maximum entropy inverse reinforcement learning, pp. 1433–1438. AAAI (2008)
12. Codevilla, F., Santana, E., Lopez, A., Gaidon, A.: Exploring the limitations of behavior cloning for autonomous driving. In: 2019 IEEE/CVF International Conference on Computer Vision (ICCV), pp. 9328–9337 (2019). https://doi.org/10.1109/ICCV.2019.00942
13. Katz, G., Huang, D.W., Hauge, T., Gentili, R., Reggia, J.: A novel parsimonious cause-effect reasoning algorithm for robot imitation and plan recognition. IEEE Trans. Cognitive Developm. Syst. **10**(2), 177–193 (2018). https://doi.org/10.1109/tcds.2017.2651643
14. Srivastava, N.: Dropout: a simple way to prevent neural networks from overfitting. J. Mach. Learn. Res. **15**(1), 1929–1958 (2014)
15. Yun, S., Han, D., Chun, S., Oh, S.J., Yoo, Y., Choe, J.: Cutmix: regularization strategy to train strong classifiers with localizable features. In: 2019 IEEE/CVF International Conference on Computer Vision (ICCV), pp. 6022–6031 (Oct 2019). https://doi.org/10.1109/ICCV.2019.00612
16. Zhong, Z., Zheng, L., Kang, G., Li, S., Yang, Y.: Random erasing data augmentation. In: Proceedings of the AAAI Conference on Artificial Intelligence, vol. 34, pp. 13001–13008 (2020)
17. van den Oord, A., Vinyals, O., kavukcuoglu, k.: Neural discrete representation learning. In: Guyon, I. (eds.) Advances in Neural Information Processing Systems, vol. 30. Curran Associates, Inc. (2017)
18. Gleave, A., et al.: imitation: Clean imitation learning implementations. arXiv:2211.11972v1 [cs.LG] (2022). https://arxiv.org/abs/2211.11972
19. Anzalone, L., Barra, S., Nappi, M.: Reinforced curriculum learning for autonomous driving in carla. In: 2021 IEEE International Conference on Image Processing (ICIP), pp. 3318–3322 (2021). https://doi.org/10.1109/ICIP42928.2021.9506673

Proposal of Timestamp-Based Dynamic Context Features for Music Recommendation

Yasufumi Takama(✉), Lin Qian, and Hiroki Shibata

Tokyo Metropolitan University, 6-6 Asahigaoka, Hino, Tokyo 191-0065, Japan
ytakama@tmu.ac.jp
https://krectmt3.sd.tmu.ac.jp/en/index.html

Abstract. This paper proposes timestamp-based dynamic context features for music recommendation. Recently, we can easily enjoy a vast array of music anytime, anywhere through online services of music streaming such as Spotify, Amazon Music, and Apple Music. It causes a gap between the volume of accessible items and our ability of accessing music items and As context information including time, weather, location, country, and emotional state are known to play an important role, context-aware music recommendation has been widely studied. While conventional music recommendation methods considering context have focused on the static context of listeners, which include their nationality and languages, this paper proposes new dynamic context features that depend on the sequence of listening events and timestamp information. The proposed method does not classify the timestamp data into weekdays, weekends, morning, etc. Instead, the method uses timestamp information to analyze the listening behavior of users in chronological order. The effectiveness of the proposed features is evaluated through offline experiments using MMTD (Million Musical Tweets Dataset) dataset.

Keywords: music recommendation · context-aware recommendation

1 Introduction

In recent years, streaming services have changed the way we consume music. As communication technologies and audiovisual devices have evolved, and even audiophiles who are stubborn about physical formats must admit that it's hard to completely resist the temptation of online platforms when it comes to tackling music, especially now that the quality of streaming is getting better and better. As the history of music enters the streaming era, with access to tens of millions of songs at the touch of a touchscreen, finding old favorites and discovering new bands and artists has never been easier. People can enjoy their favorite music anytime and anywhere through music streaming services such as Spotify[1],

[1] https://open.spotify.com/.

B. Xin et al. (Eds.): IWACIII 2023, CCIS 1931, pp. 215–225, 2024.
https://doi.org/10.1007/978-981-99-7590-7_18

Amazon Music[2], and Apple Music[3]. Take Spotify as an example, it has more than 100 million tracks available to users to stream.

As a result, there is a large gap between the ability of users accessing music items and the number of available music items. This means that it is hard for users to search for suitable items in a large number of music resources. Therefore, it has become particularly important to utilize music recommendation systems and enhance the user experience. In such a context, improving the performance of music recommendation systems has been the subject of extensive research.

Similar to other domains, CF (Collaborative Filtering) [1] and CBF (Content-based Filtering) [2] are also commonly used in music recommendation systems. CF predicts ratings based on user-item interactions, which are used to find users with similar interests to the target user or items similar to those preferred by him/her. CBF recommends items to users based on item descriptions and profiles of user preferences.

Because of the characteristics of listening behaviors and music items' nature, it is hard to get an explicit feedback from listeners, including the rating values to their listened music items. To solve this problem, not only recommendation algorithms, but also the information used for the prediction is also important: context information which includes weather, location, time, country, and emotions of listeners are often used in music recommendation. Research on context-aware music recommender systems that use listeners' context information as supplementary information has been emphasized [3].

In the field of music recommendation, the user's interaction with the items (tracks) was referred to as LEs (Listening Events), which contain information about the item which the user listens to, the time, the location, and so on. The context information was classified into static information and dynamic information by determining whether or not the information changes over time. Conventional context-aware music recommendation algorithms have focused on the static context of users, which include their nationality [4] and language [5].

However, in the music domain, temporal information has a greater impact on recommendation accuracy than in other domains, especially in the current environment where music streaming services are prevalent. This is mainly because music is consumed in a different way compared to other products: the same user will usually play the same song repeatedly in a short period of time, whereas we rarely consume products such as movies or books multiple times. Also, music is consumed much more than other products in a short period of time: people will only watch a movie in a two-hour period, but listen to dozens of songs in the same period of time. Therefore, the study of dynamic context information should also be emphasized [6].

This paper proposes four new dynamic features that are calculated from timestamp information, considering two assumptions as follows: (1) 'last_track_id' (LTID) and 'since_last_track_time' (SLTT), which hold the assumption that listeners will enjoy music in the same context for a short period of time

[2] https://www.amazon.co.jp/music/prime.
[3] https://www.apple.com/jp/apple-music/.

but will enjoy music in a different context over time. (2) 'is_listened' (IL) and 'since_last_listen _time' (SLLT), which hold the assumption that listeners will regularly enjoy the same music items under the same context for a short period of time, and that their preferences will change over time.

The proposed dynamic context features are evaluated through offline experiments. Those are incorporated into the FMs (Factorization Machines) [7] using negative sampling approaches. Experiments are conducted on MMTD (Million Musical Tweets Dataset) [8] dataset. Experimental results showed that: (1) the proposed features perform better than the condition without context information or directly using timestamp information. (2) whereas previous methods used a fixed threshold to determine whether to recommend or not [5], this paper also examines the setting of an appropriate threshold.

2 Related Work

2.1 Music Recommender System

In music recommendation, because of the characteristics of listening behaviors and music items' characteristics, getting explicit feedback from listeners, e.g. the ratings to their listened music items, is difficult. On the other hand, the interaction of the user to the item is recorded as Listening Events (LE). Furthermore, the importance of information relating to context including time, weather, user's location, demographic information, and the emotions in the field of music recommendation has been reported [9].

Meanwhile, with the development of signal processing techniques, elements such as tone, tempo, and signal information of music can be extracted. However, there is still a semantic gap between those primitive data and the concept perceived by listeners [10]. That is, content features including melody, tempo, and harmony of a music item and the features relating to signal levels such as the mode, valence, duration, and energy do not accurately reflect the listener's preferences. For example, listeners would not like a song just because its acousticness is too high. Instead, they would like a song that gives them pleasure. Accurately representing such kinds of listeners' preferences in terms of content and signal features is difficult. Therefore, context information is also crucial in the music recommendation domain.

Of course, at this stage, there is still room for improvement as well as challenges in the music recommendation domain. CF can recommend suitable items to users without any domain knowledge including items' attributes. However, the data sparsity problem [11] and cold start problem [12] are still not well avoided. Due to the excessive number of items in the music domain, each user experiences a very limited number of items. As a result, the data sparsity problem is inevitable and essential for music recommendation systems.

The cold start problem occurs when the system is just starting up or when new users or new items are added to the system. As there is no interaction between these users and items, it is difficult to get enough information to make appropriate recommendations. All these problems have also arisen in music recommender systems and have received widespread attention [13].

2.2 Context-Aware Music Recommender System

Conventional recommendation systems including CBF and CF-based ones tend to employ simple user models. For instance, user-based CF describes users as a vector consisting of their ratings to items. However, this approach ignores the concept of "situated action": users interact with the system in a particular "context," and preferences for items in one context may differ from those in another [3].

The concept of context-aware computing is originated from the paradigm of the ubiquitous computing. Weiser [14] defined the concept of ubiquitous computing as "the method of enhancing computer use by making many computers available throughout the physical environment but making them effectively invisible to the user." The idea that users can access ubiquitous computing at any time and under different situations is the basis for leveraging users' context of interacting with devices. Advanced technologies such as wearable computers and smart devices have provided researchers with the means to collect/use contextual data for supporting user-computer interactions. Context information such as physical environment, time, emotion, existence of other users, and events in the past/future make it possible for systems to better understand the current needs of users [9].

For utilizing multimedia information, Martin et al. [15] have proposed a pre-filtering method that classifies users into several clusters based on their playlist names for recommending music items for users in each cluster using nearest-neighbor-based CF. But it was reported that different clusters achieved different recommendation accuracy, which was caused by the difference of cluster size. Aiming for improving the performance, they proposed to use the labels of the clusters as the feature of FMs. No negative LE is sampled by their method: instead, it supposes users give rating 1 to a track in a cluster. For the combination of <user, item, cluster> that does not exist, the rating is supported to be –1.

Schedl and Schnitzer [16] have introduced multiple types of context information, which includes the music content, the music context, and the user context, into hybrid music recommendation. Those context information is used for location-aware weighting of similarities. Factorization models are also widely used in context-aware music recommender systems, such as matrix factorization models, tensor factorization models, specialized factorization models that take non-categorical variables into account, and factorization machines (FMs) [7]. FMs supports both high prediction ability of factorization models and the flexibility of feature engineering at the same time. The input data of FMs are represented as real-valued data, which is the same as ordinary machine learning methods. FMs handle factorized interactions between variables, which is suitable for leaning in sparse settings like in recommender systems [7]. Takama et al. have developed a context-aware music recommender system based on FMs

[5]. As this paper extends this system by introducing proposed context features, this system is described in detail in Sect. 3.2.

3 Proposed Method

3.1 Dynamic Context Features

This paper proposes dynamic context features for music recommendation, which are based on timestamp information. The proposed method does not use timestamp information by simply classifying it into weekdays/weekends or morning/afternoon/night. Instead, it analyzes timestamp data for extracting the listeners' behavior in chronological order. The following assumptions about user preferences for listening to music items are supposed in this paper.

1. The users will repeatedly listen to their favorite music items.
2. The users will listen to music items continuously for a certain period of time.
3. The users will tend to listen to same/similar music items at similar times.

Based on these assumptions, the following four dynamic context features are proposed: where LE_a is a current listening event in which a user $user_a$ listened to a music item $item_a$.

- **is_listened** (IL): 1 if $user_a$ has listened to $item_a$ before, else 0.
- **last_track_id** (LTID): the id of music item that $user_a$ listened to just before LE_a. If $user_a$ has not listened to any music item before LE_a, it takes –1.
- **since_last_track_time** (SLTT): an elapsed time from the last LE before LE_a to LE_a. If LE_a is the first LE of $user_a$, it takes 315569260 (100 years).
- **since_last_listen_time** (SLLT): an elapsed time from the last time $user_a$ listened to $item_a$ to LE_a. If $user_a$ has not listened to $item_a$ before LE_a, it takes 315569260 (100 years).

The correspondence of each feature with the assumptions is as follows: IL is based on assumption 1, LTID and SLLT are based on assumption1 and 3, and SLTT is based on assumption 2 and 3.

Table 1 shows examples of LEs from the same user (id: 23493), which consists of 'user_id,' 'track_id,' 'track_title,' and timestamp information (unix timestamp) when a user listened to a music item. Table 2 shows the proposed dynamic features calculated for LEs in Table 1. Unix timestamp is a time representation defined as the total number of seconds from 00:00:00 GMT on 01/01/1970. This paper represents the features of the time interval between LEs in seconds.

3.2 Recommendation System

The proposed dynamic context features are incorporate into the context-aware music recommendation system proposed by Takama et al. [5], which uses FMs

Table 1. Examples of listening events

LE_id	user_id	track_id	track_title	timestamp
001	23493	6702647	Summit	1328650522
002	23493	7395589	How You Like Me Now	1337935712
003	23493	4193697	Mrs. Cold	1338052594
004	23493	917097	Eyes on Fire	1339407728
005	23493	917097	Eyes on Fire	1341533810
006	23493	7694530	Intro	1344333931

Table 2. Examples of dynamic features

LE_id	IL	LTID	SLTT	SLLT
001	0	-1	315569260	315569260
002	0	6702647	9285190	315569260
003	0	7395589	116882	315569260
004	0	4193697	1355134	315569260
005	1	917097	2126082	2126082
006	0	917097	2800121	315569260

and negative sampling approaches. Factorization machines (FMs) [7] are a generalized model: they can mimic most factorization models only by feature engineering. FMs estimate the interactions between categorical variables by selecting appropriate features. Factorization models are trained with those features. Since FMs can add any features other than those representing users and items, they are suitable for considering context information.

After applying label encoding for converting categorical variable to numerical one and one-hot encoding, LEs with the proposed dynamic context features such as shown in Table 1 and 2 are represented as a design matrix $E \in \{0,1\}^{s \times t}$, where s is the number of LEs ad t is the number of features a LE has. Given $x \in \{0,1\}^t$ as a LE, which corresponds to a row of the matrix, the model of FMs (degree=2) is defined as:

$$\hat{r}(x) = w_0 + \sum_{i=1}^{t} w_i x_i + x_i \sum_{i=1}^{t-1} \sum_{j=i+1}^{t} \hat{w}_{i,j} x_i x_j, \quad (1)$$

$$\hat{w}_{i,j} = \sum_{f=1}^{k} v_{i,f} v_{j,f}, \quad (2)$$

where $\hat{r}(x)$ is the predicted value, x_i is x's i-th feature, w_i is the weight of x_i, and w_0 is the global bias. The $\hat{w}_{i,j}$ shows the interaction between x_i and x_j. $v_{i,f}$

represents the f-th latent factor of x_i. Introducing latent factors can efficiently capture all pairwise interactions between features.

Negation is a naive approach as a negative sampling method: it treats interaction as negative samples if they are not positive ones. Although it effectiveness for matrix factorization models such as wALS [5] has been shown, increasing the number of negative samples could lead to high computational cost, which is serious problems when using FMs. For mitigating this problem, Takama et al. [5] selected a portion of the music items that users have never listened to as negative samples.

This paper employs the Random Sampling approach, which has proven to be more effective than Top Discount Popularity Sampling and Priority Popularity Sampling [5]. This approach holds the assumption that if listeners do not listen to a certain music item in some contextual scenario, they would not be interested in it under that context. In the proposed method, a music item is randomly selected from those not in the target LE and used for generating a negative sample for that LE. The same context information including the proposed dynamic context features as the target LE is added to the context information of the negative sample: as IL and SLLT are determined by 'track_id,' a negative sample could have different values for those features from its corresponding actual LE. However, this paper uses the same feature values for reducing the computational cost for generating negative samples. Considering that negative samples mean samples other than actual ones in general, we think this simplification does not affect the learning of FMs so much. As the number of negative samples generated is the same as the number of actual LEs, the users who have more LEs will get more negative samples.

4 Experiments

4.1 Outline

In the experiment, the MMTD dataset[4] is used to evaluate the proposed dynamic context features. The MMTD dataset [8] consists of microblog-based music listening histories, which was crawled via the Twitter Streaming API between September 2011 and April 2013. It includes timestamp, other context information (e.g., country and location), and 11 content information of music items: among them, this paper uses only timestamp as the aim of this paper is to evaluate the proposed dynamic context features that relate with time. The MMTD is generated from 1,086,808 tweets and includes 133,968 music items and 25,060 artists. The tweets were posted by 215,375 users.

We employs a 5-fold real-life split strategy [17] for evaluation because it is said to be closer to an actual situation than ordinary cross-validation. The train data and test data (TestSet) in the experiments contain both positive samples (actual LEs) and negative samples. Positive samples in the TestSet is 20% of

[4] http://www.cp.jku.at/datasets/MMTD/.

the total dataset. As explained in Sect. 3.2, the number of negative samples in TestSet is the same as that of positive samples.

MPR (Mean Percentage Ranking) is used for evaluating the performance of the top-N recommendation task. MPR is the average of the percentile ranking of all positive samples in the recommendation list over TestSet: a smaller MPR indicates better performance. RMSE (Root Mean Squared Error) and Accuracy are used for the performance of the rating prediction task.

$$RMSE = \sqrt{\frac{\sum_{x \in TestSet}(r(x) - \hat{r}(x))^2}{|TestSet|}} \tag{3}$$

$$Accuracy = \frac{|TP + TN|}{|TestSet|} \tag{4}$$

$$TP = \sum_{x \in TestSet} \delta(\hat{r}(x) \geq \theta \wedge r(x) = 1) \tag{5}$$

$$TN = \sum_{x \in TestSet} \delta(\hat{r}(x) < \theta \wedge r(x) = 0) \tag{6}$$

where $\hat{r}(x)$ is the predicted value for x (Eq. 1): $r(x)$ is 1 for actual LE (x), 0 for negative samples. $\delta(x)$ is used to calculate the number of correct prediction: $\delta(x) = 1$ if x is true, else 0. This experiment assumes that a music item in x (LE) is recommended when $\hat{r}(x)$ exceeds the threshold θ. A smaller RMSE and a larger Accuracy mean better performance.

4.2 Results

The dynamic context features proposed in this paper are evaluated by incorporating them to FMs-based context-aware music recommendation system. The experimental result when one of the proposed features is incorporated at a time is shown in Table 3. It also includes the results when no context feature is used (non-context) and when timestamp is directly used as a feature (timestamp). Bold text in the table indicate a better results than non-context and timestamp. The threshold θ is set to 0.5.

Table 3. Performance of proposed dynamic context features in MMTD.

Feature	MPR	RMSE	Accuracy
non-context	0.259816	0.910645	0.694200
timestamp	0.239146	0.341877	0.723673
IL	**0.237669**	**0.333056**	**0.882118**
SLLT	**0.233185**	**0.316119**	**0.890841**
LTID	0.263842	1.083306	**0.769608**
SLTT	0.304796	0.860436	0.677508

Table 3 shows IL and SLLT achieved better performance than non-context and timestamp, whereas the performance of LTID and SLTT got worse.

As IL and SLLT, LTID and SLTT are respectively related to each other, the effect of combining those features is investigated, of which the results are shown in Table 4. It was confirmed that the combination of IL and SLLT, both of which outperformed non-context and timestamp, also achieved better performance. However, those performance was not improved from when using each feature alone. Although a detailed analysis will be needed in future, one possible reason might be the settings of threshold: as shown below, a threshold value that achieves the best performance could be different between the model using each feature alone and that using multiple features.

Table 4. Result of combining context features in MMTD.

Features	MPR	RMSE	Accuracy
IL+ SLLT	**0.236042**	**0.321293**	**0.887843**
LTID+SLTT	0.342273	1.315931	0.696765

In the above experiments, the threshold value was set to 0.5, which means that the item should be recommended to the user when the predicted rating is greater than or equal to 0.5. This paper also investigates the effect of the threshold value on the recommendation performance when only the feature SLLT, which outperformed other features as shown in Table 3, is used. The result is shown in Fig. 1 (MPR) and Fig. 2 (Accuracy). It is observed that setting threshold around 0.7 achieved the best performance regardless of the evaluation metrics.

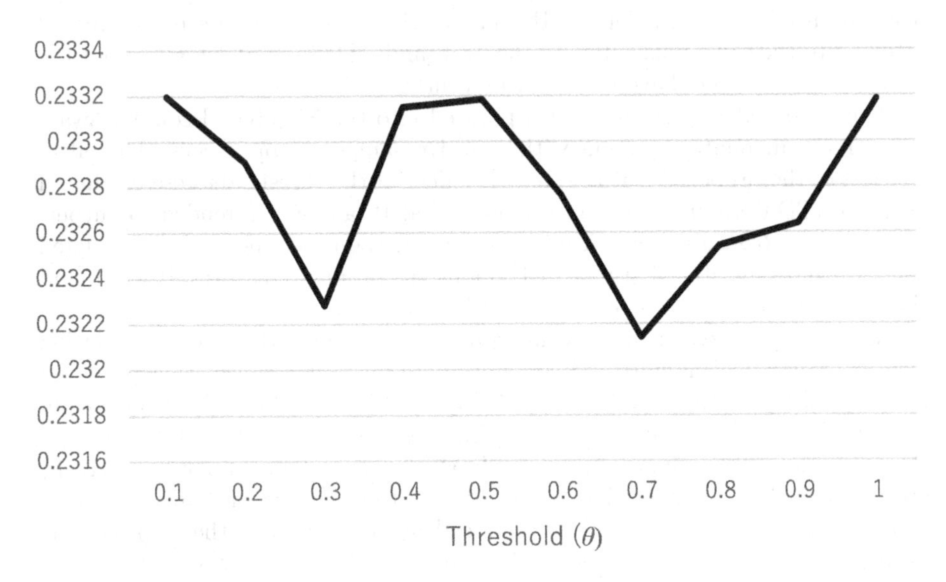

Fig. 1. Influence of threshold using SLLT (MPR).

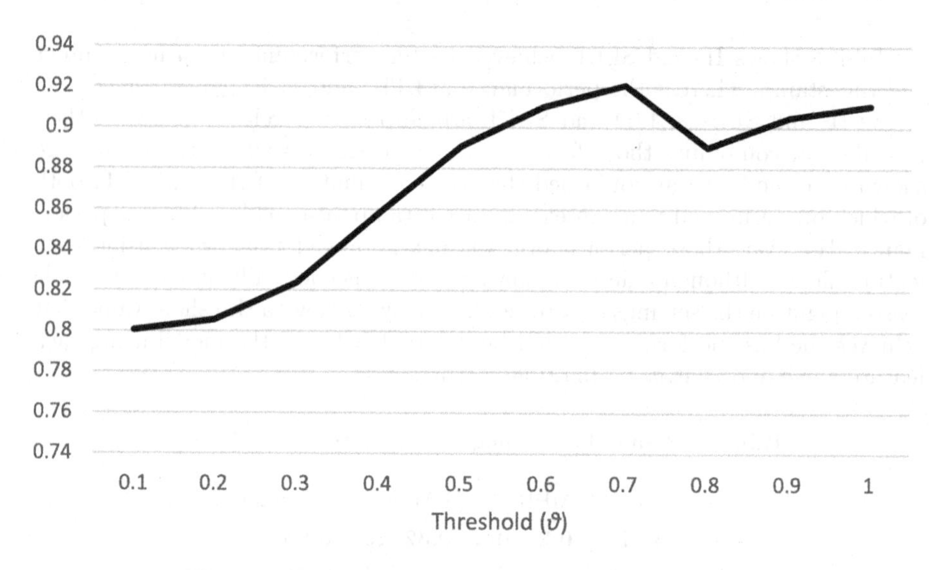

Fig. 2. Influence of threshold using SLLT (Accuracy).

5 Conclusion

This paper proposed dynamic context features, which are calculated from the timestamp information of users' listening events. The proposed features are based on the assumptions about user listening behavior. The proposed method does not use timestamp information by simply classifying it into weekdays/weekends or morning/afternoon/night. Instead, it is analyzed for extracting the listeners' behavior in chronological order. IL and SLLT are about user's experience of listening to the same music item in the past, and LTID and SLTT are about the music item that users have listened to recently.

The proposed features were incorporated into the FMs-based context-aware music recommendation system with negative sampling approaches. The effectiveness of the proposed features and the effect of the thresholds were evaluated using MMTD dataset. The results showed that IL and SLLT tended to outperform others. Regarding the effect of threshold, the experiment was conducted using the most effective feature SLLT among the proposed features. It was found that the threshold value around 0.7 had better results.

In future work, experiments using other datasets and other recommendation algorithms will be conducted to further evaluate the performance of the proposed features. Although the effectiveness of LTID and SLTT could not be confirmed in the experiment, the refinement of those features are also one of our future works. Possible ideas of refinements are to apply clustering to find similar song (track) groups instead of directly using 'track_id' and to discretize the elapsed time as 'within a day,' 'within a week,' etc. instead of directly using the elapsed time in seconds as a feature. It is also challenging that the users' listening behaviors assumed in this paper are applied to generate synthetic rating matrix [18].

Acknowledgements. This work was partially supported by JSPS KAKENHI Grant Numbers 21H03553, 22H03698, and 22K19836.

References

1. Resnick, P., Iacovou, N., Suchak, M., Bergstrom, P. Riedl, J.: An open architecture for collaborative filtering of netnews. In: ACM Conference on Computer Supported Cooperative work, pp. 175–186 (1994)
2. van Meteren, R., van Someren, M.: Using content-based filtering for recommendation. In: Proceedings of the Machine Learning in the New Information Age: MLnet/ECML2000 Workshop, vol. 30, pp. 47–56 (2000)
3. Adomavicius, G., Tuzhilin, A.: Context-Aware Recommender Systems. In: Ricci, F., Rokach, L., Shapira, B. (eds.) Recommender Systems Handbook, pp. 191–226. Springer, Boston, MA (2015). https://doi.org/10.1007/978-1-4899-7637-6_6
4. Schedl, M., Bauer, C., Reisinger, W., Kowald, D., Lex, E.: Listener modeling and context-aware music recommendation based on country archetypes. Front. Artif. Intell. **3**, 508725 (2021)
5. Takama, Y., Zhang, J., Shibata, H.: Context-aware music recommender system based on implicit feedback. Trans. Jpn. Soc. Artif. Intell. **36**(1), 1–10 (2021)
6. Sánchez-Moreno, D., Zheng, Y., Moreno-García, M.N.: Time-aware music recommender systems: modeling the evolution of implicit user preferences and user listening habits in a collaborative filtering approach. Appl. Sci. **10**(15), 5324 (2020)
7. Rendle, S.: Factorization machines with libFM. ACM Trans. Intell. Syst. Technol. **3**(3), 1–22 (2012)
8. Hauger, D., Schedl, M., Košir, A., Tkalcic, M.: The million musical tweet dataset: what we can learn from microblogs. In: Proceedings of the 14th Conference of the International Society for Music Information Retrieval, pp. 189–194 (2013)
9. Kaminskas, M., Ricci, F.: Contextual music information retrieval and recommendation: state of the art and challenges. Comput. Sci. Rev. **6**, 89–119 (2012)
10. Celma, O.: Music Recommendation, pp. 43–85. Music Recommendation and Discovery, Springer (2010)
11. Lee, S., Jihoon, Y., Park, S.: Discovery of hidden similarity on collaborative filtering to overcome sparsity problem. In: Proceedings of the 7th International Conference on Discovery Science, pp. 396–402 (2004)
12. Bobadilla, J., Ortega, F., Hernando, A., Gutierrez, A.: Recommender systems survey. Knowl.-Based Syst. **46**, 109–132 (2013)
13. Raza, S., Ding, C.: Progress in context-aware recommender systems -an overview. Comput. Sci. Rev. **31**, 84–97 (2019)
14. Weiser, M.: Some computer science issues in ubiquitous computing. Commun. ACM **36**(7), 75–84 (1993)
15. Pichl, M., Zangerle, E., Specht, G.: Towards a context aware music recommendation approach: what is hidden in the playlist name. In: IEEE International Conference on Data Mining Workshop, pp. 1360–1365 (2015)
16. Schedl, M., Schnitzer, D.: Hybrid retrieval approaches to geospatial music recommendation. In: 36th International ACM SIGIR Conference on Research and Development in Information Retrieval, pp. 793–796 (2013)
17. Wang, C., Lin, S., Yang, H.: Evaluating music recommendation in a real-world setting: on data splitting and evaluation metrics. In: IEEE International Conference on Multimedia and Expo, pp. 1–6 (2015)
18. Moriyoshi, K., Shibata, H., Takama, Y.: Proposal of generation of rating matrix based on rational behaviors of users, ISCIIA2022, No. C1–4 (2022)

Method to Control Embedded Representation of Piece of Music in Playlists

Hiroki Shibata(✉), Kenta Ebine, and Yasufumi Takama

Graduate School of Tokyo Metropolitan University, Hachioji, Japan
{hshibata,ytakama}@tmu.ac.jp, kenta-ebine@ed.tmu.ac.jp

Abstract. This paper proposes a new method to control embedded representation of a piece of music provided as a set of playlists. To recommend an appropriate piece to a user, numerical representation of music has been studied so far. This study does not focus on explicit representation like signal data, rather implicit representation called embeddings obtained from users' playlists in order to avoid issues around copyright. In the previous work, naive method was proposed and raw dataset is provided to learn the model of embedding for pieces of music, however, it is still not clear the raw dataset is appropriate for the model of music recommender system. Actually, this paper shows there is bias in a raw dataset and it makes the representation tends to provide trivial result, i.e., clearly same element that can be implied only by shallow knowledge like that pieces of music composed by the same artist are similar each other. This study shows the problem quantitatively and proposes a new method to reduce self-evident feature from embedded representation.

Keywords: Music Recommendation · Learning Representation · Distributed Representation

1 Introduction

Nowadays, music streaming services such as Spotify[1] and Apple Music[2] have enabled people to listen to vast amount of pieces of music. In this situation, demand to music recommender system to provide appropriate contents from the sea of music is increasing. Music reommender system is expected to help people find what they most prefer with practical effort, without checking entire data. In addition, not only taking each piece of music one by one, playlist has been getting users' attention as the way to listen to music. Playlists are also used to share the favorite pieces of music among users. Spotify mentioned above actually provide a service to create and share the playlist. Soundcloud[3] is another services that provide the way to share playlists by the users.

[1] https://www.spotify.com/jp/.
[2] https://www.apple.com/jp/apple-music/.
[3] https://soundcloud.com.

B. Xin et al. (Eds.): IWACIII 2023, CCIS 1931, pp. 226–240, 2024.
https://doi.org/10.1007/978-981-99-7590-7_19

To recommend an appropriate piece of music to a user by computing, numerical representation of a piece of music has been studied so far, such as explicit representation like signal data, and statistical data of the signal. However, this study does not focus on such explicit data in order to avoid issues arround copyright. Instead, the study focuses on implicit representation (numerical representation) called embeddings obtained from users' playlists. As the embedding, this paper focuses on distributed representation that is often employed in a domain of Natural Language Processing (NLP) [1–3]. Distributed representation is applied to various tasks such as, obtaining a embedding vector for each item in a recommendation system [4,5], link prediction in social networking services [6]. It is also applied to music recommendation by introducing embedding representation of pieces of music [13] with those methods to obtain distributed representation in NLP.

Core functionality of the distributed representation is making a vector so called embedding representation of words in a corpus that represents their words. It is archived based on the distributional hypothesis that means "the meaning of a word is generated by its neighbor words [1]." The original hypothesis is only applied to a text, however, it can be applied to any sequence of finite variety of elements if there is a rule among their arrangement similar to the syntax that defines a sentence of natural language. Actually there are studies to propose generating embeddings of pieces of music from the listening history of users [13], in which music titles are lined up in temporal axis. However, the authors suppose its investigation of obtained embedding representation is still insufficient (problem 1), and there was a problem of biases by which embeddings represent trivial feature of the dataset, in which, most pieces of music in high raked position that are placed close to a certain piece in the embedding space belong to the same artist as that of the piece (problem 2). It lacks serendipity. As for expected functionality in practical use, users want to be recommended pieces (here Y), that are similar to a piece (here x) a target user likes but not from the same artists as who composed x, because the user probably knows Y already.

To these 2 problems (problem 1, 2) above, the contributions of this paper are followings.

- Providing a quantitative investigation for the obtained embeddings
- Proposal of a new method to control the embeddings to reduce trivial feature, similar pieces of music are almost from the same artists

From these results above, the paper contributes further development of a method to obtain a meaningful embeddings and gives detailed insight to embeddings of piece of music from music playlist, towards practical implementation of music recommender system.

1.1 Notations

Notations that are supposed to uncommon is explained here. Although all notations are explained at their emerging point too, please see the section for convenience.

For positive integer is denoted as $\mathbb{N}$. For $n \in \mathbb{N}$, $[n] = \{1, 2, ..., n\}$, set of all integers from 1 to n. $\delta(a, b)$ is a Kronecker's delta that takes 1 if $a = b$ and 0 otherwise. $\chi(x; A)$ is an index function that takes 1 if $x \in A$ and 0 otherwise. $x \in A$ means x is belongs to A, this definition related to a set is follows set theory. For 2 set A, B, $A \cap B, A \cup B, A - B$ are intersection, union, and difference of A and B. $f : A \to B$ denotes a mapping, noted also as $f(a) = b$ with some example of $a \in A, b \in B$.

For given elements $a_1, a_2, ..., a_n$, a finite set A consists of these elements is denoted as $A = \{a_1, a_2, ..., a_n\}$, and sequence (or series) of these elements is denoted as $(a_1, a_2, ..., a_n)$ or $(a_i; i \in [n])$. That is, $\{\}$ is used for a set, $()$ is used for a sequence that the order is defined by its index and multiple occurrences of the same element is possible. A^n is a set of all tuples (or sequence, or pairs in the case of 2) of $a = (a_i; i \in [n])$ (Note that pairs, tuples, sequence and series are all the same concept.). $\#A$ is a number of elements in a set A.

2 Related Work

2.1 Distributed Representation

The distributed representation of word is an embedding of words in its vocabulary on a vector space (typically, numerical vector space with the dimension $N > 1$) [2,3]. Each vector of a word is calculated (learnt) under the set of restriction, called model of the distributed representation. These vectors for words are called embedding and its space to which the vectors belong is called embedding space frequently. In the model, words have similar meanings are placed closer point to each other, compared with those have distant meanings. Those vectors show additive property, such that if the model predicts $v(w), w \in W$ that is closest to $v(\text{king}) + v(\text{female}) - v(\text{male})$, it will be $v(\text{queen})$, as well known, where $v : W \to R^D$ denotes a mapping from W to a numerical vector space R^D. Here W is a set of words (vocabulary). Therefore, in this model, tasks that need the meaning of words can be done with commonly known algebraic operation of vectors in a vector space, which provide consistent and quantitative way of the task. Typical dimension of the embedding space D is $100 \leq D \leq 1000$ [2].

As the representative model and method of the distributed representation of words, there is Word2Vec [2,3]. It models a mapping from words (W) to a embedding vector space with 2 layers Neural Networks, and it is divided into 2 sub-categories called CBOW (Continuous Bag-of-Words) and Skip-gram. A model based-on CBOW is defined so that it solves a problem to predict a word (here w $(\in W)$) provided that the w's surrounding words are given. Surroundings means a neighbor relation in a sequence. More strictly, it is given as a sequence of words $(w_i; i \in [N]), w_i \in W$, where there is a $i \in [N]$ that holds $w_i = w$, and for positive integer A, $[A]$ means $[A] = \{1, 2, ..., N\}$. Note that usually $\#W < N$. the task is that for all $i \in [N]$, the model is learnt so that it predicts a word placed at i only with the sequence of words $(w_j; j \in [i + c] - [i - c - 1] - \{i\})$, where c is called the window size. Because there is multiple $i \in I(w') \subset [N]$ that $w_i = w'$ for each $w' \in W$, prediction for each place of $i \in I(w')$ cannot be done exactly, that is, there is a tradeoff among occurrences of w'.

On Skip-gram model, on the other hand, for each place i, the task predicts a word for each place $j \in [i + c] - [i - c - 1] - \{i\}$. Skip-gram is more time complicated CBOW but it usually gives better results than CBOW [2].

2.2 Music Recommendation

There are studies on music recommendation, such as [7–9,11–14]. This study focuses on music playlists and embedding for pieces of music obtained from the playlists, because playlists are everywhere on music streaming services nowadays and does not depend on the contents of music, which enables the system avoid copyright related issues. That is, it can be expected to be available in the future continuously. Moreover, it explains music in groups that express users' subjective categorization of music, which enables the system to understand not only users' preferences to the music, but also the situation that music is possibly listened to. Speaking to applicability, embedding of distributed representation has variety of advantages, therefore, music embedding from playlists is worth studying. This section introduces representative work for embeddings of music from playlists under the assumption of distributed representation.

Wang et al. [13] have proposed a method to calculate a vector for each piece of music by, regarding sequences of pieces listened to by users successively at once, as sentences dealt with in Skip-gram-based Word2Vec. Recommendation is done using user-based collaborative filtering (CF) in which similarity between any pair of users needed in CF is calculated based on the above-mentioned vectors. The main focus of this study [13] is to recommend a piece of music, rather than a playlist of music. In addition, a qualitative discussion on obtained vector is provided in which it is mentioned that vectors among the same artists are similar to each other. However, over all analysis is remained just in showing visualization without quantitative evaluation of results. In addition, there is a problem in the method that most high ranked pieces of music come from the same artists and the results of recommendation lacks serendipity (diversity).

Zhou et al. [14] have proposed a method based on CBOW-based Word2Vec to calculate a vector for each piece of music using all listening history of users. At the recommendation for a user u, 2 vector $\bar{a}, \bar{b}$ is calculated first where $\bar{a}$ is a mean vector of all vector for pieces of music in user u's listening history while $\bar{b}$ is a mean vector for recent listened to pieces of music by user u. Then a piece of music with a vector c that closest to $(\bar{a} + \bar{b})/2$ is recommended to u. This study also focuses on recommending just a piece of music, but does not provide any quantitative analysis for obtained embedding vectors. In addition, the proposed method from [14] does not provide a discussion to remove bias mentioned above either.

There are also studies that proposes to generate and recommend a playlist itself. Claudio et al. [16] have proposed a method to generate a playlist for recommendation by taking a piece of music designated by a user as an input. Flexer et al. [17] have proposed a playlist generation method that recognizing user's context at the recommended time. It employs signal characteristics and lyrics of music. Although its applicability is limited because it uses specific type of data such as lyrics and signal characteristics, it shows practical recommendation method. However, in this paper, we focus on more general formalization by employing only the playlists and seek its capability as the playlists data is ubiquitous as mentioned in the introduction.

3 Proposed Method and Investigation

This paper proposes a new method to control the embedding vectors for pieces of music obtained from playlists as dataset of which sequences are assumed to hold the distributional hypothesis that, "latent structure that characterizes a sequence of playlist exists and the structure can be employed to obtain a effective embedding representation for music recommendation [1]." As a model for embeddings, Word2vec [2] is employed following to Wang et al. [13]. To the best of our knowledge, quantitative investigation to the characteristics of learnt embedding vectors has not been conducted sufficiently, therefore, in addition to the proposal of a new method, the present paper provides a quantitative analysis to the obtained embeddings.

For dataset, Spotify Playlists Dataset[4] that is published by Martin et al. [18], is used to obtain embeddings of pieces of music. This data set consists of playlists published by users who posted tweets on Twitter[5] with hash tag "#nowplaying" via Spotify. Number of composition titles in the dataset is 12,867,130. Each entry of dataset consists of, "user_id" (user name on Spotify), "artist name," "track name," "playlist name." Playlist id is assigned at the preprocessing period combining user_id and playlist name so that each string is identical in the dataset. After that, the dataset has 188,437 playlists. 88% of playlists has less than 100 pieces of titles, while the longest playlists has 47,362 titles. Common parameter configurations are,

$$c = 9, D = 50, N_{\mathrm{s}} = 5,$$

where N_{s} is parameter related to the number of negative samplings used at the learning. Actual implementation of the model relays on Gensim[6] of Python's library.

Here is common definitions in experiments and investigation. As a similarity, here after cosine similarity denoted as $\cos(u, v)$ for vectors u, v is used. Let a set of pieces of music M, Let a set of playlist P, its element be $p = (p_i \in M; i = 1, 2, ...) \in P$, the length of p be n_p.

[4] https://dbis.uibk.ac.at/node/263.
[5] https://twitter.com.
[6] https://radimrehurek.com/gensim/index.html.

Table 1. 10 pieces of music that each $m \in M(m*)$ of them has the 10 highest similarity of $\cos(m_*, m')$ from the top.

Similar songs m' to m_*	$\cos(m_*, m')$
michael jackson - bad	0.925
michael jackson - billie jean	0.923
michael jackson - beds are burning	0.861
michael jackson - beautiful firl	0.856
michael jackson - black or white	0.856
michael jackson - ben	0.844
the jacksons - beat it	0.839
meat loaf - bat out of hell	0.835
the jacksons - blame it on the boogie	0.834
u2-beautiful day	0.829

3.1 Investigation on Embeddings

In distributed representation, embeddings for similar words will be placed near by each other, however, it is not clear that property also holds for the case of dataset on pieces of music sequences (playlists). In this section, it is investigated.

An qualitative investigation about similar pieces of music is provided first. Letting $m_* =$ "michael jackson - beat it", and $M(m*)$ be a set of those $m' \in M$ with top 10 highest $\cos(m_*, m')$, Table 1 shows the list of artist-titles $m' \in M(m_*)$ with corresponding value $\cos(m_*, m')$. From the table, most of artists are the same. That means, estimated with distributed representation of pieces of music tend to be the same artists with the focused piece. This tendency coincides with that reported by Wang at el. [13].

Next the paper provide an investigation into similarity of pairs of pieces occur in the same playlists. Additional mathematical formulations are needed for statistics. So those are defined first. Subset M^γ of M is defined as,

$$M^\gamma = \left\{ m; \sum_{i \in n_p} \delta(p_i, m) \leq 1, p \in P, m \in M \right\},$$

where δ is Kronecker's Delta. The investigation is conducted on this subset M^γ. Let the frequency of co-occurrence be $f : M^2 \to \mathbb{N}$. f is defined for $m, m' \in M$ as,

$$f(m, m') = \sum_{p \in P_m} \sum_{i \in [n_p]} \delta(m', p_i), P_m = \{p; \exists i \in [n_p] \, [p_i = m], p \in P\}.$$

Then defining a sorted sequence $(l_i^m; i \in [\#M^\gamma], l_i^m \in M^\gamma), i \neq j \Rightarrow l_i^m \neq l_j^m$ so that $l_i^m \geq l_{i+1}^m$ with the order $\geq$: $m' \geq m'' \Leftrightarrow f(m, m') \geq f(m, m'')$, letting a set

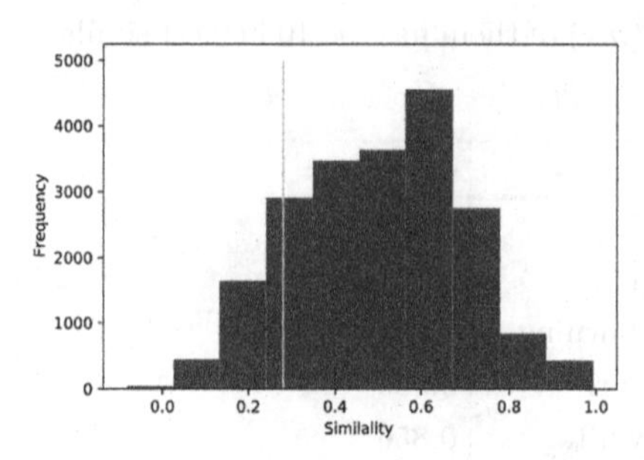

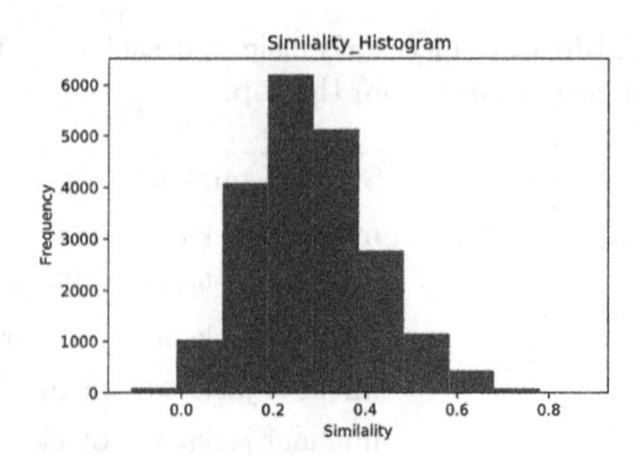

Fig. 1. (a): A histogram of cosine similarity between similar pairs in $\mathfrak{M}$. Vertical line in the middle denotes a mean position of the histogram in (b). (b): A histogram of cosine similarity between un-similar paris in $\mathfrak{M}^*$.

of similar piece to m be S_m defined as $S_m = \{l_i^m; i \in [\alpha]\}$, a set of similar pairs $\mathfrak{M}$ is defined as,

$$\mathfrak{M} = \{(x, x'); g(x, x') \geq \beta, x, x' \in M^\gamma\}, g(x, x') = \sum_{m \in M^\gamma} \chi(x; S_m)\chi(x'; S_m), \tag{1}$$

where $\chi(x; X) \in [0, 1], x \in X \Leftrightarrow \chi(x; X) = 1$ is an index function, and $\alpha = 30, \beta = 3$ is used in this experiment. Note that the higher $g(x, x')$, the more similar the sets to which x, x' belong. In analogy in word sequence (sentence), if $g(x, x')$ take higher value, this pair (x, x') will occur in the same sentences frequently, that is, these has the similar or related meanings each other. Such words should placed in the near position on embedding space of models based on the assumption of distributed representation [1]. Let non-similar set of pairs of pieces of music be $\mathfrak{M}^* \subset (M^\gamma)^2 - \mathfrak{M}$. $\mathfrak{M}^*$ is defined by collecting pairs $(x, x') \in (M^\gamma)^2 - \mathfrak{M}$ at random.

Histogram of cosine similarity between pairs for each $\mathfrak{M}, \mathfrak{M}^*$ is prepared in Fig. 1(a), Fig. 1(b) respectively. Parameters are $\gamma = 100, \alpha = 30, \beta = 3$. A vertical line in the middle of Fig. 1(a) denotes a mean position of the histogram in Fig. 1(b) to help the comparison. From the figures, it is confirmed that value of the similarity distributes in higher position of the axis for $\mathfrak{M}$, while in lower position for $\mathfrak{M}^*$ than for $\mathfrak{M}$. Average similarities for $\mathfrak{M}, fM^*$ are $0.5, 0.28$ respectively indeed. These results indicate similar piece of music has closer embedding vectors each other than those of un-similar.

In the next investigation about similarity in sequence 2 piece of music, m_1 = "daft punk-get lucky", m_2 = "m83-midnight city" $\in M^\gamma, \gamma = 1000$ is selected. In addition, 2 playlist $p^1, p^2 \in P$ is selected that contain these 2 piece respectively, that is, $\exists i \in [n_{p^j}] \left[m_j = p_i^j\right], j = 1, 2$. Graphs of cosine similarity, $\cos\left(m_j, p_i^j\right), i \in [n_{p^j}]$ are shown for each $j = 1, 2$ in Fig. 2(a), Fig. 2(b) respectively. i corresponds to the horizontal axis named song order in the graph. The position with similarity 1 corresponds to the index i of the

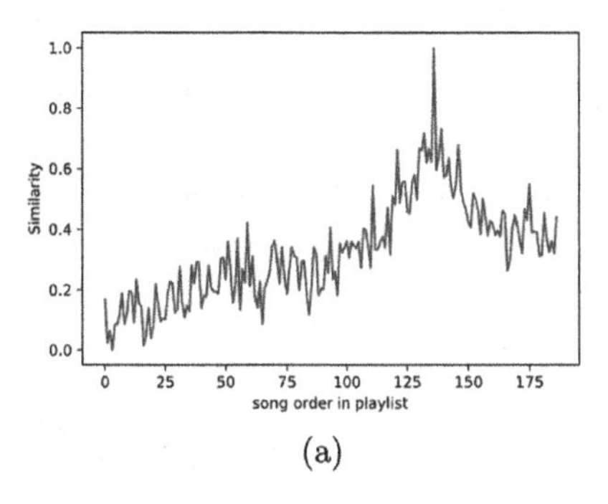

(a)

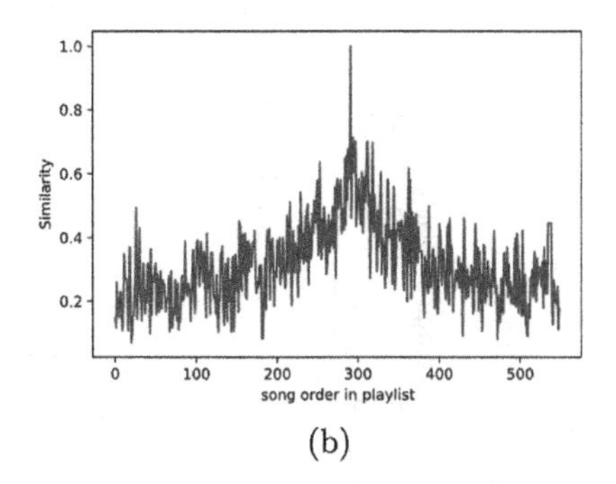

(b)

Fig. 2. Inner playlist similarity. (a): $m_1 =$ "daft punk-get lucky" and $p_i^1, i \in [n_{p^1}]$. (b): $m_2 =$ "m83-midnight city" and $p_i^2, i \in [n_{p^2}]$, i corresponds the axis named song order in the graph.

Table 2. Most 5 similar pairs of pieces (m, m'), and cosine similarity $\cos(m, m')$

similar pair(m, m')	$\cos(m, m')$
christophe beck – the north mountain christophe beck – wolves	0.727
christophe beck – sorcery christophe beck – the trolls	0.851
jack white - entitlement jack white – want and able	0.611
childish gambino – playing around before the party start childish gambino – death by numbers	0.774
red hot chili peppers – behind the sun red hot chili peppers – knock me down	0.452

pieces $m_1 = p_i^1, m_2 = p_i^2$. From these figures, piece p_k^j close to m_j has higher similarity than the distant piece p_k^j.

In the next, over all tendency about similarity is investigated. Picking up a set of 100 playlist (E), in which each $p \in E$ holds $n_p \geq 5, p \in P$ at random, the following average similarity μ_d with distance $d = 1, 2, ..., 49$,

$$\mu_d = \frac{1}{\sum_{p \in E} \#I(d,p)} \sum_{p \in E} \sum_{i \in I(d,p)} \cos(p_d, p_{i+d}), I(d,p) = \{i; i \in [n_p], i + d \leq n_p\}$$

is calculated, where $\#X$ for a set X is the number of elements of X. Fig. 3 visualize the relation between the distance d at the horizontal axis and μ_d at the vertical axis. It can be confirmed that until the distance d is less than 20, μ_d is decreasing along with d. The reason in the region $d > 20$, no decreasing tendency of μ_d along with d, can be considered as that this experiment set the windows size of the Word2Vec to 9 that less than 20. In this case, it is supposed that occurrence of pairs with distance larger than 9 within the same window

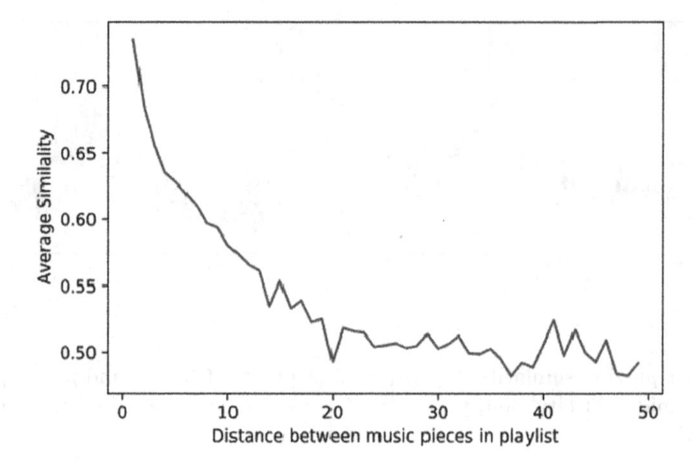

Fig. 3. Relation of distance between pieces of music in a playlist $d = 1, 2, ..., 49$, and average similarity μ_d.

become rare. Lower bound of μ_d seems to be around 0.5. This is higher than 0.28 of average similarity in Fig. 1(b). In this result, a pair in the same window or same playlist has high similarity between each other of the pair.

Those results so far are the intended features of distributed representation. So it is confirmed that models of distributed representation for words can be applied to the sequence of pieces of music.

To see in more detail, Table 2 is prepared. It shows most 5 pairs (m, m') sorted with descending order of the value $g(m, m')$ in Eq. (1). From this table, all the pairs of elements consist the same artist. The reason of this result is considered to be that many users include pieces of music with the same artists within the same playlist. From the results, there is strong tendency that pieces with the same artists have high similarity. Because this is trivial results, the tendency must be changed any time it is needed. To mitigate this problem and change the tendency, this paper proposes a new method to control the characteristics of embedding with respect to their similarity among each other.

3.2 Proposed Method to Reduce Bias

To reduce the chance the same artists appear, the proposed method removes the redundant playlists from the dataset, i.e., playlists that are similar to another with respect to the kind of artists with in them are all removed except one of them. The similarity to measure this redundancy is introduced as following. Letting P be a set of playlists, A be a set of artists in dataset, and $f_p : A \rightarrow \in [0, \infty), p \in P, a \in A$ be the realized frequency of a in playlists p, the measure is the following modified Simpson coefficient $t^2 : P \times P \rightarrow \mathbb{R}$ defined as,

$$t^2(p, q) = \frac{\sum_{a \in A} \min \{f_p(a), f_q(a)\}}{\min \{\sum_{a \in A} f_p(a), \sum_{a \in A} f_q(a)\}}. \quad (2)$$

Table 3. Top 10 similar pieces to $m =$ "michael jackson - thriller" with original dataset (W).

artist name and title of piece	similarity
michael jackson - they don't care about us	0.91
michael jackson - this is it	0.88
michael jackson - this time around	0.87
michael jackson - threatened	0.87
michael jackson - the way you make me feel	0.85
cyndi lauper - time after time	0.83
michael jackson - this place hotel	0.81
ke$ha - tik tok	0.80
donna summer - this time i know it's for real	0.79
michael jackson - todo mi amor eres tu	0.79

If $\forall a \in A\, [f_p(a) \leq 1, f_q(a) \leq 1]$, and letting $A_{p'} = \{a; f_p(a) = 1\}$ for $p' \in P$, then t^2 become the following ordinal simpson coefficient t^1,

$$t^1(p,q) = \frac{\#\,(A_p \cap A_q)}{\min\{\#A_p, \#A_q\}}. \tag{3}$$

The procedure of the proposed method is as follows.

$$\hat{P} = \left\{ p; p \in P, \exists a' \in A \left[\sum_{i \in n_p} \delta(a(p_i), a') \geq \frac{n_p}{2} \right] \right\} \tag{4}$$

$$Q_p = \left\{ p'; t^k(p', p) \geq 0.5, p \in \hat{P} \right\}, \hat{Q} = \left\{ \underset{p' \in Q_p}{\operatorname{argmax}}\,(n_{p'})\,; p \in \hat{P} \right\} \tag{5}$$

Then, the proposed method takes $P^* = \hat{Q} \cup (P - \hat{P})$ as the dataset of playlists. It is assumed $\operatorname{argmax}_a f(a)$ represents identical element, that is, its result never change. To the best of our knowledge, there is no study to try to overcome the problem of the existence of the same artists in the nearby vectors by controlling the embedding with modified dataset. Hereafter, results using original dataset P and modified dataset P^* with t^1, and P^* with t^2 are investigated. For simplicity, these results are designated with W, W1, W2 respectively in the paper.

In Table 3, 4, 5, pieces m' similar to $m =$ "michael jackson - thriller" is listed with corresponding similarity $\cos(m, m')$ between embedding vectors m, m'. From these tables, when comparing W1 with W, it can be found that the same artist (michael jackson) in W1 becomes less than half of W. At the same time, although W2 is only different in 3 with respect to the number of the same artists, similarities are decreased from the pieces in W. Because there was not so significant difference in the results W1 and W2, only the results of W and W1 are investigated in the following part.

Table 4. Top 10 similar pieces to m = "michael jackson - thriller" with modified dataset P^* using original simpson coefficient t^1 (W1).

artist name and title of piece	similarity
michael jackson - this is it	0.85
ke$ha - tik tok	0.85
michael jackson – they don't care about us	0.84
cyndi lauper - time after time	0.84
the rock masters - thunderstruck	0.84
richard o'brien - time warp	0.83
donna summer – this time i know it's for real	0.83
maroon 5 - this love	0.82
timbaland - throw it on me	0.81
michael jackson – the way you make me feel	0.80

Table 5. Top 10 similar pieces to m = "michael jackson - thriller" with modified dataset P^* using modified simpson coefficient t^2 (W2).

artist name and title of piece	similarity
michael jackson – they don't care about us	0.89
michael jackson - threatened	0.83
michael jackson – this time around	0.82
michael jackson – the way you make me feel	0.82
maroon 5 - this love	0.82
eurythmics – thorn in my side	0.82
cyndi lauper – time after time	0.82
michael jackson - this is it	0.82
ke$ha - tik tok	0.81
the jacksons – this place hotel	0.81

In Table 6, 7, pieces m' similar to m = "madonna-like a virgin" is listed with corresponding similarity $\cos(m, m')$. The same tendency as the case in michael jackson is observed. Moreover, the piece "david bowie" listed in the high position in the result of W1 is similar with respect to the fact that they are both singer song writer. Therefore, while decreasing similarity among the same artists, it may increase serendipity. Therefore it may suitable to the practical recommendation.

Table 6. Top 10 similar pieces to $m =$ "madonna-like a virgin" with original dataset (W).

artist name and title of piece	similarity
madonna - like a prayer	0.98
madonna - like it or not	0.87
madonna - live to tell	0.85
madonna - la isla bonita	0.82
deniece williams - let's hear it for the boy	0.84
bon jovi - livin' on a prayer	0.84
roxette - listen to your heart	0.82
madonna - like a prayer 2008	0.82
madonna - like a virgin/hollywood	0.81
fleetwood mac - little lies	0.81

Table 7. Top 10 similar pieces to $m =$ "madonna-like a virgin" with modified dataset P^* using modified simpson coefficient t^1 (W1).

artist name and title of piece	similarity
madonna - like a prayer	0.98
roxette - listen to your heart	0.89
deniece williams - let's hear it for the boy	0.83
fleetwood mac - little lies	0.83
david bowie - let's dance	0.83
bon jovi - livin' on a prayer	0.82
madonna - la isla bonita	0.82
thompson twins - lies	0.81
madonna - like it or not	0.81
prince - little red corvette	0.81

The results for $m =$ "james blunt-i'll be your man" is shown in Table 8, 9. For these results, significantly, the number of the same artists decreased from W to W1. Its ranking of the original artist (james blunt) decreased as well. It is supposed that the proposed method has bigger effect to decrease the similarity of unfamiliar artists.

Table 8. Top 10 similar pieces to m = "james blunt-i'll be your man" with original dataset (W).

artist name and title of piece	similarity
james blunt - i'll take everything	0.87
james blunt - if time is all i have	0.87
james morrison - i won't let you go	0.81
jason mraz - i won't give up	0.79
james blunt - i really want you	0.79
krezip - i would stay	0.79
jason mraz - i'm yours	0.79
lenny kravitz - i'll be waiting	0.78
noel gallagher's high flying birds – if i had a gun	0.78
joshua radin - i'd rather be with you	0.78

Table 9. Top 10 similar pieces to m = "james blunt-i'll be your man" with modified dataset P^* using modified simpson coefficient t^1 (W1).

artist name and title of piece	similarity
olly murs - i've tried everything	0.85
james morrison - i won't let you go	0.83
gavin mikhail – i will follow you into the dark	0.83
lenny kravitz - i'll be waiting	0.82
jason mraz - i'm yours	0.82
joshua radin - i'd rather be with you	0.82
jason mraz - i won't give up	0.81
owl city - i'll meet you there	0.81
james blunt - if time is all i have	0.79
james arthur - impossible	0.79

4 Conclusion

This paper shows quantitative analysis of embedding vectors for a piece of music obtained with the model based on the assumption of distributed representation, with the implementation on Word2Vec. The intended feature of embedding vectors was observed, in which pieces of music are placed in closer position with the similar pieces and distant position from the un-similar pieces on the embedded vector space, where the similarity among pieces of music is defined by the position in the playlist. It was confirmed the analogy replacing a concept of sentence in the original Word2Vec with a playlist of pieces of music is valid from quantitative investigation in the paper. Furthermore, this paper discussed the problem especially exists in employing the raw playlist as a training dataset, i.e., most of the pieces of music close to each other in the embedding space come from the same artist. To overcome the problem, the paper proposed a method to control generated embedding vector by manipulating the dataset, which reduce the redundant playlists that consist of almost one artist. The way of manipulation was intuitive and easy to understand. Moreover, the results was consistent with the expected effect, and shows suitable characteristics when it is considered to be applied in practical recommendation, as it increases the serendipity of recommended list of pieces.

As the remaining work, investigation must be done more concretely. For the aim, selection of suitable measure to evaluate the statistics for music recommendation is needed. By applying proposed method as the component of a recommender system, it is expected that recommendation with high user's satisfaction and attention can be archived.

Acknowledgements. This work was partially supported by JSPS KAKENHI Grant Numbers 21H03553, 22H03698, and 22K19836.

References

1. Harris, Z.S.: Distributional Structure. WORD **10**(2–3), 146–162 (1954)
2. Mikolov, T., Chen, K., Corrad, o G., Dean, J.: Efficient estimation of word representations in vector space. In: Proceedings of International Conference on Learning Representations Workshops Track (2013)
3. Mikolov, T., Sutskever, I., Chen, K., Corrado, G., Dean, J.: Distributed representations of words and phrases and their compositionality. Adv. Neural. Inf. Process. Syst. **26**, 3111–3119 (2013)
4. Krishnamurthy, B., Puri, N., Goel, R.: Learning vector-space representations of items for recommendations using word embedding models. Procedia Comput. Sci. **80**, 2205–2210 (2016)
5. Yoon, Y., Lee, J.: Movie recommendation using metadata based Word2Vec algorithm. In: 2018 International Conference on Platform Technology and Service, pp. 1–6 (2018)
6. Amiri, M., Shobi, A.: A link prediction strategy for personalized tweet recommendation through Doc2Vec approach. Res. Econ. Manag. **2**(4), 63–76 (2017)
7. Baltrunas, L., et al.: InCarMusic: context-aware music recommendations in a car. In: Huemer, C., Setzer, T. (eds.) EC-Web 2011. LNBIP, vol. 85, pp. 89–100. Springer, Heidelberg (2011). https://doi.org/10.1007/978-3-642-23014-1_8
8. Yapriady, B., Uitdenbogerd A.: Combining demographic data with collaborative filtering for automatic music recommendation. In: 9th International Conference on Knowledge-Based and Intelligent Information and Engineering Systems, pp. 201–207 (2005)
9. Radhika, N.: Music recommendation system based on user's sentiment. Int. J. Sci. Res. 383–384 (2015)
10. Jawaheer, G., Szomszor, M., Kostkova, P.: Comparison of implicit and explicit feedback from an online music recommendation service. In: International Workshop on Information Heterogeneity and Fusion in Recommender Systems (2010)
11. Bogdanov, D., Haro, M., Fuhrmann, F., Gómez, E., Herrera, P.: Content-based music recommendation based on user preference examples. In: Workshop on Music Recommendation and Discovery, Colocated with ACM RecSys 2010 (2010)

12. Cano, P., Koppenberger, M., Wack, N.: Content-based music audio recommendation. In: Proceedings of the 13th ACM International Conference on Multimedia (2005)
13. Wang, D., Deng, S., Xu, G.: Sequence-based context-aware music recommendation. Inf. Retrieval J. **21**, 230–252 (2018)
14. Zhou, Y., Tian, P.: Context-aware music recommendation based on Word2Vec. In: International Computer Science and Applications Conference, pp. 50–54 (2019)
15. Köse, B., Eken, S., Sayar, A.: Playlist generation via vector representation of songs. In: Angelov, P., Manolopoulos, Y., Iliadis, L., Roy, A., Vellasco, M. (eds.) INNS 2016. AISC, vol. 529, pp. 179–185. Springer, Cham (2017). https://doi.org/10.1007/978-3-319-47898-2_19
16. Baccigalupo, C., Plaza, E.: Case-Based Sequential Ordering of Songs for Playlist Recommendation, pp. 286–300. Advances in Case-Based Reasoning, European conference (2006)
17. Flexer, A., Schnitzer, D., Gasser, M., Widmer, G.: Playlist generation using start and end songs. In: International Conference on Music Information Retrieval, pp. 173–178 (2008)
18. Pichl, M., Zangerle, E., Specht, G.: Towards a context-aware music recommendation approach: what is hidden in the playlist name?. In: Proceedings of 15th IEEE International Conference on Data Mining Workshops, pp. 1360–1365 (2015)

Design and Implementation of ANFIS on FPGA and Verification with Class Classification Problem

Moegi Utami[1(✉)], Yukinobu Hoshino[1], and Namal Rathnayake[2]

[1] Kochi University of Technology, 185 Miyanokuchi, Tosayamada-cho, Kami-shi Kochi 782-8502, Japan
275047e@gs.kochi-tech.ac.jp, hoshino.yukinobu@kochi-tech.ac.jp
[2] University of Tokyo, 7-3-1, Hongo Bunkyo-ku, Tokyo 113-8655, Japan

Abstract. In this research, we implemented ANFIS (Adaptive-Network-Based Fuzzy Inference System) on an FPGA(Field Programmable Gate Array) as a System-on-Chip (SoC) system and evaluated its performance using iris flower classification. Additionally, we developed and tested an IEEE 754 16-bit floating-point format tailored for ANFIS operations. The evaluation focused on computational accuracy and speed. The results demonstrate that this FPGA-based ANFIS implementation, coupled with the 16-bit floating-point format, yields high efficiency and superior computational performance.

Keywords: FPGA · ANFIS

1 Introduction

In recent years, there has been a notable paradigm shift in the domain of Internet of Things (IoT) technology, with increasing emphasis on a distributed architecture termed"edge computing". Diverging from the conventional approach, which predominantly centers on centralizing data processing within the cloud, edge computing leverages edge devices or servers embedded in IoT devices to conduct localized data processing. Consequently, only essential data is transmitted and shared with the cloud. This novel approach significantly enhances the real-time responsiveness of information processing while concurrently alleviating the cloud's computational burden by reducing data transmission and processing demands. An area of particular significance where edge devices demonstrate considerable promise is disaster prediction. Through strategic deployment of sensors near water bodies, accurate forecasts of water levels can be attained, enabling the proactive identification of floods and other potential hazards well in advance. This proactive strategy is pivotal in mitigating the impact of disasters, safeguarding communities, and preserving critical infrastructure from potential harm. However, because real-time performance is required, they must operate at

Supported by organization x.

B. Xin et al. (Eds.): IWACIII 2023, CCIS 1931, pp. 241–252, 2024.
https://doi.org/10.1007/978-981-99-7590-7_20

high speeds. Additionally, in the event of a disaster, they must be able to operate on standby power when the primary power supply is unavailable. Therefore, FPGAs are utilized in control units for designing edge devices in embedded systems due to their ability to operate with low power consumption, minimize heat generation during operation, and reduce the overall device footprint. Gomez-Pulid et al. conducted a performance comparison between FPGAs, CPUs, and GPUs, with their experimental results demonstrating that FPGAs outperform CPUs and GPUs in terms of speed and power efficiency when operated continuously for 24 h at an equivalent cost when clustered [1]. FPGAs offer the advantage of being suitable for outdoor installation due to their low heat generation, and their low power consumption contributes to reduced long-term operating costs. Moreover, FPGAs have a compact footprint when integrated into embedded systems, enabling installation in more locations compared to CPUs. Thus, FPGAs surpass CPUs as superior edge device controllers, capable of prolonged operation while occupying minimal space. In this study, we developed and implemented a program based on the learned ANFIS in both FPGA and CPU environments. Our primary focus was on designing the hardware logic circuit for ANFIS in FPGA, emphasizing power consumption and processing speed. Furthermore, we conducted a thorough verification process, comparing the results with CPU processing, to ensure the accuracy of our FPGA-based implementation falls within an acceptable range.

2 Applying AFIS to the Iris Classification

In this paper, we apply a 4-input, 3-MF (membership function) Adaptive-Network-Based Fuzzy Inference System [2,3] to create a program for AI in edge devices. The objective of this program is to identify three types of iris based on four characteristics. The dataset used for machine learning and iris identification is the iris-dataset. It consists of 150 data items, including measurements of petal and sepal length (cm) and width (cm) for three species of iris (setosa, versicolor, and virginica), with 50 data items per variety. These four data items are input into the ANFIS, and the processing results determine the classification of each data set. By assigning a unique value to each iris variety, the ANFIS classification calculation produces the output based on those distinct values. ANFIS for iris identification in this study was designed by using "Neuro-FuzzyDesigner" in MATLAB R2017a and loading the teacher data into the ANFIS automatic design function. The dataset prepared for machine learning is a modified version of iris-dataset. The input data was prepared as the teacher data, which consisted of a total of 150 rows of five items, four of which were the length (cm) and width (cm) of the petals and sepals of the iris, and the eigenvalues of each cultivar were recorded. The eigenvalues for each variety were set as follows. In this study, ANFIS for iris identification was designed using the "Neuro-Fuzzy Designer" in MATLAB R2017a. The teacher data was loaded into the ANFIS automatic design function. For machine learning, a modified version of the iris-dataset was prepared. The input data was arranged as teacher data, comprising a total of 150 rows, with five items each. Among these, four items represented the length (cm) and width (cm) of the petals and sepals of the iris, while the remaining item recorded the eigenvalues for each cultivar (Fig. 1).

The eigenvalues for each variety were set as follows:
setosa : 100
versicolor : 200
virginica : 300

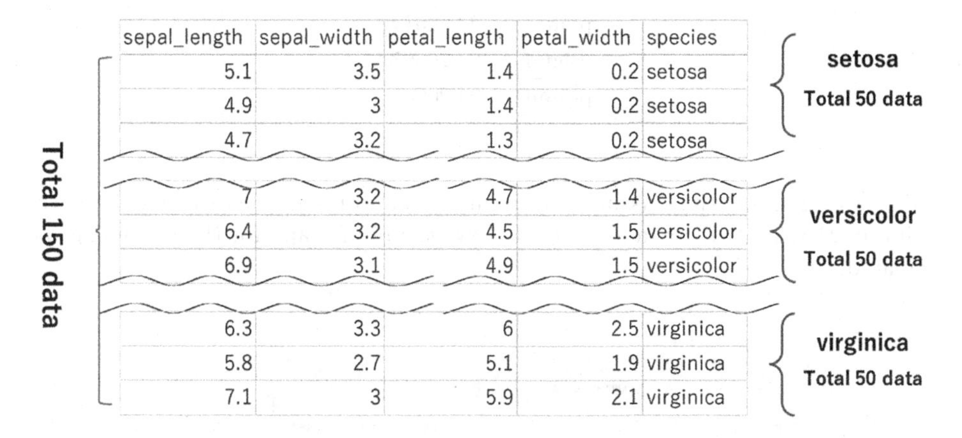

sepal_length	sepal_width	petal_length	petal_width	species
5.1	3.5	1.4	0.2	setosa
4.9	3	1.4	0.2	setosa
4.7	3.2	1.3	0.2	setosa
7	3.2	4.7	1.4	versicolor
6.4	3.2	4.5	1.5	versicolor
6.9	3.1	4.9	1.5	versicolor
6.3	3.3	6	2.5	virginica
5.8	2.7	5.1	1.9	virginica
7.1	3	5.9	2.1	virginica

Fig. 1. irisdataset

During machine learning, the MFs are set and learnt for the previous case section. The error values of each MF are shown in Table 1 after learning until the error value of each MF is minimised. The function with the smallest error value that could be designed using only four arithmetic operations was pimf, so the ANFIS was designed with pimf (Pi-shaped membership function) as the MF for the precondition. The function of pimf is as in Eq. (1).

$$f(x,a,b,c,d) = \begin{cases} 0 & (x \leq a) \\ 2(\frac{x-a}{b-a})^2 & (a \leq x \leq \frac{a+b}{2}) \\ 1-2(\frac{x-b}{b-a})^2 & (\frac{a+b}{2} \leq x \leq b) \\ 1 & (b \leq x \leq c) \\ 1-2(\frac{x-c}{d-c})^2 & (c \leq x \leq \frac{c+d}{2}) \\ 2(\frac{x-d}{d-c})^2 & (\frac{c+d}{2} \leq x \leq d) \\ 0 & (d \leq x) \end{cases} \tag{1}$$

x:input value
a,b,c,d: constant parameter

Table 1. Minimal error value due to the MF of the previous case.

MF	error value
trimf	1.0951
trapmf	3.7536
gbellmf	0.021761
gaussmf	0.15818
gauss2mf	0.13541
pimf	0.66251
dsigmf	0.026067
psigmf	0.026067

Figure 2 shows the classification results after educating the iris dataset with pimf in MATLAB R2017a and Fig. 3 shows the pimf graph for the petal lengths created.

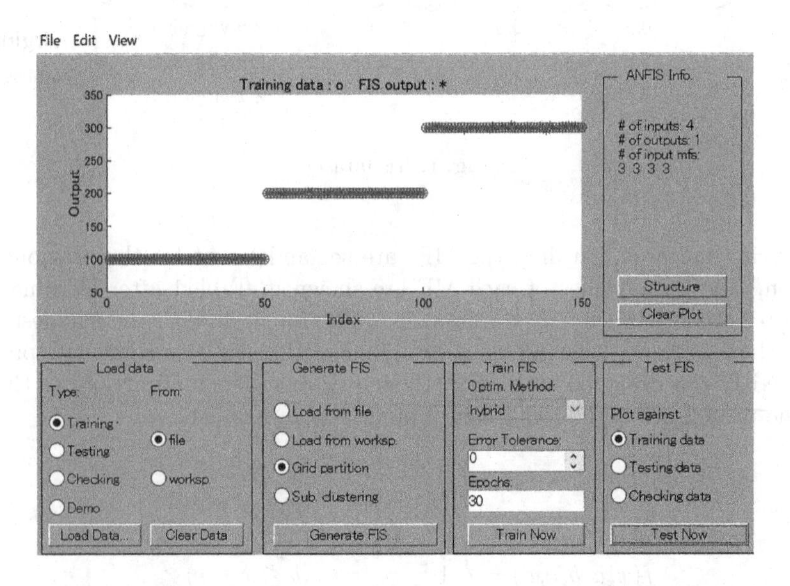

Fig. 2. Post-training classification results for iris-dataset by MATLAb 2017a

ANFIS is structured as a hierarchical network with an input layer and five processing layers. As illustrated in Fig. 4, which represents a 2-input, 2-MF ANFIS, the processing at each layer is summarized as follows. In the input layer, two data items are input to each neuron. Operations are then conducted in the processing layers, spanning from layer 1 to layer 5, following a series of steps.

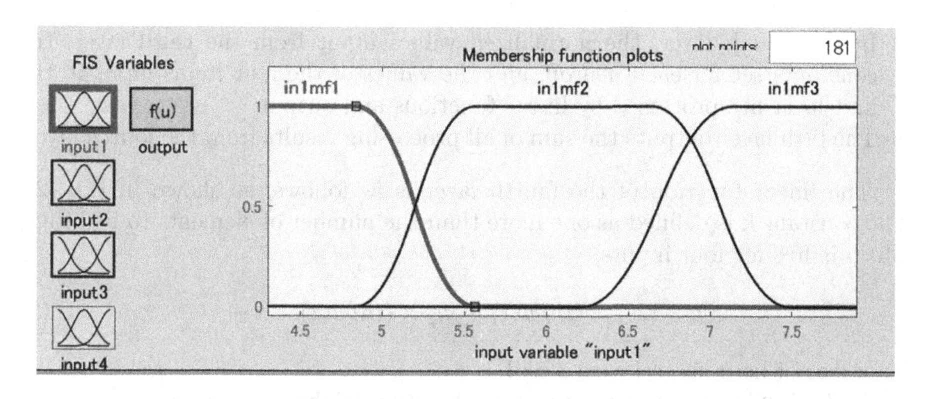

Fig. 3. Graph of pimf for petal length created in MATLAB R2019a.

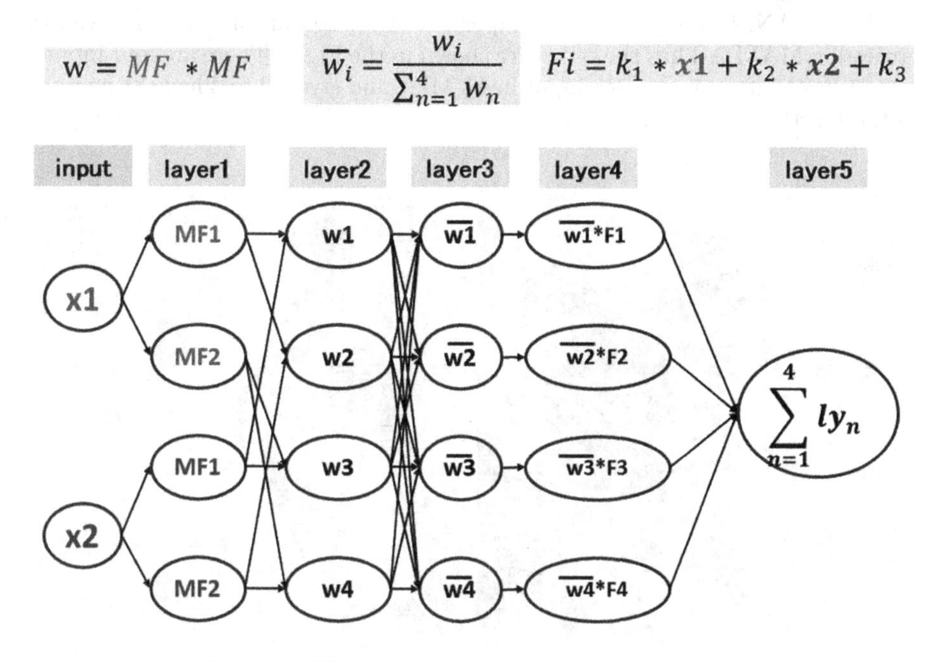

Fig. 4. ANFIS structure with 2 inputs and 2 MFs

1. In the first layer, the values of each neuron input from the input layer are processed by three neurons that store the membership functions (MFs) to compute the degree of truth for each classification.
2. In the second layer, the results of processing by the MF of each neuron in the first layer are combined and multiplied by one from each classification.
3. In the third layer, the result obtained in the second layer is divided by the sum of all neurons in the second layer, and the normalized result is output to each neuron in the next layer.

4. In the fourth layer, the normalized values input from the third layer, the constants set for each neuron, and the values of the four items input in the first layer are processed by linear functions and output.
5. The fifth layer outputs the sum of all processing results from the fourth layer.

The linear function of the fourth layer is as follows, as shown in Eq. (2). The constant k is defined as one more than the number of elements to be input, which is five for four inputs.

$$\overline{w_i} \times (k_{i1} \times (input1) + k_{i2} \times (input2) + k_{i3}) \tag{2}$$

i:Number of neurons in layers 3 and 4

k_1, k_2, k_3:Constants in the posterior function in the "i" neuron

input1,input2: values of the input layer

As the ANFIS created in this paper has 4 inputs and 3 MFs, the ANFIS created in MATLAB has the structure shown in the Fig. 5. There are four inputs, layer 1 has 12 neurons as each input has 3 MFs, and layers 2 to 4 have 81 neurons in the fourth power of 3.

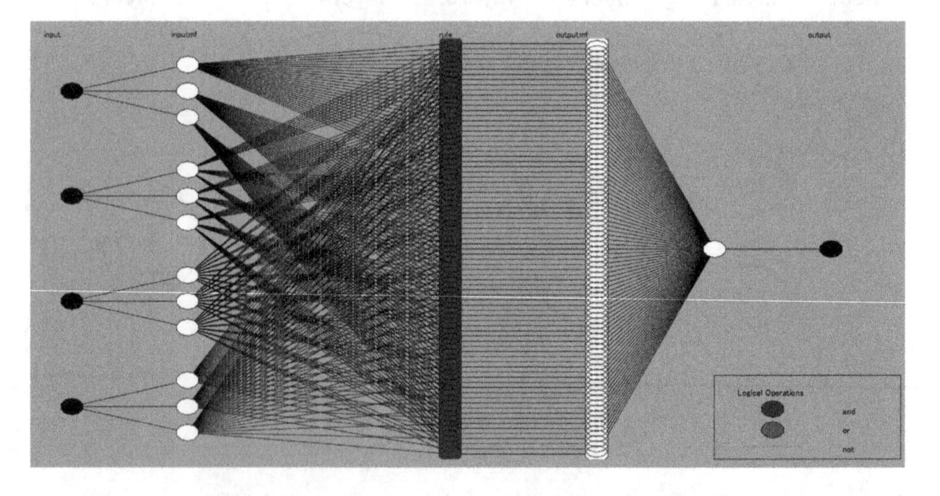

Fig. 5. Post-training classification results for iris-dataset by MATLAb 2017a

3 Hardware Program Design for 16bit ANFIS

The logic circuit was designed using VerilogHDL as the hardware description language and Quartus II 13.1 for development. When designing a hardware logic circuit for ANFIS to identify iris, it is essential to convert floating-point values into a format that can be processed in VerilogHDL. In this paper, the floating-point data handled in the logic circuit is converted into a 16-bit binary single-precision floating-point data type (IEEE754) for arithmetic operations. The decision to choose fp16, which offers lower precision compared to fp32 in processing results, was driven by the emphasis on cost and processing speed in the hardware logic circuit's design. Although fp16 may yield lower accuracy compared to CPU's 32-bit floating-point operations, its implementation is expected to provide sufficient accuracy for class classification. Additionally, fp16 requires fewer resources in terms of memory usage, making it more efficient for the overall logic circuit. Achieving accurate class classification allows for increased processing speed and reduced power consumption despite the slight reduction in processing precision compared to the CPU's fp32 operations (Fig. 6).

S	E	M
1bit	5bit	10bit

S:Sign (0:plus 1:minus)
E:Exponent (Bias +15)
F:Fraction

Fig. 6. IEEE 754 16bit floating point

Hardware-software co-design was employed to input the four items of the dataset into the hardware logic circuit. After converting the decimal point values to fp16, the four items are fed into the hardware logic circuit by outputting them to the physical addresses of the pins on the FPGA board. Subsequently, the processing results are retrieved from the pin's virtual address by the software program, converted back to 32-bit floating point, and verified.

The operations executed in layers 1 through 5 of ANFIS were designed by incorporating the logic circuits for the four arithmetic operations and comparison operations of fp16. The logic circuits for the four arithmetic operations of fp16 were designed with reference to Reference [4]. The structure of the ANFIS logic circuitry for layers 1 through 5 is illustrated in Fig. 7. The four pin-input items are labeled as x1 to x4, and the ANFIS processing results are output to the pins labeled as output.The FPGA is always running, but a reset signal is generated when the input x1 x4 is changed. The software programme then retrieves the processing results from the virtual address of the pin, converts them to 32-bit floating point and outputs the processing results for each individual Iris in a csv file. Each layer processing is handled by a pipelined data-flow type calculation [5].

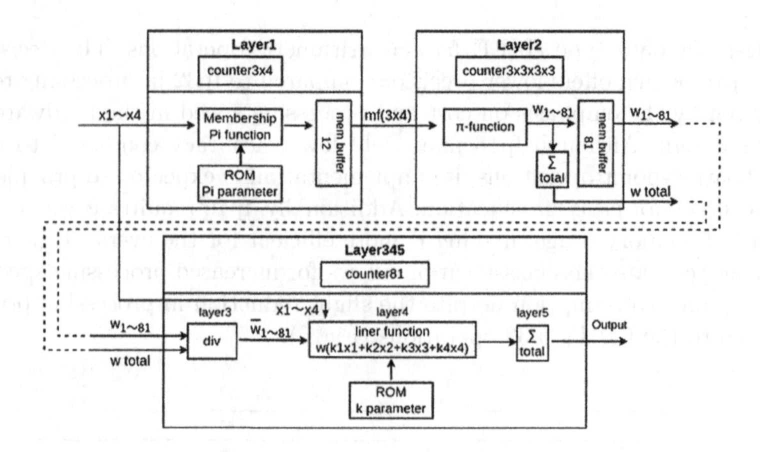

Fig. 7. Structure of the ANFIS logic circuit with 4 inputs and 3 MFs.

The number of pins and logic elements used are shown in Table (2).

Table 2. Hardware Logic Circuit Functions

clock frequency	50 MHz
logic utilization	9,038/32,070 (28%)
Total number of registers	13245
Total number of pins	368/457 (81%)
Total number of block memory bits	103,262/4,065,280 (3%)
DSP block total	16/87 (18%)

Table 3 provides a summary of the processing times for each layer circuit and the entire logic circuit. The simulation results clearly demonstrate that the processing speed approximately doubles when the frequency of the FPGA-implemented logic circuit is doubled. The busy signals of each layer of ANFIS are as shown in Fig. 8, the processing time of each layer is the time from the rise to fall of the busy signal, and the operating end time of the logic circuit is simultaneous with the fall of the busy signals of ly345.

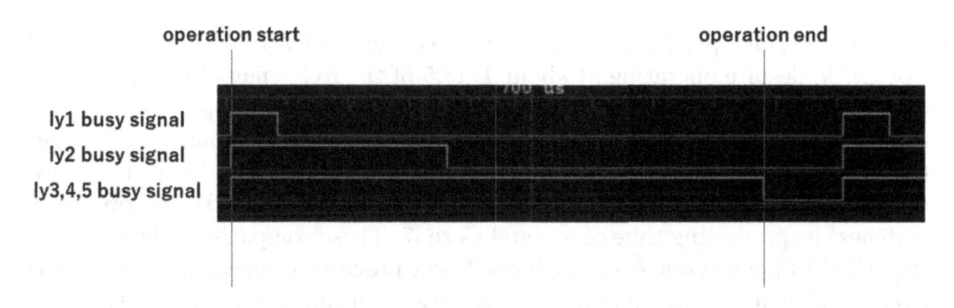

Fig. 8. Busy signals at each layer of ANFIS

Table 3. ANFIS layers of hardware logic circuits and overall processing time [us]

frequency	layer1	layer2	layer3,4,5	entire circuit
50[MHz]	7.717	27.588	59.630	87.218
100[MHz]	3.863	13.789	25.950	43.602

In order to compare the results of hardware AFIS, we designed an AFIS software program for iris identification using C language. This software program was implemented on two different devices: a PC equipped with a high-speed CPU mounted on an SoC FPGA and an ARM processor. The purpose of this comparison is to evaluate the processing performance and computational speed of the 16-bit AFIS designed for FPGAs. Additionally, we developed a software program to read four data items from the iris dataset and perform iris identification. The hardware logic was implemented on the two different devices with varying frequencies and numbers of CPU chips. The average processing time for 150 data points from the 1st to the 5th layers is presented in Table 4.

Table 4. Processing speed by implementation on CPU

CPU	frequency	Number of cores	processing time
ARM®CortexA9	925[MHz]	2	160.431[us]
Intel Core i7	2.8[GHz]	8	12.738[us]

4 Results and Comparison

Based on the data presented in Tables 3 and 4, we observe that Intel Core i7, Model-sim, Cyclone V SEA5 SoC, and ARM CortexA9, in that order, exhibit faster processing speeds. Comparing the Cyclone V SEA5 SoC with the ARM CortexA9, despite operating at about 1/18.5 of the frequency of the ARM CortexA9, the Cyclone V SEA5 SoC achieves a processing speed approximately 1.84 times faster. Additionally, Model-sim's processing speed is around 0.292 times that of an Intel Core i7, even though it operates at 1/28th of the frequency. Interestingly, an FPGA with an operating frequency of 400 [MHz] theoretically matches the processing time of an Intel Core i7. These comparisons demonstrate that FPGAs possess one core but boast faster processing speeds in ANFIS logic circuits, allowing them to handle data operations within the data input wait time required for an edge device. Furthermore, the FPGA cluster incurs lower costs for cluster design and electricity consumption when operating 24 h a day. Consequently, the FPGA cluster designed in this study, implementing the ANFIS hardware logic circuitry, exhibits sufficient processing speed for edge device control and proves to be a more cost-effective alternative compared to the CPU cluster. We conducted a comparison of software programs and hardware logic circuits for iris identification using ANFIS. Table 5 presents a comparison of correlation coefficients and accuracy between the software program executed on a Dual ARM CortexA9 (CPU) and the hardware logic circuit executed on a Cyclone V SEA5 SoC (FPGA) with the results processed by ANFIS using eigenvalues for each iris variety. The correlation coefficients demonstrate that the CPU identification results exhibit a very strong correlation with each eigenvalue, and similarly, the FPGA identification results also show a strong correlation with each eigenvalue. However, it is worth noting that the mean relative error between the identification results and the eigenvalues is approximately 3.4 times higher for FPGA than for CPU. Additionally, the standard deviation is approximately 3 times higher for FPGA than for CPU.

Table 5. CPU Execution Result

Control devices	correlation coefficient	Average of relative error	standard deviation
CPU	1.0000	0.12[%]	0.21
FPGA	0.9998	0.41[%]	0.61

5 Conclusions and Future Work

In this study, we designed and implemented an ANFIS system for iris identification to compare FPGAs and CPUs as edge device controllers. The results demonstrate that FPGAs can achieve faster processing speeds than CPUs, even at lower operating frequencies suitable for edge device control. Moreover, the FPGA implementation of ANFIS for iris recognition outperforms CPUs with multiple cores and similar frequencies, all at a lower cost. For future work, we aim to design a logic circuit with higher precision using 16-bit arithmetic and implement it in a robot. While our current hardware logic circuit utilizes 16fp, research by Henry et al. indicates that a design employing bfloat16 (bf16) can attain accuracy and speed equivalent to fp32 in deep learning [6]. By modifying the logic circuits for quadrature and comparison operations, which currently operate with fp16, to work with bf16, we aspire to implement a hardware logic circuit that matches the precision and speed of a 32-bit floating-point CPU. Additionally, we operated FPGA ANFIS by reading a dataset from a SoC-FPGA, but this does not allow us to verify real-time performance. As a future step, we plan to validate the FPGA's performance as a controller by implementing the hardware logic circuit on a smaller FPGA board and executing robot control. This will enable us to assess the FPGA's capabilities for real-time applications effectively.

Furthermore, while ANFIS for iris identification was implemented in FPGAs in this paper, previous work has included a photovoltaic panel emulator [7], a heater plate Algorithm for adjusting the pulse-width modulation (PWM) duty cycle [8], bilateral teleoperation system [9]. We believe that it is possible to apply ANFIS to classification and learning problems as there have been successful simulations using ANFIS implemented on FPGAs, and we intend to develop a new ANFIS for the future. We plan to verify whether the logic circuit created in this paper can be applied to ANFIS with other data sets. Integrating these technologies into IoT devices for disaster monitoring holds vast potential. AI camera devices can create an efficient disaster monitoring network, enabling real-time data and early detection. The AI accelerator designed for disaster prediction can revolutionize management practices, improving accuracy and response strategies. Our commitment to pushing technological boundaries for disaster management stems from the desire to safeguard lives and communities globally. With IoT devices and AI accelerators, we envision swift and effective disaster responses, minimizing casualties and environmental impact. We're dedicated to making the world more resilient and will persist in advancing science and technology toward this goal.

Acknowledgment. This work was supported by JSPS KAKENHI Fostering Joint International Research(B) 22KK0160.

References

1. Gomez-Pulido, J.A., Vega-Rodriguez, M.A., Sanchez-Perez, J.M., et al.: Accelerating floating-point fitness functions in evolutionary algorithms: a FPGA-CPU-GPU performance comparison. Genet. Program Evolvable Mach. **12**, 403–427 (2011)
2. Takagi, T., Sugeno, M.: Fuzzy identification of systems and its applications to modeling and control. IEEE Trans. Syst. Man Cybern. SMC-15(1), 116–132 (1985)
3. Jang, J.S.: ANFIS: adaptive-network-based fuzzy inference system. IEEE Trans. Syst. Man Cybern. **23**(3), 665–685 (1993)
4. Shirazi, N., Walters, A., Athanas, P.: Quantitative analysis of floating point arithmetic on FPGA based custom computing machines. In: Proceedings IEEE Symposium on FPGAs for Custom Computing Machines, pp. 155–162. IEEE (1995)

5. Kim, H., Choi, K.I.: A pipelined non-deterministic finite automaton-based string matching scheme using merged state transitions in an FPGA. PloS one **11**(10), e0163535 (2016)
6. Henry, G., Tang, P.T.P., Heinecke, A.: Leveraging the bfloat16 artificial intelligence datatype for higher-precision computations. In: 2019 IEEE 26th Symposium on Computer Arithmetic (ARITH), pp. 69–76 (2019)
7. Gómez-Castañeda, F., Tornez-Xavier, G.M., Flores-Nava, L.M., Arellano-Cárdenas, O., Moreno-Cadenas, J.A.: Photovoltaic panel emulator in FPGA technology using ANFIS approach. In: 2014 11th International Conference on Electrical Engineering, Computing Science and Automatic Control (CCE), Ciudad del Carmen, Mexico, pp. 1–6 (2014). https://doi.org/10.1109/ICEEE.2014.6978289
8. Huang, C.W., Pan, S.T., Zhou, J.T., Chang, C.Y.: Enhanced temperature control method using ANFIS with FPGA. Sci. World J. 2014, 8 (2014). Article ID 239261. https://doi.org/10.1155/2014/239261
9. Khati, H., et al.: Neuro-fuzzy control of bilateral teleoperation system using FPGA. Iran. J. Fuzzy Syst. **16**(6), 17–32 (2019)

Intelligent Optimization and Decision-Making

Beacon Localization Method Based on Flower Pollination-Fireworks Algorithm

Zhaofeng Du, He Huang, and Bin Xin(✉)

School of Automation, Beijing Institute of Technology, Beijing 100081, China
brucebin@bit.edu.cn

Abstract. Urban warfare environments are complex, highly concealed, and dangerous, particularly with the strong unknowns present in indoor environments. Utilizing mobile robots for searching inside buildings can significantly reduce casualties and yield significant military benefits. However, indoor environments lack support from the Global Positioning System (GPS), making it impossible for mobile robots to utilize GPS for positioning during mission execution. Achieving precise autonomous indoor localization for mobile robots has become a current hot topic of great concern. To address this issue, this paper proposes the use of Ultra Wide Band (UWB) beacons with Received Signal Strength Indication (RSSI) for achieving precise indoor localization. An RSSI-based localization method using the Flower Pollination-Fireworks Algorithm (FPFA) is introduced to solve the indoor positioning problem for mobile robots. Compared to traditional RSSI localization algorithms, the proposed algorithm effectively tackles factors such as signal attenuation and multipath effects, thereby enhancing the accuracy and stability of localization. Additionally, simulation experiments and error analysis were conducted on the algorithm, demonstrating its superiority in terms of accuracy and robustness.

Keywords: Beacon positioning · Fireworks Algorithm · Flower Pollination-Fireworks Algorithm

1 Introduction

With the development of positioning technology, the demand for positioning has expanded from outdoor to indoor and underground spaces. However, traditional positioning technologies such as GPS face many challenges in these environments, including multipath effects, signal obstruction, and signal attenuation [1]. Therefore, in order to meet the demand for indoor positioning services, researchers have begun to explore positioning technologies that are more suitable for indoor environments.

1.1 Wireless Sensor Positioning Technology

Visual positioning technology offers high positioning accuracy but requires stable motion of the agent and faces challenges in complex environments or environments with poor

B. Xin et al. (Eds.): IWACIII 2023, CCIS 1931, pp. 255–269, 2024.
https://doi.org/10.1007/978-981-99-7590-7_21

visibility [2]. The basic principle of wireless sensor positioning involves deploying in the environment and placing signal receivers at the target points to be measured. By measuring the relative position relationship between the receivers and transmitters, the position of the target can be calculated for localization, as shown in Fig. 1.

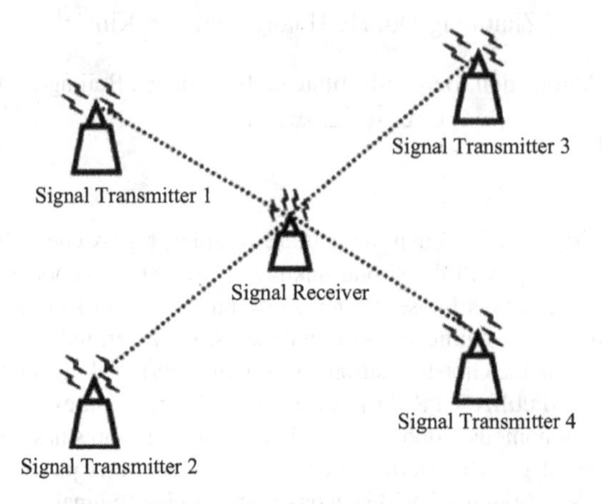

Fig. 1. Wireless Sensor Positioning Principle

Currently, sensor positioning technologies mainly include the following methods:

a) Time Difference of Arrival (TDOA): This method calculates the position of the transmitter by measuring the time difference between receiving the same signal at different receiver locations.
b) Time of Arrival (TOA): This method determines the target position by measuring the time difference of signal arrival at different sensors.
c) Angle of Arrival (AOA): This method utilizes the differences in signal arrival angles at different antennas to determine the target position.
d) Received Signal Strength Indication (RSSI): This method evaluates the communication quality by measuring the strength of the received signal at the receiver.

In addition, new positioning technologies have been proposed. For example, Venkatraman et al. introduced a new technique that estimates the true position or line-of-sight (LOS) distance in TOA measurements, even in the presence of non-line-of-sight (NLOS) distances [3]. The algorithm effectively utilizes constraints and nonlinear methods to extract the relationship between NLOS errors and the true range, enabling precise positioning of the target with only three base station measurements.

Zhang et al. proposed a new localization tracking algorithm in 2013 [4]. It utilizes Sigma-point Kalman smoothing, incorporates human dynamic walking models, and combines low-cost RSSI sensors to localize and track the target by perceiving the target's position and estimating its velocity. In 2019, Zheng et al. utilized an approximately effective substitute method to estimate and evaluate the accuracy of Angle of Arrival (AOA) measurements [5]. They designed a novel algorithm that improves the positioning performance by weighting all factors related to distance estimation accuracy in the AOA measurements.

These new techniques contribute to advancements in positioning technology and offer improved accuracy, robustness, and performance in various positioning applications.

Beacon, a kind of wireless signal positioning devices, is primarily used for localization, with RSSI technology being the main technique. Beacons consist of signal receivers and transmitters, where the devices emitting signals are referred to as base stations, and the devices determining the position of signal receivers are called tags. The signal types currently used include infrared, Bluetooth, WiFi, Ultra-Wideband (UWB), and the fusion of WiFi and vision technologies [6–11]. These technologies are the current main research directions, with UWB signals offering relatively high positioning accuracy, improved distance resolution, and strong resistance to multipath effects.

1.2 Main Research Content

The main objective of this article is to address the indoor positioning problem using UWB beacons and the RSSI technique. After the beacons receive the signals, the RSSI algorithm, improved by the Flower Pollination-Fireworks Algorithm, is employed to process the signal data and solve the beacon positioning problem. The research findings of this article are extensively validated through experiments, demonstrating the effectiveness of the proposed algorithm. By utilizing the algorithm proposed in this article, accurate positioning can be achieved in indoor environments.

2 Beacon Positioning Model

2.1 UWB Beacon

Beacons are lightweight, weighing around a few grams, and are widely employed due to their portability and low cost. Beacons measure the relative distance between a target point and the transmitter source. However, the accuracy of propagation varies among these different technologies. Optical transmission, for example, has lower positioning accuracy and is highly susceptible to environmental factors, making stability challenging to ensure. Technologies like Bluetooth and WiFi are not highly sensitive to the RSSI value changes with distance, resulting in relatively lower positioning accuracy.

In contrast, UWB technology utilizes non-sinusoidal narrow pulses and exhibits strong resistance to multipath effects. Therefore, in this article, UWB beacons are selected to investigate their application and performance in indoor positioning.

2.2 Basic Principles of Beacon Positioning

The basic principle of beacon positioning is to calculate the relative distance between the target point and the transmitter source by measuring the signal attenuation during propagation.

During the positioning measurement process, the initial signal from the transmitter source is known, while the received signal strength is a function of the propagation distance. As the propagation distance increases, the received signal strength gradually weakens. In the ideal scenario, the received signal strength is logarithmically related to the distance between the target point and the transmitter center, as described by (1) [12], which considers the received signal power, propagation distance, path loss, and transmit signal power.

Therefore, based on the received signal strength, it is possible to calculate the relative Euclidean distance between the target point and the transmitter source.

$$Pr = I - 10u\lg d + \eta \tag{1}$$

In (1), the received signal power is represented by Pr, where I represents the initial signal power. u is the signal fading factor, which represents the environmental loss and typically ranges from 2 to 4. d is the Euclidean distance between the target point and the transmitter source. η is the environmental interference factor, which is influenced by the current environment of the beacon and obstacles in the propagation path. It can be approximated as a Gaussian random signal with a mean of 0 and a variance of σ, measured in dBm. Both I and η reflect the parameters of signal transmission quality between beacons and are influenced by the environment.

2.3 Factors Affecting Beacon Positioning

In two-dimensional positioning, at least three non-collinear beacon signals are required to determine the position of the target point. By using the positions of the beacons as the centers of circles with radii equal to the measured distances, three circles are constructed, and the intersection of these circles represents the position of the target point.

In three-dimensional positioning, the number of beacons needs to be expanded to at least four non-coplanar beacons. Spheres are constructed with the beacon positions and measured distances, and the intersection of these four spheres represents the position of the target point.

From this, it is evident that the number and accuracy of beacon positions are crucial for achieving accurate positioning. When conducting beacon-based positioning, having an adequate number of well-positioned beacons is essential to ensure precise and reliable location determination. The applicability of beacons is limited by the requirements for practical accuracy [13]. Since the measurement error of a single beacon is positively correlated with the signal propagation distance, we use r to represent the effective distance of the beacon, which is the maximum range where the signal strength meets the measurement error requirement. Considering the small indoor space and the presence of numerous obstacles, using beacons with high transmit power can adversely affect the positioning results. Therefore, the selected beacons have low transmit power and a relatively small effective propagation distance.

In practical applications, obstacles can significantly interfere with signal propagation. If a beacon cannot obtain stable measurement results due to obstacles between the beacon and the target point, the measurement data from that beacon is discarded. Therefore, in the actual measurements, we define effective beacons as those within the line of sight range of the target point, while beacons in other situations are considered invalid. Figure 2 illustrates the concept of invalid/valid beacons.

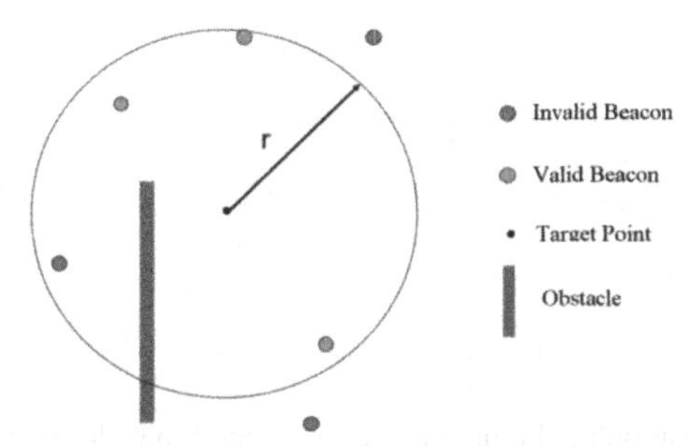

Fig. 2. Invalid/Valid Beacon Illustration.

To ensure the reliability and practicality of this research, the LD150 self-powered positioning beacon module was used for experimentation, as shown in Fig. 3. This module can be used as both a beacon transmitter (base station) and a signal receiver device (tag). In this study, the beacon module's model and parameter settings were obtained by measuring the physical beacon to ensure the accuracy of the experimental results.

Fig. 3. LD150 Beacon

3 RSSI Localization Algorithm Based on Flower Pollination-Fireworks Algorithm

3.1 RSSI Localization Algorithm

RSSI is a target localization technology based on signal strength. As the distance from the signal source increases, the received signal strength gradually weakens. By using the signal propagation model and establishing a functional relationship between signal strength and distance, it is possible to determine the target's position by treating the signal source as the center and the signal strength as concentric circles. RSSI technology enables the acquisition of indoor position information for targets, which has significant practical value in various applications.

As shown in Fig. 4, using (1), the distances from point O to the three transmitters can be calculated. This yields a system of (2), which can be solved simultaneously to obtain the coordinates of point O.

$$\begin{cases} (x_a - x_o)^2 + (y_a - y_o)^2 = {d_{oa}}^2 \\ (x_b - x_o)^2 + (y_b - y_o)^2 = {d_{ob}}^2 \\ (x_c - x_o)^2 + (y_c - y_o)^2 = {d_{oc}}^2 \end{cases} \quad (2)$$

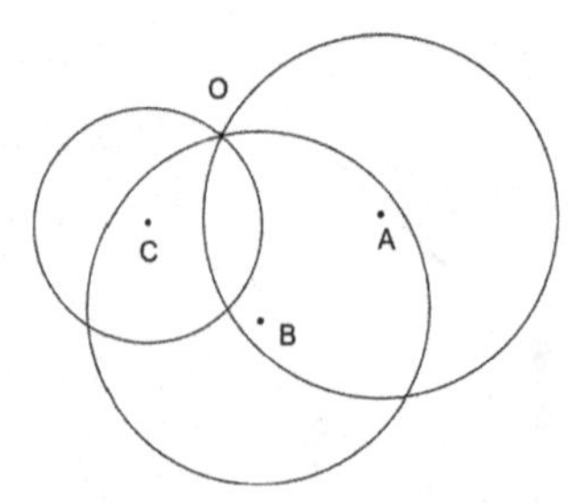

Fig. 4. Three points positioning. Point O is the target point to be located, while A, B, and C are three signal transmitters. Point O measures the signal strength from each of the three transmitters.

However, in practical applications, the signal strength is influenced by environmental interference, leading to errors in the calculated Euclidean distances. As a result, the positioning accuracy may not meet the requirements. To address this issue, this paper proposes an RSSI positioning method based on the Flower Pollination - Fireworks Algorithm (FP-FWA).

3.2 Fireworks Algorithm

The Fireworks Algorithm (FWA) is a swarm intelligence algorithm. It was proposed by Professor Ying Tan from Peking University in 2010, inspired by the observation of fireworks exploding and producing sparks to illuminate the surrounding area [14]. In the algorithm, each exploding firework represents a feasible solution in the search space, and the process of searching the neighborhood is achieved through the generation of new fireworks from the explosion. The optimization process of the Fireworks Algorithm can be summarized into the following steps:

i. Initialization: Randomly generate initial fireworks in the feasible solution space, where each firework represents a feasible solution.
ii. Fitness Evaluation: Evaluate the fitness value of each firework based on the optimization objective function. This determines the quality of each firework, which is used to generate a different number of explosion sparks and Gaussian mutation sparks at different explosion radii. The explosion sparks are primarily used to search the neighborhood of the old fireworks. Fireworks with higher fitness values have smaller explosion ranges but generate a large number of new fireworks. Fireworks with lower fitness values have larger explosion ranges but generate only a few new fireworks, as shown in Fig. 5.
iii. Termination Check: Check if the termination condition is met. If the condition is satisfied, the algorithm stops the search. Otherwise, select a subset of individuals from the old fireworks, explosion sparks, mutated old fireworks, and Gaussian mutation sparks to form the next generation of fireworks for the next iteration.

These steps are repeated until the termination condition is met, optimizing the solutions in the search space.

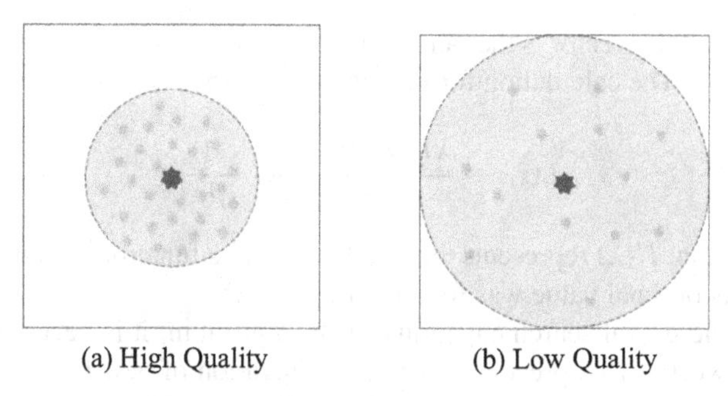

Fig. 5. The Effects of Different Quality Fireworks

3.3 Improved Fireworks Algorithm Based on Flower Pollination

The FWA is a swarm intelligence algorithm known for its simple algorithmic structure and powerful global search capability. However, in cases where the solution space is large or the objective function is complex, the ignition and diffusion operations of the FWA can significantly increase computational complexity and easily lead the algorithm to fall into local optimal solution. Therefore, in this paper, an improved version of the Fireworks Algorithm is proposed by incorporating the cross-pollination mechanism from the Flower Pollination Algorithm. This is achieved by using Levy flight to generate new fireworks during explosion, and selecting the next generation fireworks based on the density of exploded fireworks.

Levy flight is a special type of random process characterized by its significant discontinuity and non-stationarity in its path. Levy flight does not have a fixed step size but instead undergoes random variations according to a certain probability distribution. This probability distribution is often referred to as the Levy distribution, which exhibits heavy-tailed properties, meaning that there is a significant increase in probability density in the tail of the distribution. This implies that in Levy flight, there are numerous small step sizes and a few extremely large step sizes.

In the original algorithm, the explosion radius and the number of sparks generated by an old firework are calculated based on the fitness values of the current firework and the rest of the population. For the firework x_i, (3) and (4) are used to calculate the explosion radius A_i and the number of sparks generated S_i, respectively.

$$A_i = \widehat{A} \times |L| \times \frac{f(x_i) - y_{\min} + \varepsilon}{\sum_{i=1}^{N} (f(x_i) - y_{\min}) + \varepsilon} \tag{3}$$

$$S_i = M \times \frac{y_{\max} - f(x_i) + \varepsilon}{\sum_{i=1}^{N} (y_{\max} - f(x_i)) + \varepsilon} \tag{4}$$

In the equations, $y_{\min} = \min(f(x_i)), (i = 1, 2, ..., N)$ is the firework with the minimum fitness value in the current population. $y_{\max} = \max(f(x_i)), (i = 1, 2, ..., N)$ is the firework with the maximum fitness value. $\widehat{A}$ is the coefficient for adjusting the explosion radius. M is the maximum number of sparks. ε is a constant to avoid division by zero. And L is an explosion vector whose dimension is the same as that of firework x_i, and each dimension of the vector is the random number which follows the Levy distribution Levy distribution. The calculation formula is given as (5).

$$L(x_i) \sim \frac{\lambda \Gamma(\lambda) \sin(\frac{\pi\lambda}{2})}{\pi} \cdot \frac{1}{{x_i}^{1+\lambda}} \tag{5}$$

In the equation, $\Gamma(\lambda)$ represents the standard gamma function, and through multiple experiments, its optimal value was found to be $\lambda = 1.5$.

To ensure the global search capability of the algorithm, it is necessary to limit the number of fireworks. This prevents excessive generation of new fireworks at positions with good fitness values or the generation of a small number of new fireworks at positions with lower fitness values, which can weaken the local search and prevent the algorithm from getting stuck in local optima. Equation (6) imposes a constraint on the number of fireworks.

$$S_i = \begin{cases} round(a * M),\ S_i < aM \\ round(b * M),\ S_i > bM, 0 < a < b < 1 \\ round(S_i), \quad else \end{cases} \tag{6}$$

By imposing constraints, the number of fireworks generated by the fireworks algorithm can be controlled within different fitness value ranges, thereby avoiding excessive or insufficient fireworks and improving the search efficiency of the algorithm.

The fireworks algorithm introduces Gaussian mutation fireworks to increase the diversity of the explosion fireworks population. In this process, the old selection operator is replaced by the mutation operator, and the new generated is the mutation fireworks. The generation process of Gaussian mutation fireworks in the algorithm is as follows: the firework x_i is randomly selected from the old fireworks population. Then, a dimension is randomly chosen for the selected firework x_i, and Gaussian mutation is performed on that dimension. The old fireworks generate the mutation fireworks. Assume that the dimension selected for mutation in firework x_i is denoted as k. The Gaussian mutation operation for the old firework can be represented as (7):

$$\hat{x}_{ik} = x_{ik} \times e \tag{7}$$

where $e \sim N(1, 1)$, $N(1, 1)$ represents a Gaussian distribution with a mean of 1 and a variance of 1.

When generating new fireworks and Gaussian mutation fireworks from the old operators and selected mutation operator, it is possible for the new fireworks to exceed the boundaries of the feasible domain, represented as boundary Ω. If the new fireworks exceed the boundaries in dimension k, the algorithm ensures that the generated fireworks outside the boundaries are mapped to a new position within the feasible domain based on the mapping rule defined by (8). Specifically, for fireworks x_i that exceed the boundaries, they are mapped as follows:

$$\hat{x}_{ik} = x_{LB,k} + \hat{x}_{ik} \% (x_{UB,k} - x_{LB,k}) \tag{8}$$

In this equation, $x_{UB,k}$, $x_{LB,k}$ represent the upper and lower bounds of the solution space, respectively. % is the symbol used for the operation of calculating the remainder.

To ensure that excellent individuals in the firework population can be passed on to the next generation, after the explosion of old fireworks and the generation of explosion sparks and Gaussian mutation sparks, the algorithm forms a candidate set, including old fireworks, new fireworks generated by explosions, and mutated particles and Gaussian mutation sparks. Then, a subset of individuals is selected from the candidate set as the next generation of fireworks for further explosions. The candidate set is denoted as C, and the size of the firework population is denoted as N. The candidate set will directly select the particle with the highest current fitness value as the elite particle for the next generation of fireworks. As for the remaining $N - 1$ fireworks, the algorithm uses the roulette wheel method to select the next generation fireworks from the candidate set. In the algorithm, the calculation formulas for the selection probability of candidate individuals $p(x_i)$ are given by (9) and (10):

$$p(x_i) = \frac{R(x_i)}{\sum_{x_j \in K} R(x_j)} \tag{9}$$

$$R(x_i) = \sum_{x_j \in K} d(x_i - x_j) = \sum_{x_j \in K} \|x_i - x_j\| \tag{10}$$

In the equations, $R(x_i)$ represents the sum of distances between the current firework and all other fireworks in the candidate set K, excluding firework x_i. If the current firework is located in an area of high firework density within the candidate set, meaning there are many other fireworks surrounding it, the probability of selecting the current firework will decrease. This is because regions with high firework density may have already been sufficiently explored, and further strengthening the selection probability of individuals in that area would weaken the algorithm's global search capability and increase the likelihood of getting trapped in local optima.

3.4 The Algorithm Flow of FP-FWA

This subsection proposes an improvement to the fireworks algorithm for indoor beacon localization by utilizing hetero-flower pollination. In the algorithm, the beacon positions and the Euclidean distances between each beacon and the target point are considered as decision variables. The objective is to optimize the coordinates of the target point. The constraint requires that the target coordinates remain within the localization range. The termination condition is set as the maximum number of iterations. The optimization process involves operations such as explosion, mutation, dispersion, and others, aiming to achieve more accurate localization coordinates with lower errors. The pseudocode for the Flower Pollination-Fireworks Algorithm, as shown in Algorithm 1, is as follows:

Algorithm 1: RSSI Localization Algorithm based on FP-FWA

Input: Beacon positions, Euclidean distance from beacons to the target, algorithm parameters, termination condition
Output: Fitness value, coordinates of the estimated point
1: Randomly generate a certain number of individuals' positions within the solution space.
2: Calculate the fitness value for each individual.
3: Perform roulette wheel selection based on the firework density.
4: Calculate the explosion radius and brightness for the selected individuals.
5: Randomly select fireworks for Gaussian mutation.
6: Check if the fireworks exceed the boundaries; if so, apply a mapping operation.
7: Calculate the fitness value for all individuals in the updated population.
8: Check if the termination condition is met. If satisfied, end the algorithm and proceed to step 9. If not, return to step 3 and continue the algorithm.
9: Output the results. The algorithm terminates.

4 Simulation Experiments and Results Analysis

4.1 Preparations Before the Algorithm Experiments

The experiments were conducted using a Dell DESKTOP-CGBH3VN computer and MATLAB 2021b simulation software. In both experiments, the same beacon deployment environment was used, and a comparison was made with the localization algorithm in reference [15] to evaluate the performance of the Flower Pollination-Fireworks Algorithm.

In the first experiment, the improved RSSI localization algorithm based on the FP-FWA was used to randomly select five positions in an open area and near the corners of the wall in the space that needs to be searched. The localization errors were calculated for each position, and the overall error was observed to demonstrate the effectiveness of the RSSI localization algorithm.

In the second experiment, 20 positions were randomly selected within the locatable area of the space. The RSSI localization algorithm based on the Flower Pollination-Fireworks Algorithm and the localization algorithm described in reference [16] were applied separately for localization. The superiority of the algorithms was determined based on the magnitude of the errors.

Within line-of-sight range, UWB positioning beacons based on RSSI technology have high measurement accuracy. However, when measuring beyond the line-of-sight range, there are significant errors and unstable data issues in the localization calculations. Therefore, in this study, the beacons were chosen to be within the line-of-sight range. To facilitate simulation experiments and improve experimental reliability, low-power UWB beacons with built-in DW1000 chips were used for error experiments.

The experiments were conducted in an indoor environment where a beacon transmission module and a receiving module were placed on the ground without any obstacles between them. The beacon transmission module remains stationary, while the receiving module changes its distance by moving to complete the measurements. Figure 6 shows the experimental setup. The experiment captures the measurements and errors of the beacons at different distances, as shown in Table 1, and displays them on the user interface of the upper computer. Figure 7 illustrates the display on the upper computer during a 20 m ranging measurement.

According to the measured values and actual distances, the least squares method is used to fit the ranging measurements. The fitting equation can be expressed as follows: $Y = aX^3 + bX^2 + cX + d$, where a, b, and c are all constants, and the calibration (11) is obtained through calculations.

$$y = 3.3026 * 10^{-5}x^3 - 0.0022x^2 + 1.0426x - 0.0726 \tag{11}$$

In the equation, x represents the true value, and y represents the measured value within the line-of-sight range of the beacon. The fitting curve is shown in Fig. 8. In subsequent simulations, the measured values are obtained based on the ranging calibration equation of the beacons for experimental purposes.

Table 1. Measurement data.

Index	True value (m)	Measured value (m)	Error (m)
1	1.5	1.49	+0.01
2	3	3.06	−0.06
3	5	5.10	−0.10
4	7	7.13	−0.13
5	9	9.17	−0.17
6	10	10.19	−0.19
7	20	20.19	−0.19
8	30	30.20	−0.20

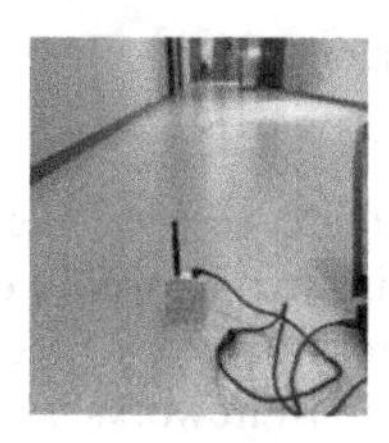

(a) Signal Transmitter (b) Signal Receiver

Fig. 6. Actual Testing of the Beacon

4.2 Localization Algorithm Experiment

This section conducts experiments to verify the effectiveness and superiority of the improved RSSI localization algorithm. In the first set of experiments, five positions

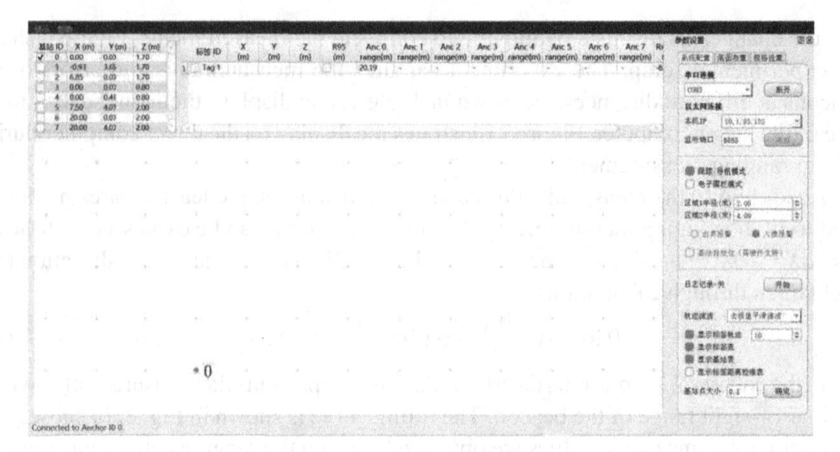

Fig. 7. Display of Measurement Software on the Upper Computer in Actual Testing

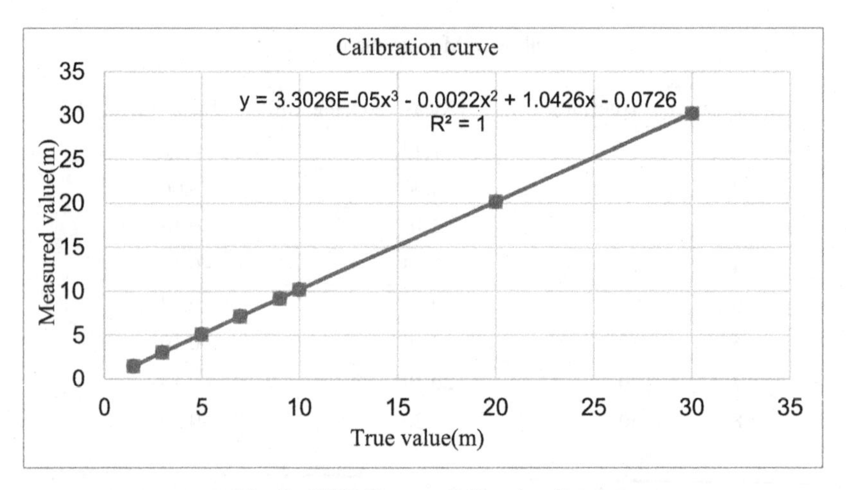

Fig. 8. UWB Beacon Calibration Curve

were randomly selected in an open area and near the corners of the wall within the space to calculate the localization errors. The localization algorithm errors are shown in Table 2. From the analysis of the experimental results, it can be observed that the RSSI localization algorithm based on Flower Pollination-Fireworks is able to effectively locate the test positions in three-dimensional space, with small errors and small error fluctuations. The localization errors differ between areas near obstacles and open areas, with larger errors near obstacles and smaller errors in open areas.

Location 1 from the case study is selected for analysis. The convergence curve of the algorithm for Location 1 is shown in Fig. 9.

The convergence curve reveals that the algorithm exhibits a fast convergence rate in the early stages of iteration. By the 50th iteration, it is already very close to the optimal solution attainable by the algorithm, indicating strong optimization capability.

Table 2. Simulated error.

Area	Index	Error (m)
Open area	1	0.1395
	2	0.1592
	3	0.0699
	4	0.1797
	5	0.2091
Near obstacles	1	0.3670
	2	0.3384
	3	0.3582
	4	0.3088
	5	0.3541

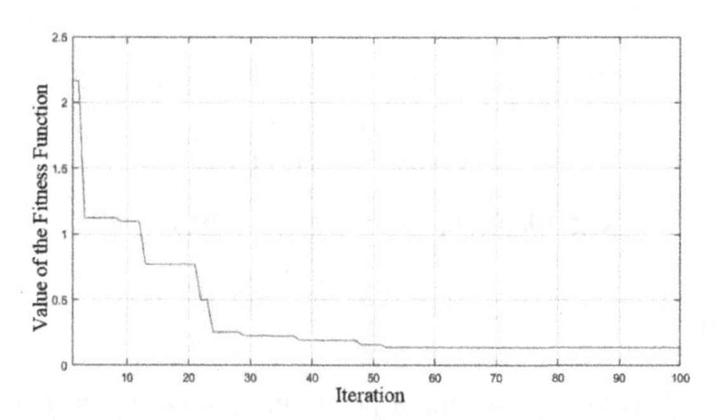

Fig. 9. Positioning Error Iteration

The simulation experiments and data analysis provide evidence of the effectiveness of the RSSI localization algorithm based on the Flower Pollination-Fireworks Algorithm.

In order to demonstrate the superiority of the algorithm, 20 random positions were selected in both open areas and areas near obstacles within the space. The positioning calculations were performed simultaneously using the Weighted Centroid Localization Algorithm (WCLA) described in reference [15] and the RSSI localization algorithm based on the Flower Pollination-Fireworks Algorithm. The obtained absolute error results are shown in Fig. 10. The results indicate that the RSSI localization algorithm based on the Flower Pollination-Fireworks Algorithm can improve the indoor positioning accuracy by 10 cm. Through comparative experiments, the superiority of the RSSI localization algorithm based on the Flower Pollination-Fireworks Algorithm is demonstrated.

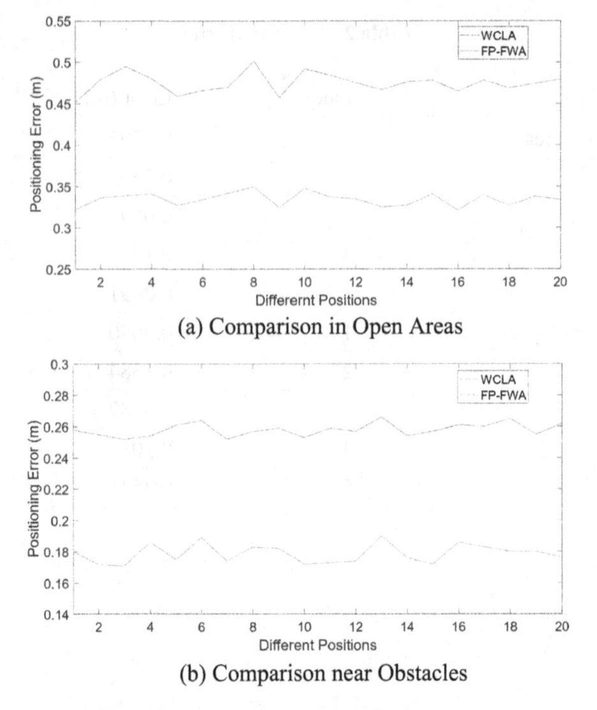

(a) Comparison in Open Areas

(b) Comparison near Obstacles

Fig. 10. Comparison of Positioning Errors

5 Conclusion

The aim of this study is to address indoor positioning issues by utilizing UWB beacons based on RSSI technology. Building upon this, a new localization algorithm is proposed in this paper. The objective of this algorithm is to improve accuracy of the positioning calculations. The algorithm demonstrates excellent optimization performance, as it avoids getting stuck in local optima and is capable of finding the global optimum. The experimental results demonstrate that compared to other existing RSSI localization algorithms, the RSSI localization algorithm based on the Flower Pollination-Fireworks Algorithm achieves higher positioning accuracy. It effectively addresses the indoor beacon positioning problem. These findings highlight the algorithm's significant practical value and its ability to provide an effective solution for indoor positioning challenges.

References

1. Li, C.T., Cheng, J., Chen, K.Y.: Top 10 technologies for indoor positioning on construction sites. Autom. Constr. **118**, 103309 (2020)
2. Cadena, C., Carlone, L., Carrillo, H., et al.: Past, present, and future of simultaneous localization and mapping: toward the robust-perception age. IEEE Trans. Robots **32**(6), 1309–1332 (2016)
3. Venkatraman, S., Caffery, J., You, H.R.: A novel To a location algorithm using LoS range estimation for NLoS environments. IEEE Trans. Veh. Technol. **53**(5), 1515–1524 (2004)

4. Zhang, S., Jiang, H., Yang, K.: Detection and localization for an unknown emitter using TDOA measurements and sparsity of received signals in a synchronized wireless sensor network. In: IEEE International Conference on Acoustics, Speech and Signal Processing, Vancouver, pp. 5146–5149. IEEE (2013)
5. Zheng, Y., Sheng, M., Liu, J., et al.: Exploiting AoA estimation accuracy for indoor localization: a weighted AoA-based approach. IEEE Wirel. Commun. Lett. **8**(1), 65–68 (2018)
6. Zhu, H., Xie, Y.H.: Research on application of beacon-based location technology in American Universities. IOP Conf. Ser. Earth Environ. Sci. **252**(5), 052–059 (2019)
7. Nuaimi, K.A., Kamel, H.: A survey of indoor positioning systems and algorithms. In: 2011 International Conference on Innovations in Information Technology (IIT), Abu Dhabi, pp. 185–190. IEEE (2011)
8. Yassin, A., Nasser, Y., Awad, M., et al.: Recent advances in indoor localization: a survey on theoretical approaches and applications. IEEE Commun. Surv. Tutor. **19**, 1327–1346 (2011)
9. Bhattd, D., Babu, S.R., Chudgar, H.S.: A novel approach towards utilizing Dempster Shafer fusion theory to enhance WiFi positioning system accuracy. Pervasive Mob. Comput. **37**, 115–123 (2017)
10. Tian, Y., Huang, B.Q., Jia, B., et al.: Optimizing AP and beacon placement in WiFi and BLE hybrid localization. J. Netw. Comput. Appl. **164**, 654–662 (2020)
11. Martins, P., Abbasi, M., Sa, F., et al.: Improving bluetooth beacon-based indoor location and fingerprinting. J. Ambient. Intell. Humaniz. Comput. **11**, 3907–3919 (2020)
12. Zhang, K., Zhang, G., Yu, X.W., et al.: Boundary-based anchor selection method for WSNs node localization. Arab. J. Sci. Eng. **46**, 3779–3792 (2021)
13. Martins, P., Abbasi, M., Sa, F., et al.: Intelligent beacon location and fingerprinting. Procedia Comput. Sci. **151**, 9–16 (2019)
14. Tan, Y., Zhu, Y.: Fireworks algorithm for optimization. In: Tan, Y., Shi, Y., Tan, K.C. (eds.) ICSI 2010. LNCS, vol. 6145, pp. 355–364. Springer, Heidelberg (2010). https://doi.org/10.1007/978-3-642-13495-1_44
15. Kim, K., Shin, Y.: A distance boundary with virtual nodes for the weighted centroid localization algorithm. Sensors **18**, 1054 (2018)
16. Panag, T.S., Dhillon, J.S.: Maximal coverage hybrid search algorithm for deployment in wireless sensor networks. Wirel. Netw. **25**, 637–652 (2019)

Parameter Identification for Fictitious Play Algorithm in Repeated Games

Hongcheng Dong[1,2] and Yifen Mu[2(✉)]

[1] School of Mathematical Sciences, University of Chinese Academy of Sciences, Beijing, China
donghongcheng@amss.ac.cn

[2] Key Lab of Systems and Control, Academy of Mathematics and Systems Science, Chinese Academy of Sciences, Beijing 100190, People's Republic of China
mu@amss.ac.cn

Abstract. In the previous works [1] and [2], we solved the optimal strategy of the human player against a machine player who makes decisions based on Fictitious Play in infinitely repeated 2×2 games, in which the information is assumed to be complete and perfect. In this paper, we consider the problem of identification when the human player does not know the initial assessment of the machine player. In this scenario, we propose an identification algorithm for the human player and prove that the process of identification will end successfully in a finite time if the machine's payoff parameter is rational. When the machine's payoff parameter is irrational, the identification process will not end, which implies some advantage for the algorithm with irrational parameters.

Keywords: Repeated games · Algorithm identification · Fictitious Play · Dynamical game systems

1 Introduction

In this paper, we will study the problem of parameter identification in repeated games between a human player and a machine player which adopts some learning algorithms to take its action. To be specific, we will try to identify the initial assessment of the Fictitious Play algorithm in repeated 2×2 games. This is a simple and starting scenario for the problem of algorithm identification, which constitutes a necessary part of the evolution and control for dynamical game systems. With the development of Artificial Intelligence (AI), such games involving learning algorithms will become more and more common and important.

In the past decade, lots of algorithms have been developed to play games with different features, from the complete information 0-sum game like Go to the incomplete information 0-sum game like poker to the multi-player incomplete information stochastic games like the electronic game StarCraft [3–8]. These AIs are based on different learning algorithms to generate the near-optimal strategy in specific games which can perform very well and even beat the top human players.

These developments make AI algorithms more and more participating in people's life and work. Hence the game involving algorithms are becoming common

B. Xin et al. (Eds.): IWACIII 2023, CCIS 1931, pp. 270–282, 2024.
https://doi.org/10.1007/978-981-99-7590-7_22

and important, which are called algorithm games in the literature [9]. Early there have been many works analyzing the repeated games between symmetric algorithms. Researchers try to investigate whether the equilibrium (Nash Equilibrium, correlated equilibrium, etc) will arise as the long-run outcome when the players adopt the same algorithm to update their actions in the game. This topic has been studied extensively and still attracts attention of researchers in game theory [10–15], control theory [16,17] and computer sciences [18,19]. In fact, the convergence results provide theoretical basis for the training to optimal strategies in the construction of game AI we stated at the beginning.

On the other hand, from the point of opponent exploitation, we can provide a different perspective to understand the learning algorithms by considering the asymmetric algorithm game, i.e. the game between an algorithm (called a machine player) and a perfect opponent (called a human player). Such games also happen in many different situations. For example, many people play chess game or computer game with an AI for entertainment, the unmanned autonomous systems for different tasks would interact with each other or the human. Besides, algorithms may be forced to play a game by being attacked or fooled, which may cause surprising even serious consequences, see [20] and [21] in which scientists find that some tiny perturbations may cause the algorithms to make mistakes in classifying an image. This technical vulnerability may bring on attacks to medical learning systems, or the face recognition system designed for crime detection, which further lead to uneconomical or even dangerous consequences for the society [22–25].

Thus, in order to apply these algorithms in practice in a correct way, to understand and handle the system involving algorithms, to design better algorithms to play specific games, the analysis for the human-machine game system is necessary and urgent. However, to the best of our knowledge, research on such systems is not sufficient and related works are not much in the literature. Previously, [26,27] studies the optimal strategy against an opponent with finite memory and gives the theoretical results. Recently, [28,29] and [30] use myopic best response or look-ahead strategy to fight against the opponent which is approximated by Recurrent Neural Network (RNN). Also, [31] presents safe strategy and proposes an algorithm of exploiting sub-optimal opponents under the condition of ensuring safety, and [32] presents an exact algorithm in imperfect information games to exploit the opponent using the Dirichlet prior distribution.

In this paper we assume that the machine player uses the classical Fictitious Play (FP) algorithm to update its actions. FP algorithm was the first learning algorithm to achieve Nash equilibrium [33]. When the players adopt FP algorithm, convergence has been proved for repeated games with two players or zero-sum payoffs [11,34,35]. However, even for simple 3×3 general-sum game, the convergence does not hold [36]. This implies the complexity of the dynamical game systems driven by learning algorithms. So far, Fictitious Play and many variants have been studied, wherein stochastic FP [37] considers the perturbed payoff in the game and the players choose a distribution on the best response according to the private information about the perturbation, in weakened FP

[38] and generalised weakened FP [39] players take the $\epsilon-$best response as his action. Recently, [40] proposes the Full-Width Extensive Form Fictitious Self-Play (XSP) based on reinforcement learning and supervised learning for extensive form games, and [41] further proposes Neural Fictitious Self Play (NFSP) which uses neural networks to approximate the mapping of FP.

In [1] and [2], we have proved and solved the optimal strategy against the machine adopting FP under the assumption of complete and perfect information. In this paper, we will assume that some parameters in the algorithm is unknown to the human player and consider the identification problem. Specifically, we assume that the human player know that the machine adopts the FP algorithm but does not know the initial assessment of the algorithm. In order to get the near-optimal averaged utility over the infinite time, one natural idea for the human player is to identify the unknown parameters of the machine. Since the human can infer the inequality of the assessment parameters from the stage action of the machine, this seems very possible given enough probes. In this paper, we will give a simple and natural algorithm to identify the assessment-parameter in the FP algorithm and prove that the identification process will stop successfully in finite time if the machine's payoff parameter is a rational number. However, by an example, we will show that the identification process can not stop if the machine's payoff parameter is irrational. This finding implies some advantage of the learning algorithms with irrational parameters and may help design better algorithms.

The paper is organized as below: Sect. 2 gives the problem formulation; Sect. 3 gives the results when the machine's payoff parameter is rational and illustrates the case when the machine's payoff parameter is irrational by giving an example; Sect. 4 concludes the paper with some remarks and the future work.

2 Problem Formulation

Consider a 2×2 general-sum strategic-form game. Player 1 and Player 2 are called the machine player and the human player. The machine has two actions, denoted by A, B. The human has two actions, denoted by a, b. Thus there are 4 different possible outcomes (equally, the action profile) of the game: Aa, Ab, Ba, Bb. Given any outcome, the machine and the human have their individual utility $q_i, w_i, i = 1, 2, 3, 4$. We describe the game by the bi-matrix below.

Player 1	Player 2	
	a	b
A	q_1,w_1	q_2,w_2
B	q_3, w_3	q_4, w_4

The mixed strategy of the player is a probability distribution over the pure action set {A,B} or {a,b}.

Consider the repeated game. Denote the action of the machine and the human at time t by α_t^1, α_t^2. The machine player will choose its action α_t^1 according to the Fictitious Play (FP) algorithm, i.e.,

$$\alpha_t^1 = BR((\frac{\kappa_t(a)}{\kappa_t(a)+\kappa_t(b)}, \frac{\kappa_t(b)}{\kappa_t(a)+\kappa_t(b)})) \tag{1}$$

where the BR function denotes the best response of the machine player against his assessment $\kappa_t(a), \kappa_t(b)$ to the human's behavior and $\kappa_t(a), \kappa_t(b)$ are non-negative real numbers which are updated by

$$\kappa_t(i) = \kappa_{t-1}(i) + \begin{cases} 1, & if\, \alpha_{t-1}^2 = i; \\ 0, & if\, \alpha_{t-1}^2 \neq i. \end{cases} \tag{2}$$

where $i = a, b$. Here we let Player 1 choose A when both actions A, B are the best response of the machine player.

Obviously, once the initial assessments $\kappa_0(a), \kappa_0(b)$ are fixed, the updating rule of the machine is totally determined. Then how the system evolve will be determined by the human's action sequence. If the human takes his action sequence to be $\{\alpha_t^2\}, t = 1, 2, \ldots$, then at each time t, the human will get an instantaneous utility $u_t = u_t(\alpha_t^1, \alpha_t^2) \in \{w_1, w_2, w_3, w_4\}$.

Define the averaged utility of the human over the infinite time to be

$$U_\infty = \limsup_{T\to\infty} \frac{\sum_{t=1}^T u_t(\alpha_t^1, \alpha_t^2)}{T}, \tag{3}$$

which always exists.

In [1] and [2], by assuming the complete and perfect information, we have solved the optimal strategy of the human player to get the optimal U_∞. Now we assume that the human does not know the initial assessment $(\kappa_0(a), \kappa_0(b))$, then how should the human do in order to get a bigger U_∞?

One natural idea for the human is to identify the initial parameter $(\kappa_0(a), \kappa_0(b)$. This seems possible since the human can get more information with the system running.

Before stating the related results, like we have done in the previous works, we rewrite the bi-matrix into a new one:

Player 1	Player 2	
	a	b
A	$q_3+\Delta_1, w_1$	q_2, w_2
B	q_3, w_3	$q_2+\Delta_2, w_4$

where $\Delta_1 > 0, \Delta_2 > 0$.

Then strategy updating rule of the machine is rewritten to be

$$\alpha_t^1 = \begin{cases} A, & if\, \Delta_1 \cdot \kappa_t(a) \geq \Delta_2 \cdot \kappa_t(b); \\ B, & otherwise. \end{cases}$$

3 The Identification for Parameters in the FP Algorithm

In [1] and [2], we have investigated the dynamical game systems in which the machine uses the FP algorithm to update its actions. We showed that the human's optimal strategy depends on the ratio $\frac{\Delta_2}{\Delta_1}$ and the long-run behavior of the system depends on $\frac{\Delta_2}{\Delta_1}$ being rational or irrational too.

By the explicit form of the human's optimal strategy [1,2], it is independent of the specific values of $\kappa_0(a)$ and $\kappa_0(b)$ but is solely dependent on the parameter

$$K_t = \Delta_1 \cdot \kappa_t(a) - \Delta_2 \cdot \kappa_t(b)$$

which can be computed by the initial assessment $(\kappa_0(a), \kappa_0(b))$ and the realized actions α_t^1, α_t^2. For example, if $w_2 > w_3 > max\{w_1, w_4\}$, the optimal strategy of the human is just the naive/myopic best response of the machine's action which can correctly predicted by the human. Denote the prediction of the human for the machine's action to be $\tilde{\alpha}_t^1$.

Now, assume that the machine's initial assessment $(\kappa_0(a), \kappa_0(b))$ is unknown to the human. Then for the human it is enough to identify the initial assessment parameter

$$K = \Delta_1 \cdot \kappa_0(a) - \Delta_2 \cdot \kappa_0(b)$$

in order to get his optimal strategy. This offers great convenience to the human compared to determining the precise values of $\kappa_0(a)$ and $\kappa_0(b)$.

On the other hand, according to Eq. (6), the machine takes actions according to an inequality of $(\kappa_0(a) + X_t(a), \kappa_0(b) + X_t(b))$, where $X_t(a)$ and $X_t(b)$ are the numbers of times at which the human player takes action a and action b up to time t (not included). Thus it is possible for the human to infer the feasible set of K from the machine's action.

Next we will give an algorithm to identify K. Since the system behavior is very different for rational and irrational $\frac{\Delta_2}{\Delta_1}$, we will also study the identification for rational and irrational $\frac{\Delta_2}{\Delta_1}$ respectively.

3.1 The Identification Algorithm for Assessment Parameter K

Now the goal of the human is to determine the value of the parameter $K = \Delta_1 \cdot \kappa_0(a) - \Delta_2 \cdot \kappa_0(b)$. We will take the game with the relationship $w_2 > w_3 > max\{w_1, w_4\}$ as a typical case to state the identification results. However, it is easy to see that the other cases share the same idea.

Denote the estimation of K by $\tilde{K}$. In the following, the estimation at each time in the identification process is denoted by $\tilde{K}_t$, and the subscript t is omitted when it does not lead to misunderstanding.

We give an identification algorithm as below:

Algorithm 1. Identify initial assessment parameter K

```
1: function F(κ0(a), κ0(b), M)      ▷ Identify the initial evaluation under the game
   matrix M
2:     K ← Δ1 · κ0(a) − Δ2 · κ0(b)
3:     initial K̃
4:     for t do
5:         α̃1(t) = BS(K̃, M, t, h)          ▷ BS is the optimal strategy in [1] and [2]
6:         α1(t) = BS(K, M, t, h)
7:         if α̃1(t) ≠ α1(t) then                              ▷ wrong forecast
8:             Update K̃  ▷ Update to a value that ensures the correct action before
9:         end if
10:        Update h                        ▷ Update history action sequence
11:    end for
12:    return K̃                          ▷ Output the final identification result
13: end function
```

By the identification algorithm, the estimation $\tilde{K}$ is updated as follows:

$$\tilde{K}_{t+1} = \begin{cases} \epsilon - \Delta_1 \cdot X_t(a) + \Delta_2 \cdot X_t(b), & if\, \alpha_t^1 = A, \tilde{\alpha}_t^1 = B, \\ -\epsilon - \Delta_1 \cdot X_t(a) + \Delta_2 \cdot X_t(b), & if\, \alpha_t^1 = B, \tilde{\alpha}_t^1 = A. \end{cases}$$

where $\epsilon > 0$ is small enough.

Then we have

Theorem 1. *Assume that $\frac{\Delta_2}{\Delta_1}$ is a rational number and the human player adopts the identification Algorithm 1 above. Then, for any initial identification value $\tilde{K}_0$, there exists a finite time t_f such that for all $t \geq t_f$, $\tilde{\alpha}_t^1 \equiv \alpha_t^1$, i.e., the human player can predict the machine's action correctly after t_f.*

Proof. When the human adopts the identification Algorithm 1, the evolution path of the system is definite, that is, the sequence $\{\alpha_t^1\}$, $\{\tilde{\alpha}_t^1\}$, $\{\alpha_t^2\}$, $X_t(a)$, $X_t(b)$ are determined. Denote $\eta_1 = \min_t\{K + f_t : K + f_t \geq 0\}$, $\eta_2 = \max_t\{K + f_t : K + f_t < 0\}$, where $f_t = \Delta_1 \cdot X_t(a) - \Delta_2 \cdot X_t(b)$).

We will prove this theorem in three steps.

Step 1: First, we prove that when $\tilde{K} \in [K - \eta_1, K + \eta_2)$, the human's prediction of the machine's action $\tilde{\alpha}_1^t$ can always be consistent with player 1's action α_1^t, i.e.,, $\tilde{\alpha}_t^1 \equiv \alpha_t^1$.

In this case, if $K + f_t \geq 0$, then $\tilde{K} + f_t \geq K - \eta_1 + f_t = K + f_t - \eta_1$, then from the definition of η_1 $\tilde{K} + f_t \geq 0$.

If $K + f_t < 0$, then $\tilde{K} + f_t < K + \eta_2 + f_t = K + f_t - \eta_2$, then from the definition of η_2, $\tilde{K} + f_t < 0$.

Step 2: Next, we prove that for any initial $\tilde{K}_0$, $\tilde{K}_t$ will enter $[K - \eta_1, K + \eta_2)$, and stop updating.

For the initial $\tilde{K}_0$, if $\tilde{K}_0 \in [K - \eta_1, K + \eta_2)$, then from the previous step, $\tilde{\alpha}_t^1 \equiv \alpha_t^1$, $\forall t \geq 0$. That is, the human will always predict correctly, so $\tilde{K}_t$ stops updating.

Suppose the human made a wrong prediction at time t_0, i.e., $\tilde{\alpha}_{t_0}^1 \neq \alpha_{t_0}^1$. Without loss of generality, we can set $\alpha_{t_0}^1 = B$, $\tilde{\alpha}_{t_0}^1 = A$, which corresponds to $K + \Delta_1 \cdot X_{t_0}(a) - \Delta_2 \cdot X_{t_0}(b)) < 0$ and $\tilde{K}_{t_0} + \Delta_1 \cdot X_{t_0}(a) - \Delta_2 \cdot X_{t_0}(b)) \geq 0$.

Then according to the identification Algorithm 1,

$$\tilde{K}_{t_0+1} = -\epsilon - \Delta_1 \cdot X_{t_0}(a) + \Delta_2 \cdot X_{t_0}(b) = -\epsilon - f_{t_0}.$$

Then there are three situations to be discussed below.

(1) If $\tilde{\alpha}_t^1 \equiv \alpha_t^1, \forall t \geq t_0 + 1$, i.e., the human has been predicting correctly after t_0, then the estimation stops updating.
(2) If the human still predicts wrongly after t_0, then denote $t_1 = \arg\min_{t \geq t_0+1}\{\tilde{\alpha}_t^1 \neq \alpha_t^1\}$ to be the time of the next mistake. Then, in this case, at time $t_0+1, t_0+2, \ldots, t_1$, the human will not update the estimation, i.e., $\tilde{K}_{t_0+1} = \tilde{K}_{t_0+2} = \cdots = \tilde{K}_{t_1}$.

(2.1) If $\alpha_{t_1}^1 = B$ and $\tilde{\alpha}_{t_1}^1 = A$, which means $K + f_{t_1} < 0$ and $\tilde{K}_{t_1} + f_{t_1} \geq 0$, according to the identification Algorithm 1, $\tilde{K}_{t_1+1} = -\epsilon - f_{t_1}$.

In this case, we first prove that $f_{t_1} > f_{t_0}$. If not, then

$$\tilde{K}_{t_1} + f_{t_1} = \tilde{K}_{t_0+1} + f_{t_1} \leq \tilde{K}_{t_0+1} + f_{t_0} = -\epsilon < 0,$$

contradicts with $\tilde{K}_{t_1} + f_{t_1} \geq 0$. So it must hold $f_{t_1} > f_{t_0}$.

From $f_{t_1} > f_{t_0}$,

$$\tilde{K}_{t_1+1} = -\epsilon - f_{t_1} < -\epsilon - f_{t_0} = \tilde{K}_{t_0+1}.$$

And by calculation,

$$\tilde{K}_{t_0+1} - \tilde{K}_{t_1+1} = -f_{t_0} + f_{t_1} = -\Delta_1 \cdot (X_{t_0}(a) - X_{t_1}(a)) + \Delta_2 \cdot (X_{t_0}(b) - X_{t_1}(b)).$$

Define $\eta = \min_{m,n \in \mathrm{N}+}\{|\Delta_1 \cdot m - \Delta_2 \cdot n| > 0\}$. By the rationality of $\frac{\Delta_2}{\Delta_1}$, η is a positive constant. So $\tilde{K}_{t_1+1} - \tilde{K}_{t_0+1} \leq -\eta$. That is, when the estimation $\tilde{K}$ is larger than K, the updated estimation will be smaller than the previous estimation by at least a positive constant.

(2.2) Assume that $\alpha_{t_1}^1 = A$ and $\tilde{\alpha}_{t_1}^1 = B$, that is, the prediction mistake at time t_1 is different from the prediction mistake at time t_0 and assume that there exists a future time $t_j, j \geq 2$ at which the prediction mistake is the same with the time t_0. Denote $t_j = \arg\min_{t \geq t_0+1}\{\alpha_t^1 = B, \tilde{\alpha}_t^1 = A\}$.

In this case, first of all, it holds that $K + f_{t_{j-1}} \geq 0$, i.e., $K \geq -f_{t_{j-1}}$. According to the identification Algorithm 1, $\tilde{K}_{t_{j-1}+1} = \epsilon - f_{t_{j-1}}$.

Meanwhile, at t_0, $\alpha_{t_0}^1 = B$, which requires $K + f_{t_0} < 0$, so $K < -f_{t_0}$. Thus we get $-f_{t_{j-1}} < -f_{t_0}$, i.e.,, $f_{t_{j-1}} > f_{t_0}$. By the definition of η, it must hold $f_{t_{j-1}} \geq f_{t_0} + \eta$.

Since ϵ is small enough, $f_{t_{j-1}} > 2\epsilon + f_{t_0}$. Hence we have $\epsilon - f_{t_{j-1}} < -\epsilon - f_{t_0}$, that is, $\tilde{K}_{t_{j-1}+1} < \tilde{K}_{t_0+1}$.

On the other hand, at time t_j, $\alpha_t^1 = B$, $\tilde{\alpha}_t^1 = A$, which requires $K + f_{t_j} < 0$, $\tilde{K}_{t_j} + f_{t_j} \geq 0$. According to the identification Algorithm 1, it holds $\tilde{K}_{t_j+1} = -\epsilon - f_{t_j}$.

Now we aim to prove $f_{t_j} > f_{t_0}$. If not, it holds $f_{t_j} \leq f_{t_0}$. Then according to the inequality $\tilde{K}_{t_{j-1}+1} < \tilde{K}_{t_0+1}$ we just proved and the updating formula of $\tilde{K}_{t_0+1}$, we have

$$\tilde{K}_{t_j} + f_{t_j} = \tilde{K}_{t_{j-1}+1} + f_{t_j} < \tilde{K}_{t_0+1} + f_{t_0} = -\epsilon < 0,$$

which contradicts with $\tilde{K}_{t_j} + f_{t_j} \geq 0$. So we prove that $f_{t_j} > f_{t_0}$.

From $f_{t_j} > f_{t_0}$, according to the updating formula of $\tilde{K}_{t_j+1}$,

$$\tilde{K}_{t_j+1} = -\epsilon - f_{t_j} < -\epsilon - f_{t_0} = \tilde{K}_{t_0+1},$$

i.e., it holds $\tilde{K}_{t_j+1} < \tilde{K}_{t_0+1}$.

Then by the same way with in (2.1), we get $\tilde{K}_{t_0+1} - \tilde{K}_{t_j+1} \geq \eta$, which means that if the estimation value of the human is larger than K for more than once, i.e., the human will make the same prediction mistake at least twice, then there is a good property between the adjacently updated estimates, i.e.,, the updated estimation is smaller than the previous estimation by at least a positive constant.

(2.3) If $\alpha_{t_1}^1 = A$ and $\tilde{\alpha}_{t_1}^1 = B$, and there is no time $t_p > t_1$ making $\alpha_{t_p}^1 = B$, $\tilde{\alpha}_{t_p}^1 = A$, i.e., the human will never make the same mistake as at time t_0 after time t_0, then it is only necessary to analyze the prediction mistakes corresponding to $\alpha_t^1 = A$ and $\tilde{\alpha}_t^1 = B$. And this analysis is symmetric with the above.

To sum up, if the estimation is not "sufficiently" correct, then the human must make a mistake at some time, so the updated value of $\tilde{K}_t$ will move towards the correct direction at a speed greater than a positive constant until the estimation is "sufficiently" correct. And then the human will never make mistakes. Obviously this process will end after only a finite time.

That proves the theorem. ∎

Remark 1: When $t \geq t_f$, although the human's prediction is always correct, $\tilde{K}$ and K can still be different. This is because that the FP algorithm only requires the inequality of K holds.

Remark 2: Through the proof of Theorem 1, it can be computed that after at most $\lceil \frac{\tilde{K}_{max} - \tilde{K}_{min}}{\eta} \rceil + 2$ updates, $\tilde{K}_t$ will be "sufficiently" correct. If the evaluation range $[\tilde{K}_{min}, \tilde{K}_{max}]$ about K is given at the initial time, then the initial estimation $\tilde{K}_0$ can be set to be any number in the interval.

Below we will give an example to illustrate how the identification is carried out.

Consider the following game:

Player 1	Player 2	
	a	b
A	2, 2	1, 5
B	0, 4	4, 3

where $\Delta_1 = 2, \Delta_2 = 3$.

Assuming the machine's initial assessment of the human is $\kappa_0(a) = 5$, $\kappa_0(b) = 4$, then $K = \Delta_1 \cdot \kappa_0(a) - \Delta_2 \cdot \kappa_0(b) = -2$. Let the initial estimation be $\tilde{K}_0 = 1$ and take $\epsilon = 0.1$. Then under the identification Algorithm 1, the evolution of the game system is as follows in Table 1, where the values in () represent the values of K and $\tilde{K}$ respectively at the current moment.

Table 1. The evolution process of system parameters under the Algorithm 1

t	$K + \Delta_1 X_t(a) - \Delta_2 X_t(b)$	$\tilde{K}_t + \Delta_1 X_t(a) - \Delta_2 X_t(b)$	α_t^1	$\tilde{\alpha}_t^1$	α_t^2
0	$(-2) + 0$	$(1) + 0$	B	A	b
1	$(-2) - 3$	$(-0.1) - 3$	B	B	a
2	$(-2) - 1$	$(-0.1) - 1$	B	B	a
3	$(-2) + 1$	$(-0.1) + 1$	B	A	b
4	$(-2) - 2$	$(-1.1) - 2$	B	B	a
5	$(-2) + 0$	$(-1.1) + 0$	B	B	a
6	$(-2) + 2$	$(-1.1) + 2$	A	A	b
7	$(-2) - 1$	$(-1.1) - 1$	B	B	a
8	$(-2) + 1$	$(-1.1) + 1$	B	B	a
9	$(-2) + 3$	$(-1.1) + 3$	A	A	b
10	$(-2) + 0$	$(-1.1) + 0$	B	B	a
11	$(-2) + 2$	$(-1.1) + 2$	A	A	b
12	$(-2) - 1$	$(-1.1) - 1$	B	B	a
13	$(-2) + 1$	$(-1.1) + 1$	B	B	a
14	$(-2) + 3$	$(-1.1) + 3$	A	A	b

The results in Table 1 are represented by Figs. 1 and 2 as follows.

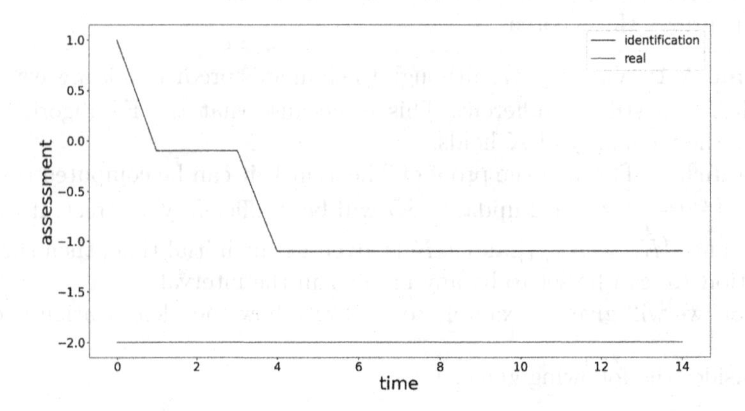

Fig. 1. The change of $\tilde{K}$ along time t

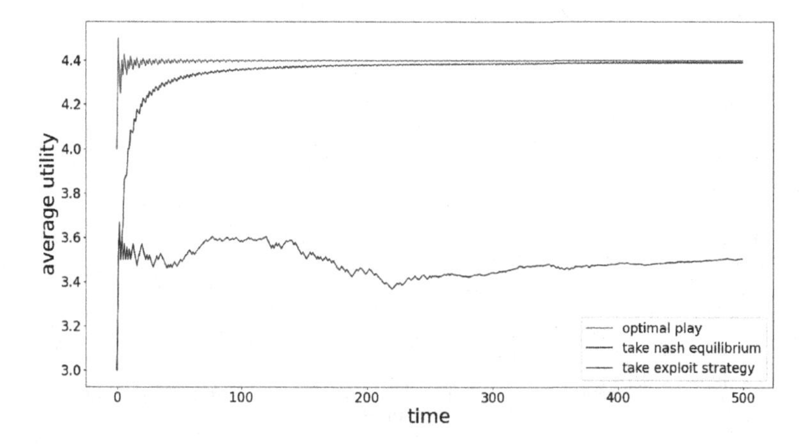

Fig. 2. The human's averaged utility along time t

As can be seen from the Table 1 and Figs. 1 and 2, if the prediction of the machine's action at time t is wrong, the estimation $\tilde{K}$ will be updated at time $t+1$. In this example, after $t \geq 5$, the predictions are accurate, implying that the estimation is "sufficiently" accurate. By "sufficient" accuracy, we find that the precise identification of $\tilde{K}$ is not necessary and it suffices for $\tilde{K}$ to approximate the value of K "closely".

Theorem 1 gives the result on identification for rational $\frac{\Delta_2}{\Delta_1}$. When $\frac{\Delta_2}{\Delta_1}$ is irrational, the situation will be different as shown in the next subsection.

3.2 The Identification for Irrational $\frac{\Delta_2}{\Delta_1}$

For the irrational $\frac{\Delta_2}{\Delta_1}$, consider the following game matrix:

Player 1	Player 2	
	a	b
A	$\sqrt{2}$, 2	1, 5
B	0, 4	4, 3

where $\Delta_1 = \sqrt{2}$, $\Delta_2 = 3$.

Assume Player 1's initial assessment of Player 2 is $\kappa_0(a) = 5$, $\kappa_0(b) = 4$, then $K = \Delta_1 \cdot \kappa_0(a) - \Delta_2 \cdot \kappa_0(b) = -2$. Let the initial estimate be $\tilde{K}_0 = 1$, take $\epsilon = 10^{-9}$.

Then the estimation $\tilde{K}_t$ changes with time as shown in Fig. 3 below where the vertical ordinate being 1 means that the estimation is updated at this moment.

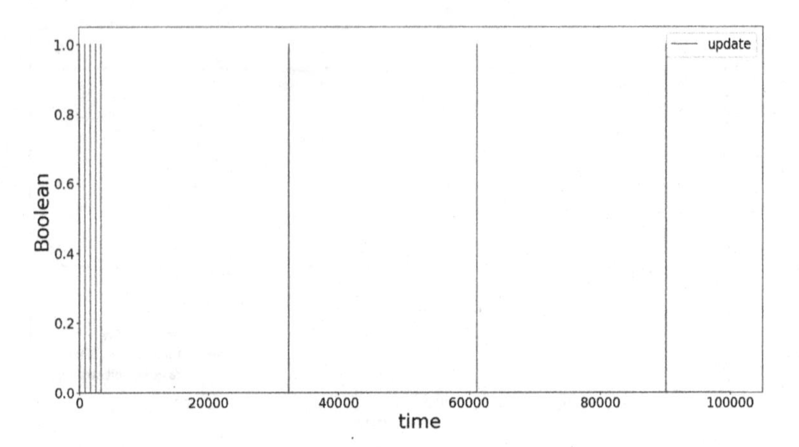

Fig. 3. Estimated value $\tilde{K}_t$ over time

From Fig. 3, we can see that for the irrational $\frac{\Delta_2}{\Delta_1}$, the identification will not end in any finite time. Along with time, the estimation error becomes more and more smaller. However, since in the FP algorithm, the parameter K is irrational, the inequality will never stop changing its signs. Thus for almost all the initial estimation and games (with measure 1), the identification process will never stop and the human will never get the total prediction of the machine. This might help us to design better algorithms.

4 Conclusions and Future Work

In this paper, we considered the repeated human-machine games where the machine uses the Fictitious Play algorithm to update its action at eat time. In the previous works [1] and [2], we solved the optimal strategy of the human player against the machine under assumption of complete and perfect information. In this paper, we assume that the human player does not know the initial assessment of the machine player and consider the identification problem of the human. We propose an identification algorithm for the human player and prove that the identification can end successfully in a finite time if the machine's payoff parameter is rational. When the machine's payoff parameter is irrational, the identification process will not end, which implies some advantage for the algorithm with irrational parameters. The results in this paper are rigorous and might shed some light on general games and algorithms.

This paper can be regarded as the first step to solve the problem of algorithm identification, which is a necessary part to exploit an algorithm in repeated games in the future application of some AI. The algorithm can also be regarded as an approximation of the real human behavior, thus the algorithm identification is the necessary step to find the pattern of the opponent's behavior. We will leave these general problems as future work.

Acknowledgement. This work was supported by the National Key Research and Development Program of China under grant No.2022YFA1004600, the Natural Science Foundation of China under Grant T2293770, the Strategic Priority Research Program of Chinese Academy of Sciences under Grant No. XDA27000000, the Major Project on New Generation of Artificial Intelligence from the Ministry of Science and Technology (MOST) of China under Grant No. 2018AAA0101002.

References

1. Dong, H., Mu, Y.: The optimal strategy against fictitious Play in infinitely repeated games. In: Proceedings of the 41st Chinese Control Conference, pp. 6852–6857 (2022)
2. Dong, H., Mu, Y.F.: The optimal strategy against the opponent adopting fictitious play algorithm in infinitely repeated 2×2 games. SSRN Electron. J. (2022). https://doi.org/10.2139/ssrn.4201849
3. Silver, D., Huang, A., Maddison, C.J., et al.: Mastering the game of Go with deep neural networks and tree search. Nature **529**(7587), 484–489 (2016)
4. Silver, D., Hunert, T., Schrittwieser, J., et al.: A general reinforcement learning algorithm that masters chess, shogi, and go through self-play. Science **362**(6419), 1140–1144 (2018)
5. Moravčík, M., Mchmid, M., Burch, N., et al.: DeepStack: expert-level artificial intelligence in heads-up no-limit poker. Science **356**(6337), 508–513 (2017)
6. Brown, N., Sandholm, T.: Superhuman AI for heads-up no-limit poker: libratus beats top professionals. Science **359**(6374), 418–424 (2017)
7. Brown, N., Sandholm, T.: Superhuman AI for multiplayer poker. Science **365**(6456), 885–890 (2019)
8. Vinyals, O., Babuschkin, I., Czarnecki, W.M., et al.: Grandmaster level in StarCraft II using multi-agent reinforcement learning. Nature **575**(7782), 350–354 (2019)
9. Bouzy, B., Métivier, M., Pellier, D.: Hedging algorithms and repeated matrix games. arXiv preprint arXiv:1810.06443 (2018)
10. Brown, G.W.: Some Notes on Computation of Games Solutions. RAND Corp., Santa Monica (1949)
11. Robinson, J.: An iterative method of solving a game. Ann. Math., 296–301 (1951)
12. Monderer, D., Sela, A.: A 2 × 2 game without the fictitious play property. Games Econ. Behav. **14**(1), 144–148 (1996)
13. Monderer, D., Shapley, L.S.: Fictitious play property for games with identical interests. J. Econ. Theory **68**(1), 258–265 (1996)
14. Christian, E., Valkanova, K.: Fictitious play in networks. Games Econ. Behav. **123**, 182–206 (2020)
15. Fudenberg, D., Drew, F., Levine, D.K., et al.: The Theory of Learning in Games. MIT press, Cambridge (1998)
16. Yuan, S., Guo, L.: Stochastic adaptive dynamical games. Sci China Math **46**, 1367–1382 (2016)
17. Hu, H.Y., Guo, L.: Non-cooperative stochastic adaptive multi-player games. Control Theory Appl. **35**(5) (2018)
18. Littman, M.L.: Markov games as a framework for multi-agent reinforcement learning. In: Machine Learning Proceedings, pp. 157–163. Morgan Kaufmann (1994)
19. Hu, J., Wellman, M.P.: Nash Q-learning for general-sum stochastic games. J. Mach. Learn. Res. **4**, 1039–1069 (2003)

20. Szegedy, C., et al.: Intriguing properties of neural networks. In: Proceedings of the International Conference on Learning Representations (2014)
21. Nguyen, A., et al.: Deep neural networks are easily fooled: high confidence predictions for unrecognizable images. In: Proceedings of the IEEE Conference on Computer Vision and Pattern Recognition, pp. 427–436 (2015)
22. Finlayson, S.G., et al.: Adversarial attacks on medical machine learning. Science **363**(6433), 1287–1289 (2019)
23. Sharif, M., et al.: Accessorize to a crime: real and stealthy attacks on state-of-the-art face recognition. In: Proceedings of the 2016 ACM SIGSAC Conference on Computer and Communications Security, pp. 1528–1540 (2016)
24. Shamma, J.S.: Game theory, learning, and control systems. Natl. Sci. Rev. **7**(7), 1118–1119 (2020)
25. Cao, M.: Merging game theory and control theory in the era of AI and autonomy. Natl. Sci. Rev. **7**(7), 1122–1124 (2020)
26. Mu, Y., Guo, L.: Towards a theory of game-based non-equilibrium control systems. J. Syst. Sci. Complex. **25**(2), 209–226 (2012)
27. Mu, Y., Guo, L.: Optimization and identification in a non-equilibrium dynamic game. In: The 48th IEEE Conference on Decision and Control, Shanghai, China, pp. 5750–5755 (2009)
28. Deng, X., et al.: Exploiting a no-regret opponent in repeated zero-sum games, personal communication
29. Tang, Z., Zhu, Y., Zhao, D., et al.: Enhanced rolling horizon evolution algorithm with opponent model learning. IEEE Trans. Games (2020)
30. Deng, Y., Schneider, J., Sivan, B.: Strategizing against no-regret learners. In: Advances in Neural Information Processing Systems, vol. 32 (2019)
31. Ganzfried, S., Sandholm, T.: Safe opponent exploitation. ACM Trans. Econ. Comput. **3**(2), 1–28 (2015)
32. Ganzfried, S., Sun, Q.: Bayesian opponent exploitation in imperfect-information games. In: 2018 IEEE Conference on Computational Intelligence and Games, pp. 1–8. IEEE (2018)
33. Brown, G.W.: Iterative solution of games by fictitious play. Act. Anal. Prod. Allocat. **13**(1), 374–376 (1951)
34. Miyasawa, K.: On the convergence of learning processes in a 2×2 non-zero-person game, Technical Report Research Memorandum No. 33, Econometric Research Program, Princeton University
35. Sayin, M.O., Parise, F., Ozdaglar, A.: Fictitious play in zero-sum stochastic games. SIAM J. Control. Optim. **60**(4), 2095–2114 (2022)
36. Shapley, L.: Some topics in two-person games. Adv. Game Theory **52**, 1–29 (1964)
37. Fudenberg, D., Kreps, D.M.: Learning mixed equilibria. Games Econ. Behav. **5**(3), 320–367 (1993)
38. Van der Genugten, B.: A weakened form of fictitious play in two-person zero-sum games. Int. Game Theory Rev. **2**(04), 307–328 (2000)
39. Leslie, D.S., Collins, E.J.: Generalised weakened fictitious play. Games Econ. Behav. **56**(2), 285–298 (2006)
40. Heinrich, J., Lanctot, M., Silver, D.: Fictitious self-play in extensive-form games. In: International Conference on Machine Learning. PMLR (2015)
41. Heinrich, J., Silver, D.: Deep reinforcement learning from self-play in imperfect-information games. arXiv:1603.01121 (2016)

An Improved Hypervolume-Based Evolutionary Algorithm for Many-Objective Optimization

Chengxin Wen[1,2], Lihua Li[1], and Hongbin Ma[1,2(✉)]

[1] School of Automation, Beijing Institute of Technology, Beijing 100081, China
{3120205445,hua}@bit.edu.cn
[2] National Key Lab of Autonomous Intelligent Unmanned Systems, Beijing Institute of Technology, Beijing 100081, China
mathmhb@bit.edu.cn

Abstract. The hypervolume indicator is commonly utilized in indicator-based evolutionary algorithms due to its strict adherence to the Pareto domination relationship. However, its high computational complexity in high-dimensional objective spaces limits its widespread adoption and application. In this paper, we propose a fast and efficient method for approximating the overall hypervolume to overcome this challenge. We then integrate this method into the basic evolutionary computation framework, forming an algorithm for solving many-objective optimization problems. To evaluate its performance, we compared our proposed algorithm with six state-of-the-art algorithms on WFG and DTLZ test problems with 3, 5, 10, and 15 objectives. The results demonstrate that our proposed method is highly competitive in most cases.

Keywords: Evolutionary algorithms · Overall hypervolume approximation · Many-objective optimization

1 Introduction

The purpose of optimization is to find the optimal value of a function. The problem becomes complicated when there are multiple objectives for optimization, which are called multi-objective optimization problems (MOPs). For MOPs, the ultimate goal is to solve the Pareto Front (PF) of the entire problem, and the solutions in PF are mutually non-dominant. When there are more than three optimization objectives in the problem, MOPs are generally called Many-object problems (MaOPs) [8]. In solving MaOPs, evolutionary multi-objective optimization algorithms (EMOAs) have achieved good results [12]. EMOAs can be

This work was partially funded by the National Key Research and Development Plan of China (No. 2018AAA0101000) and the National Natural Science Foundation of China under grant 62076028.

B. Xin et al. (Eds.): IWACIII 2023, CCIS 1931, pp. 283–297, 2024.
https://doi.org/10.1007/978-981-99-7590-7_23

divided into three categories according to their different mechanisms: 1) Pareto-based, 2) Decomposition-based, and 3) Indicator-based.

Pareto-based EMOAs use the Pareto dominant relationship to calculate individual fitness, so as to achieve the purpose of environmental selection. For example, NSGA-II [5] uses the distance value of adjacent individuals in objective space to calculate the crowding distance, and SPEA2 [21] calculates the K-order nearest neighbor distance. Pareto-based EMOAs have been well applied in the optimization of low-dimensional space due to their simple calculation method. However, Paretobased EMOAs are not very effective in solving MaOP problems. Because as the number of optimization objects increases, the proportion of non-dominant individuals in the population also increases, resulting in Pareto-based environmental selection becoming extremely inefficient.

Decomposition-based EMOAs use decomposition strategies to decompose a MOPs into several scalar optimization subproblems, and optimize them simultaneously in a collaborative manner. For example, MOEA/D [19] uses weighted sum, Chebyshev, and PBI to decompose MOPs. Finally, the convergence of the population is judged by the scalar function criterion. However, the weight vector has a strong influence on the decomposition-based EMOAs. When the Pareto front has an irregular shape, running the algorithm with a fixed weight vector will cause the population to be unevenly distributed on the Pareto front [11]. Therefore, for decomposition-based EMOAs, the weight vector adjustment during algorithm operation is still the current research focus [1,15].

Indicator-based EMOAs use various indicators as the criteria for environmental selection of populations, and reflect the diversity and convergence of populations as the value of the indicators at the same time. For example, IGD [4] evaluates the convergence performance and diversity performance of the algorithm by calculating the sum of the minimum distances between each individual on the real Pareto front surface and the collection of individuals obtained by the algorithm. Hypervolume [22] evaluates the population quality by calculating the Lebesgue measure of the area enclosed by the population individual and the reference point. Generally speaking, the setting of indicator parameters will affect the size of the indicator to a large extent, such as the reference set in IGD and the reference point of hypervolume. By setting the parameters reasonably, the indicator-based EMOAs can achieve better performance.

Among the many population evaluation indicators, hypervolume is adopted by most indicators-based EMOAs (for example, SMS-EMOA [3,7], MOPSOhv [9] and HypE [2]) because it strictly follows the Pareto principle. The main idea of hypervolume-based EMOAs is to transform MOPs into single-objective problem by maximizing the hypervolume of population. However, for hypervolume-based environmental selection, most algorithms use the following steps:

- First, the hypervolume contribution of each solution in the solution set is computed.
- Then, the solution with the smallest hypervolume contribution is discarded.
- Finally, judge whether the size of the solution set meets the requirements, if not, return to the first step.

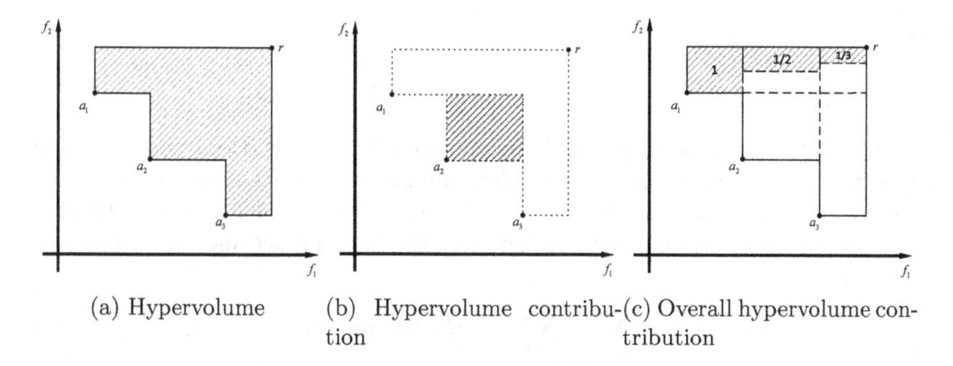

(a) Hypervolume (b) Hypervolume contribution (c) Overall hypervolume contribution

Fig. 1. Illustration of Basic Concepts

This method is called a greedy algorithm. For a solution set of size 2N, if N solutions are retained through environmental selection, it is necessary to calculate the hypervolume contribution of the solution set N times. Although a variety of hypervolume approximation methods have been proposed (such as IWFG [15], R2HCA [16], etc.), each environmental selection needs to run the approximation algorithm multiple times, which causes the algorithm to be time-consuming. To solve this problem, this paper designs an environment selection mechanism to avoid multiple calculations by screening the overall hypervolume of each solution, designs a new method to approximate the overall hypervolume contribution, and finally integrates this method into an EMOA framework. In the experimental part, we compared our method with two hypervolume-based EMOAs and four advanced EMOAs, tested on the DTLZ and WFG test suits of 3, 5, 8, 10, and 15 objectives.

The rest of this article is organized as follows. In Sect. 2, we briefly review some basic concepts. The details of the proposed algorithm are described in Sect. 3. In Sect. 4, the experiment content is introduced. In Sect. 5, the experimental results are presented and analyzed. We conclude this paper in Sect. 6.

2 Preliminary Concepts

2.1 Hypervolume

Consider a non-dominated point set S in the target space. We assume that they are all non-overlapping, then the hypervolume of this solution set is computed as follows.

$$HV(S, r) = \mathcal{L}(\bigcup_{s \in S} \{b | s \prec b \prec r\}), \tag{1}$$

where $\mathcal{L}(\dots)$ is the Lebesgue measure of a set. Another important concept is the hypervolume contribution, which reflects how well each solution contributes to the entire solution set. The hypervolume contribution of solution **s** in solution set S is expressed as

$$HVC(\mathbf{s}, S, r) = HV(S, r) - HV(S/\{\mathbf{a}\}, r), \tag{2}$$

The last concept is the overall hypervolume contribution. Because the general hypervolume contribution only considers the measure of the area dominated by the current solution alone but ignores the area dominated by other solutions. The overall hypervolume contribution considers this impact and is expressed as follows.

$$I_h(\mathbf{s}, S, r) = \sum_{i=1}^{|S|} \sum_{\mathbf{s} \in \mathcal{A}, |\mathcal{A}|=i} \frac{1}{i} HVC(worse(\mathcal{A}), S/\mathcal{A}, r). \tag{3}$$

where $worse(\ldots)$ represents the worst value in each dimension in objective point set. For example, a minimization problem represents the maximum objective value in each dimension, while a maximization problem does the opposite. Figure 1 shows the geometric interpretation of the three concepts respectively.

2.2 R2-Based Hypervolume Contribution Approximation

The basic idea of the R2-based hypervolume contribution approximation(R_2^{HCA}) method is to use different line segments to estimate the hypervolume contribution only in the exclusive dominanced area. This method can directly approximate the hypervolume contribution and use all the direction vectors of the hypervolume contribution area. Compared with the Monte Carlo sampling method, the R_2^{HCA} further reduces the time complexity of the hypervolume calculation, and has achieved a significant improvement in the estimation accuracy. The literature shows that the calculation time of the R_2^{HCA} only increases linearly with the increase of the number of objectives, and when the number of direction vectors does not change, as the number of objects increases, the accuracy of the hypervolume estimation decreases slowly. However, the current R_2^{HCA} only estimates the exclusive hypervolume indicator value of each solution. This is feasible under the $(\mu+1)$ strategy EMOAs, but applied to the $(\mu+\mu')$ strategy EMOAs It may lead to eliminating individual sets of high-combination hypervolume indicators in the environmental selection stage.

3 Proposed Method

3.1 General Framework

The general framework of proposed method is roughly same as HypE. Considering that only one offspring in each generation will have a higher probability of premature convergence to the local optimum, we adopts the $(\mu + \mu')$ evolutionary strategy. Algorithm 1 gives the pseudo-code of the whole body framework of the proposed method. In each algorithm generation, the same number of offspring as the parent is produced, then the parent and offspring are integrated into a population. Finally, environmental selection reduces the entire population

Algorithm 1. Proposed Method

Require:
Population Size, N;
Direction Vector Number, V_N;
Maximum Function Evaluations Number, FEs^{max};
Ensure:
Final Population, P;
1: Initialize Population P, Direction Vector Set Λ, reference point $\mathbf{r}$;
2: FEs $\leftarrow$ N;
3: **while** FEs $<$ FEs^{max} **do**
4: $P' \leftarrow \textit{GenerateOffspring}(P)$;
5: $P \leftarrow P \cup P'$;
6: $P \leftarrow \textit{EnvironmentalSelection}(P, N, \Lambda, \mathbf{r})$;
7: FEs $\leftarrow$ FEs + N;
8: **end while**
9: **return** P;

to a prescribed number. The specific steps of environment selection have been explained in the previous section. For the generation of offspring, the proposed method adopts a method that randomly selects two individuals, performs binary simulation crossover and polynomial mutation, generates two offspring individuals, and then repeats the above steps repeatedly until the number of offspring individuals is increased to N.

3.2 Environmental Selection

The environmental selection mechanism in proposed method is different from other hypervolume-based EMOAs. First, perform a non-dominated sort on the solution set and determine the number of solutions that need to be removed (lines 1-2). Then determine the rank number after non-dominated sorting. If it exceeds 1, randomly select DelNum solutions of the last rank with probability *selratio* and delete them (lines 4); otherwise, choose DelNum solutions of the last rank with a minor overall hypervolume contribution and delete them (lines 6). Note that the smaller the *selratio* value, the faster the calculation speed of the algorithm, but the more the convergence speed is weakened. In this article, the value of *selratio* is set to 0.5.

3.3 Overall Hypervolume Contribution Approximation

R2 DataArray Computation. To calculate the overall hypervolume contribution, we improved the hypervolume contribution estimation method proposed in [16] so that all direction vectors can be used to estimate the exclusive area and the common dominant area with other solutions.

The principle of the R_2^{HCA} is shown in Fig. 2a. In the initial stage, a randomly generated set of uniformly distributed unit vectors is utilized. To evaluate the

Algorithm 2. EnvironmentalSelection

Require:
Primitive Population, P;
Population Size, N;
Direction Vector Set, Λ;
Reference Point, $\mathbf{r}$;

Ensure:
Updated Population, P;

1: $\{F_1, F_2, ..., F_l\} \leftarrow NondominatedSort(P, N)$;
2: DelNum $\leftarrow |\{F_1, F_2, ..., F_l\}| - N$;
3: **if** $l \neq 1$ **and** randnum $<$ *selratio* **then**
4: Randomly delete DelNum solutions at the last rank;
5: **else**
6: Delete the DelNum solutions with the smallest overall hypervolume contribution in the last rank;
7: **end if**
8: **return** P;

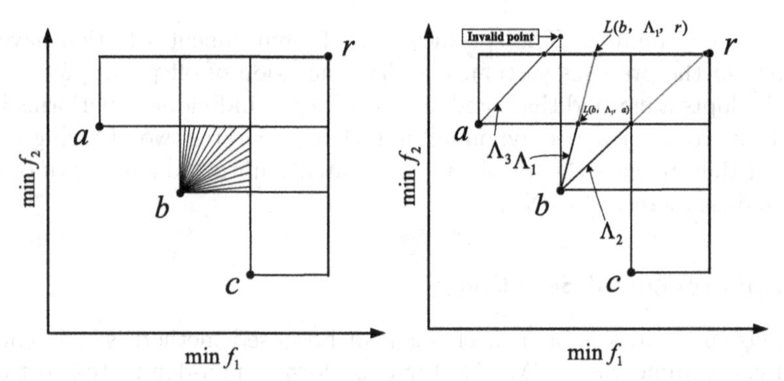

(a) The geometric meaning of R_2^{HCA} (b) Illustration of the mechanism of $Improved - R_2^{HCA}$

Fig. 2. The geometric meaning of concepts in hypervolume contribution approximation

hypervolume contribution of each solution, the length of all line segments is computed by considering the solution as the starting point and projecting into the direction of every vector in the vector set. The line segment stops at the dominant region of the nearest other solution where it intersects, serving as the endpoint. The logarithmic average of the lengths of all line segments is computed based on the objective space dimensionality, and this resulting value is utilized as an estimated hypervolume contribution. Afterward, this estimated contribution is applied to the calculations.

Due to the algorithmic mechanism, R_2^{HCA} can only estimate the exclusive dominant area of each solution. However, based on the idea of using the mean value of line segment distance to estimate hypervolume in R_2^{HCA}, we found that if the directional vector in R_2^{HCA} continues to extend, it will intersect with the dominant regions of other solutions in turn. At each focal point intersection, the number of shared solutions in the region will increase by one. By considering the length of the line segments in each region, it is possible to estimate the hypervolume of the co-dominant region with different numbers of solutions. The use of overall hypervolume calculation for each solution is proposed and implemented in HypE, and this technique has demonstrated an improvement in solution set diversity. $L(b, \Lambda_1, a)$ is used to denote the length of the line segment that starts from solution **b**, moves in the direction of Λ_1, and ends at the intersection of the dominant region of solution **a**.

The specific principle is shown in Fig. 2b. For solution **b**, the vector direction Λ_1 first intersects the dominant region of the solution **a**, then intersects the corresponding line segment of reference point **r**, and finally intersects the dominant region of the solution **c** (not shown in the figure). In the line segment marked along the Λ_1 vector direction in the figure, the black part represents the area dominated by solution **b** alone. After the first intersection, the red line segment denotes the area dominated by two solutions simultaneously, and so on. In Fig. 2b, the line segment in the direction of Λ_2 is a special case. The first part of the line segment represents the area dominated by solution **b** alone, while the latter part represents the area dominated by all three solutions. However, this situation can still be explained using the previous method. In the direction of Λ_2, the line segment passes through the intersection of the dominated region of solutions **a** and **c**, i.e., $L(b, \Lambda_2, a) = L(b, \Lambda_2, c)$, and the point on the line segment is the coincidence of the two points. Therefore, after that point, the number of points dominated by the line segment increases by 2. Therefore, In order to estimate the common dominant area of multiple solutions, we need to modify R_2^{HCA} using the following steps:

- Calculate the distances of all intersection points between the solution of the hypervolume to be estimated and other solutions in the solution set for each direction vector.
- Calculate the intersection distance between the solution of the hypervolume to be estimated and the reference point **r** for each direction vector.
- Sort the intersection distance data of each direction vector in ascending order.

The method of calculating the distance between the intersection of the line along the direction vector and the solution is consistent with the method of calculating g^{*2mth} and g^{mtch}. For any direction vector $\Lambda = \{\lambda_1, \lambda_2, ..., \lambda_m\}$, the $L(s, \Lambda, r)$ function is defined as:

$$L(s, \Lambda, r) = \min_{j \in \{1,...,m\}} \left\{ \frac{|s_j - r_j|}{\lambda_j} \right\}. \tag{4}$$

The $L(a, \Lambda, s)$ function is defined for minimization problems as

$$L(a, \Lambda, s) = \max_{j \in \{1,...,m\}} \left\{ \frac{a_j - s_j}{\lambda_j} \right\}. \tag{5}$$

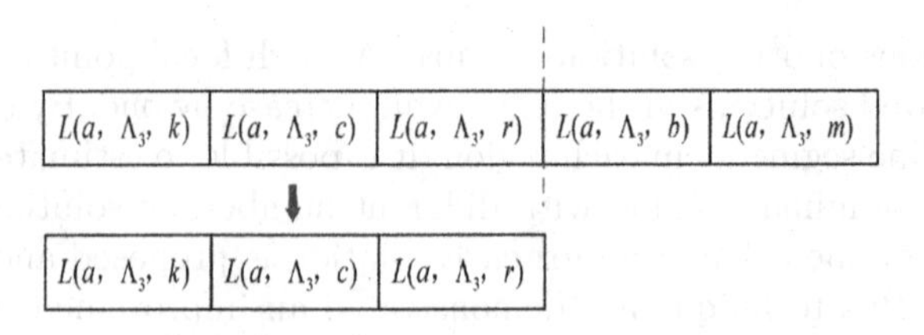

Fig. 3. The modification of R2 DataArray

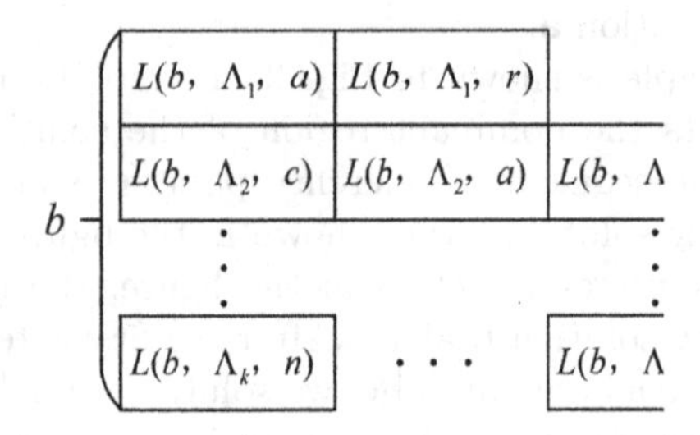

Fig. 4. The storage form of R2 DataArray

DataArray Modification. After the initialization of the R2 array is completed, there may still be invalid data in the array, as shown in Fig. 2b. An obvious problem is the intersection of solution **a** along the Λ_1 direction and solution **c**, which exceeds the calculation area of the hypervolume contribution indicator. This intersection point is invalid.

To determine whether an intersection point is invalid, we can compare the length of the line segment formed by the intersection point and the solution with the critical distance. For any point s along the direction vector Λ, the critical distance is the intersection distance between s and the reference point r. If the length of the line segment is greater than the critical distance, then the intersection point is invalid.

In the context of the R2 indicator, we can discard any intersection points that are deemed invalid and only consider the valid intersection points to avoid distortion of the hypervolume contribution estimate. The specific correction process of the R2 array is illustrated in Fig. 3. For solution **a**, we start with the intersection point along the Λ_3 direction and compare the distance of each subsequent intersection point to the reference point in ascending order. We keep the first intersection point whose distance is less than the critical distance and discard all subsequent points.

The corrected R2 array is shown in Fig. 4. Note that this is an irregular array, as some rows have more columns than others. If an irregular data structure is not supported by a certain programming environment, we can replace the discarded entries with INF.

Numerical Approximation. Once the R2 data array is calculated, we can estimate the hypervolume contribution value using the following formula:

$$HVC = \sum_{i=1}^{N} \frac{a_i}{|\Lambda_i|} \sum_{j=1}^{V_N} l_{ij}^M \tag{6}$$

Here, $|\Lambda_i|$ is the total number of vectors in the area dominated by i solutions, l_{ij} is the length of the line segment in the direction of the j-th vector in the area commonly dominated by i solutions, a_i is the hypervolume contribution coefficient of the area dominated by $i$solutions (the same definition as HypE), and M is the number of objectives.

3.4 Computational Complexity

The computational complexity of one generation of the proposed method is analyzed as follows. The initialization of the R2 data array and the process of estimating hypervolume contribution through the R2 data array occupy most of the algorithm time. In each generation, in the initialization phase of the R2 data array, $O(V_N MN^2)$ calculation is required; the sorting of each row of data in the R2 array requires $O(V_N N^3)$ calculation; in the environment selection phase, non-dominated sorting adopts The T-ENS [20] algorithm has a computational complexity of $O(MNlogN/logM)$. So the computational complexity of environment selection is $O(V_N N^3)$. In general, since m is generally much smaller than N, the computational complexity of the entire the proposed method algorithm is $O(V_N N^3)$.

4 Experimental Design

4.1 Benchmark Problems

To evaluate the performance of the proposed method, we conducted experiments on the DTLZ$\{1-7\}$ [6] and WFG$\{1-9\}$ [10] test sets. Both of these test sets allow for arbitrary adjustment of the number of objectives and decision variables.

Following the suggestions in [6], we selected test problems with PFs of different shapes, including convex, concave, linear, continuous, non-convex, and discontinuous. These different types of PFs can evaluate the performance of the algorithm in all aspects. We set the number of objectives to $m = \{3, 5, 10, 15\}$.

For the number of decision variables in the DTLZ test set, DTLZ1 is $m+4$, DTLZ2-6 is $m+9$, and DTLZ7 is $m+19$. For the WFG test set, based on the recommendation in [10], we set the number of decision variables to $n = k + l$, where the position-related variable $k = 2 * (m - 1)$ and the distance-related variable $l = 20$.

4.2 Performance Metrics

In this paper, we used two evaluation indicators - hypervolume and inverted generational distance - to evaluate the performance of the algorithm. Both of these indicators can evaluate the convergence and diversity of the solution set. The hypervolume indicator strictly follows the Pareto relationship, so it can clearly distinguish the convergence and diversity of different solution sets.

For cases where $m < 10$, we used the WFG algorithm to accurately calculate the value of hypervolume. However, when $m > 10$, the accurate calculation takes too much time, so the Monte Carlo method was used to approximate the value of hypervolume. The true ideal point r^* and the true nadir point r^{nad} were set as $(1, 1, ..., 1)$ and $(0, 0, ..., 0)$ respectively, and each solution in the solution set was normalized. The reference point used when calculating hypervolume was $\mathbf{r} = (1, 1, ..., 1)$.

The Inverted Generational Distance (IGD) [10] is another widely used metric in multi-objective scenarios, providing combined information on convergence and diversity. However, unlike hypervolume, lower IGD values indicate better solution quality. For the calculation of IGD, the selection of the reference set is critical. In our study, we used the problem reference set that came with the platEMO platform, which will be described in detail later.

Furthermore, we used the Wilcoxon rank sum test with a significance level of 0.05 to analyze the test data results of the algorithm, to determine whether one algorithm is statistically significantly different from another, as denoted by "+" (significantly better), "−" (significantly worse), or "≈" (statistically similar) compared to the proposed method.

4.3 Algorithms for Comparison

In the experimental part, two sets of comparison algorithms are adopted. The first group consists of two hypervolume-based EMOAs, namely SMS-EMOA and HypE. The number of sampling points for HypE is set to 10000, which is the same as the setting in [17]. The second group consists of four advanced EMOAs, namely: AR-MOEA [17], SPEA2+SDE [13], GFM-MOEA [18] and BCE-IBEA [14].

5 Experimental Results and Discussions

5.1 Performance Comparisons on WFG Test Suite

Table 1 presents the results of the proposed method's hypervolume indicator test on WFG1-9 with five conditions for each problem: 3, 5, 10, and 15 objectives. The proposed method achieved the optimal results in the vast majority of cases, with 35 out of 45 problems solved most efficiently, demonstrating its superior performance when compared to other state-of-the-art algorithms. HypE came in second place, producing the best results in only seven problems. Interestingly, HypE outperformed all other algorithms in the WFG3 problem, likely due to its unique approach to individual fitness calculations.

Table 1. Performance Compared on WFG Problems with Respect to the Average Hypervolume Values. The Best Average Hypervolume Values Among the Algorithms on Each Instance is Highlighted in Gray

Problem	M	ARMOEA	BCEIBEA	GFMMOEA	SPEA2SDE	HypE	SMSEMOA	Proposed Method
WFG1	3	6.9561e-1 (3.11e-2) –	7.8268e-1 (2.56e-2) +	7.7862e-1 (2.64e-2) +	7.7676e-1 (2.15e-2) +	5.5463e-1 (4.61e-2) –	7.1428e-1 (3.49e-2) –	7.3894e-1 (2.49e-2)
	5	5.1420e-1 (3.10e-2) –	7.1812e-1 (3.18e-2) +	7.2865e-1 (4.02e-2) +	6.9899e-1 (3.31e-2) ≈	6.1329e-1 (2.81e-2) –	5.5717e-1 (3.24e-2) –	6.9807e-1 (3.13e-2)
	10	3.6810e-1 (2.35e-2) –	5.0075e-1 (3.46e-2) –	5.7230e-1 (4.47e-2) ≈	4.5507e-1 (3.25e-2) –	6.1330e-1 (3.61e-2) +	4.5057e-1 (3.57e-2) –	5.7944e-1 (3.58e-2)
	15	4.2115e-1 (3.32e-2) –	4.2760e-1 (3.67e-2) –	4.8053e-1 (3.66e-2) –	3.3572e-1 (2.91e-2) –	5.5238e-1 (4.49e-2) +	3.9492e-1 (2.68e-2) –	5.0442e-1 (3.97e-2)
WFG2	3	9.0833e-1 (4.76e-3) –	9.2201e-1 (6.58e-3) –	9.2393e-1 (5.93e-3) –	9.0877e-1 (2.56e-2) –	9.2398e-1 (4.91e-3) –	9.1195e-1 (3.50e-3) –	9.2669e-1 (8.71e-3)
	5	9.3329e-1 (3.23e-2) –	9.5958e-1 (1.24e-2) –	9.6871e-1 (1.47e-2) –	9.4018e-1 (1.11e-2) –	9.7595e-1 (8.88e-3) ≈	9.5940e-1 (7.08e-3) –	9.7605e-1 (3.99e-3)
	10	8.9206e-1 (2.53e-2) –	9.4548e-1 (1.22e-2) –	9.2401e-1 (2.08e-2) –	9.2066e-1 (1.65e-2) –	9.6318e-1 (1.52e-2) ≈	9.3567e-1 (1.98e-2) –	9.6732e-1 (1.12e-2)
	15	8.5822e-1 (3.06e-2) –	9.1335e-1 (1.78e-2) –	8.6581e-1 (3.64e-2) –	8.9352e-1 (1.75e-2) –	9.3237e-1 (4.25e-2) –	8.6978e-1 (6.41e-2) –	9.5730e-1 (1.26e-2)
WFG3	3	3.4321e-1 (9.48e-3) –	3.7927e-1 (5.14e-3) –	3.6597e-1 (1.03e-2) –	3.8425e-1 (6.03e-3) ≈	4.0287e-1 (4.31e-3) +	3.6995e-1 (6.99e-3) –	3.8787e-1 (8.83e-3)
	5	1.0402e-2 (6.45e-3) –	1.2070e-1 (2.61e-2) –	5.1143e-2 (1.84e-2) –	3.9578e-2 (2.41e-2) –	2.5250e-1 (1.02e-2) +	2.0621e-1 (1.01e-2) +	1.4496e-1 (2.21e-2)
	10	0.0000e+0 (0.00e+0) ≈	0.0000e+0 (0.00e+0) ≈	0.0000e+0 (0.00e+0) ≈	0.0000e+0 (0.00e+0) ≈	2.5434e-4 (1.39e-3) ≈	0.0000e+0 (0.00e+0) ≈	0.0000e+0 (0.00e+0)
	15	0.0000e+0 (0.00e+0) ≈	0.0000e+0 (0.00e+0) ≈	0.0000e+0 (0.00e+0) ≈	0.0000e+0 (0.00e+0) ≈	0.0000e+0 (0.00e+0) ≈	0.0000e+0 (0.00e+0) ≈	0.0000e+0 (0.00e+0)
WFG4	3	5.3151e-1 (3.12e-3) –	5.4744e-1 (2.17e-3) –	5.4318e-1 (2.12e-3) –	5.4209e-1 (2.58e-3) –	5.5213e-1 (1.84e-3) –	5.2047e-1 (2.97e-3) –	5.5622e-1 (1.64e-3)
	5	6.9788e-1 (6.65e-3) –	7.2727e-1 (6.87e-3) –	7.1631e-1 (5.08e-3) –	7.1345e-1 (7.74e-3) –	7.2112e-1 (3.68e-2) –	7.4565e-1 (1.36e-2) –	7.7031e-1 (4.42e-3)
	10	8.3503e-1 (1.35e-2) –	8.4362e-1 (1.35e-2) –	7.9979e-1 (3.59e-2) –	7.9249e-1 (1.14e-2) –	6.6518e-1 (4.92e-2) –	6.4899e-1 (7.07e-2) –	8.7701e-1 (4.22e-2)
	15	6.2056e-1 (5.49e-2) –	8.6230e-1 (2.01e-2) ≈	7.6205e-1 (7.69e-2) –	7.8991e-1 (1.81e-2) –	6.3624e-1 (6.44e-2) –	5.5092e-1 (6.74e-2) –	8.7193e-1 (4.55e-2)
WFG5	3	5.1271e-1 (1.26e-3) –	5.1623e-1 (2.07e-3) –	5.1447e-1 (2.43e-3) –	5.1459e-1 (1.98e-3) –	5.1697e-1 (8.66e-4) –	5.0479e-1 (2.55e-3) –	5.2417e-1 (6.46e-4)
	5	6.8277e-1 (3.92e-3) –	7.0887e-1 (4.89e-3) –	6.9903e-1 (4.97e-3) –	6.8822e-1 (8.99e-3) –	7.3780e-1 (1.91e-3) ≈	7.2132e-1 (3.55e-3) –	7.3819e-1 (2.55e-3)
	10	8.0294e-1 (1.08e-2) –	8.2297e-1 (1.15e-2) –	7.6781e-1 (5.30e-2) –	7.4615e-1 (1.28e-2) –	7.5362e-1 (3.47e-2) –	5.4073e-1 (5.84e-2) –	8.6297e-1 (5.33e-3)
	15	3.5966e-1 (1.21e-1) –	7.9917e-1 (1.27e-2) –	7.8540e-1 (8.21e-2) –	7.1445e-1 (2.65e-2) –	7.2676e-1 (2.15e-2) –	4.8707e-1 (1.28e-2) –	8.4970e-1 (2.83e-2)
WFG6	3	5.0145e-1 (4.81e-3) –	5.1742e-1 (6.74e-3) –	5.1650e-1 (4.98e-3) –	5.1245e-1 (5.82e-3) –	5.1833e-1 (4.65e-3) –	5.0123e-1 (5.10e-3) –	5.2480e-1 (5.92e-3)
	5	6.7118e-1 (9.80e-3) –	7.0782e-1 (6.91e-3) –	7.0340e-1 (6.25e-3) –	6.8863e-1 (1.10e-2) –	7.4204e-1 (6.92e-3) ≈	7.1978e-1 (6.89e-3) –	7.3846e-1 (8.57e-3)
	10	8.4285e-1 (1.05e-2) –	8.4290e-1 (1.02e-2) –	7.6972e-1 (7.84e-2) –	7.7938e-1 (1.29e-2) –	7.3776e-1 (3.93e-2) –	7.2966e-1 (4.88e-2) –	8.7073e-1 (9.79e-3)
	15	4.5323e-1 (1.04e-1) –	8.4975e-1 (1.34e-2) –	6.7729e-1 (1.19e-1) –	7.7917e-1 (2.19e-2) –	7.1818e-1 (3.87e-2) –	6.4681e-1 (5.61e-2) –	8.5911e-1 (3.69e-2)
WFG7	3	5.3382e-1 (3.00e-3) –	5.5548e-1 (9.37e-4) –	5.5374e-1 (1.17e-3) –	5.5256e-1 (1.54e-3) –	5.5291e-1 (1.20e-3) –	5.2926e-1 (2.45e-3) –	5.6297e-1 (8.74e-4)
	5	6.8009e-1 (1.13e-2) –	7.6114e-1 (3.48e-3) –	7.4785e-1 (6.41e-3) –	7.3987e-1 (7.53e-3) –	7.8108e-1 (9.04e-3) –	7.5813e-1 (3.20e-3) –	7.8677e-1 (2.52e-3)
	10	8.6394e-1 (1.42e-2) –	8.9187e-1 (8.95e-3) –	7.7581e-1 (1.00e-1) –	8.3265e-1 (1.60e-2) –	7.4631e-1 (4.37e-2) –	7.3517e-1 (5.83e-2) –	9.1962e-1 (3.28e-2)
	15	5.1858e-1 (9.92e-2) –	8.9641e-1 (1.34e-2) –	7.2896e-1 (1.16e-1) –	8.1758e-1 (2.67e-2) –	7.2063e-1 (6.01e-2) –	6.2362e-1 (8.61e-2) –	9.1978e-1 (2.54e-2)
WFG8	3	4.8698e-1 (3.53e-3) –	4.9688e-1 (2.78e-3) –	4.9434e-1 (3.46e-3) –	4.9378e-1 (2.97e-3) –	4.9282e-1 (3.42e-3) –	4.7487e-1 (3.69e-3) –	5.0613e-1 (2.68e-3)
	5	6.3797e-1 (8.91e-3) –	6.5797e-1 (6.75e-3) –	6.6050e-1 (8.06e-3) –	6.5405e-1 (7.82e-3) –	7.0263e-1 (4.83e-3) ≈	6.7269e-1 (5.40e-3) –	7.0447e-1 (4.82e-3)
	10	7.4352e-1 (1.48e-2) –	7.2863e-1 (1.92e-2) –	7.3632e-1 (5.39e-2) –	7.6013e-1 (1.38e-2) –	7.5125e-1 (3.54e-2) –	7.1770e-1 (5.06e-2) –	8.1987e-1 (1.78e-2)
	15	4.9627e-1 (9.38e-2) –	7.6866e-1 (3.63e-2) –	4.2001e-1 (2.09e-1) –	7.7170e-1 (1.97e-2) –	7.3784e-1 (4.69e-2) –	6.8280e-1 (6.94e-2) –	8.3656e-1 (3.73e-2)
WFG9	3	4.8859e-1 (1.20e-2) –	5.2001e-1 (1.86e-2) –	5.2321e-1 (2.56e-3) –	5.0900e-1 (2.34e-2) –	5.1615e-1 (2.83e-2) –	5.0678e-1 (1.04e-2) –	5.3350e-1 (1.95e-2)
	5	5.6037e-1 (2.63e-2) –	6.6733e-1 (1.78e-2) –	6.7400e-1 (1.99e-2) –	6.6136e-1 (1.98e-2) –	7.2809e-1 (2.63e-2) ≈	7.2267e-1 (1.22e-2) –	7.3520e-1 (2.11e-2)
	10	6.0098e-1 (4.09e-2) –	6.9933e-1 (3.15e-2) –	7.1076e-1 (5.73e-2) –	6.6974e-1 (3.57e-2) –	6.9209e-1 (4.75e-2) –	5.1399e-1 (6.47e-2) –	8.2913e-1 (1.29e-2)
	15	2.6588e-1 (1.06e-1) –	6.3737e-1 (2.94e-2) –	6.5478e-1 (1.08e-1) –	6.1179e-1 (5.28e-2) –	7.0257e-1 (4.23e-2) –	4.6765e-1 (4.25e-2) –	8.4482e-1 (2.19e-2)
+/–/≈		0/43/2	2/40/3	2/39/4	1/40/4	5/30/10	1/41/3	

Table 2. Performance Compared on WFG Problems with Respect to the Average IGD Values. The Best Average IGD Values Among the Algorithms on Each Instance is Highlighted in Gray

Problem	M	ARMOEA	BCEIBEA	GFMMOEA	SPEA2SDE	SMSEMOA	HypE	Proposed Method
WFG1	3	5.3719e-1 (6.24e-2) –	3.5681e-1 (4.58e-2) +	3.8376e-1 (4.84e-2) +	3.9194e-1 (3.56e-2) +	5.2594e-1 (6.21e-2) –	1.0965e+0 (1.03e-1) –	4.5965e-1 (4.48e-2)
	5	1.2978e+0 (8.67e-2) –	7.7550e-1 (6.35e-2) +	8.3465e-1 (1.42e-1) ≈	8.2192e-1 (5.69e-2) ≈	1.2593e+0 (7.79e-2) –	1.4736e+0 (8.43e-2) –	8.1699e-1 (6.14e-2)
	10	2.2706e+0 (9.66e-2) –	1.9643e+0 (1.12e-1) –	1.8912e+0 (2.15e-1) –	2.0410e+0 (9.82e-2) –	2.2049e+0 (8.58e-2) –	2.0732e+0 (3.64e-2) –	1.6662e+0 (7.80e-2)
	15	3.0686e+0 (1.18e-1) –	3.0664e+0 (1.69e-1) –	2.6651e+0 (1.91e-1) –	3.1860e+0 (1.34e-1) –	3.1343e+0 (1.05e-1) –	2.7212e+0 (5.69e-2) –	2.5349e+0 (1.24e-1)
WFG2	3	1.6860e-1 (3.27e-3) +	1.6465e-1 (4.34e-3) +	1.7132e-1 (5.74e-3) +	2.1897e-1 (5.56e-2) +	2.3445e-1 (2.18e-2) ≈	3.0932e-1 (6.98e-3) –	2.3034e-1 (1.16e-2)
	5	5.3063e-1 (1.48e-1) +	5.1605e-1 (2.92e-2) +	5.5920e-1 (6.31e-2) +	6.0462e-1 (2.70e-2) –	9.4449e-1 (7.98e-2) –	6.1442e-1 (3.68e-2) –	5.7777e-1 (2.77e-2)
	10	1.3620e+0 (6.84e-2) –	1.2010e+0 (5.09e-2) +	1.3915e+0 (1.76e-1) –	1.3476e+0 (7.40e-2) –	2.6286e+0 (4.76e-1) –	1.3894e+0 (1.14e-1) –	1.3129e+0 (1.35e-1)
	15	2.0679e+0 (6.64e-2) –	1.8341e+0 (9.60e-2) +	2.2291e+0 (7.56e-1) –	1.9743e+0 (7.23e-2) –	5.9360e+0 (1.33e+0) –	2.8447e+0 (1.03e+0) –	1.9049e+0 (9.01e-2)
WFG3	3	1.7638e-1 (1.85e-2) –	1.0002e-1 (7.45e-3) –	1.1324e-1 (1.70e-2) –	7.2815e-2 (7.66e-3) ≈	1.2495e-1 (1.79e-2) –	4.7782e-2 (9.33e-3) +	7.6860e-2 (1.69e-2)
	5	8.8387e-1 (5.36e-2) –	4.5835e-1 (5.23e-2) –	5.7204e-1 (7.99e-2) –	7.3881e-1 (2.46e-1) –	4.6214e-1 (2.88e-1) ≈	9.1455e-2 (1.52e-2) +	3.1639e-1 (5.45e-2)
	10	3.5900e+0 (1.60e-1) –	1.2725e+0 (1.52e-1) –	8.2582e-1 (3.64e-1) –	1.9769e+0 (5.88e-1) –	2.9874e+0 (2.12e+0) –	2.2119e-1 (4.72e-2) +	3.0357e-1 (1.14e-1)
	15	6.0732e+0 (5.80e-1) –	2.2457e+0 (2.19e-1) –	8.4664e+0 (2.72e+0) –	4.3976e+0 (1.11e+0) –	6.0839e+0 (2.18e+0) –	4.0134e-1 (6.50e-2) +	7.3919e-1 (2.81e-1)
WFG4	3	2.2652e-1 (1.45e-3) +	2.1892e-1 (3.55e-3) +	2.1187e-1 (1.85e-3) +	3.2753e-1 (1.66e-2) –	3.0222e-1 (1.53e-2) –	3.3242e-1 (1.61e-2) –	2.9479e-1 (9.82e-3)
	5	1.2202e+0 (2.24e-3) +	1.1926e+0 (1.02e-2) +	1.1498e+0 (7.99e-3) +	1.4013e+0 (2.65e-2) +	1.3748e+0 (5.10e-2) +	1.6297e+0 (1.41e-1) –	1.4434e+0 (3.19e-2)
	10	6.0223e+0 (1.28e-1) –	5.1331e+0 (1.04e-1) +	4.9869e+0 (3.11e-1) +	5.4460e+0 (9.57e-2) +	9.7522e+0 (1.28e+0) –	9.5151e+0 (8.51e-1) –	5.7215e+0 (5.80e-1)
	15	1.2006e+1 (5.74e-1) –	9.1424e+0 (2.40e-1) +	1.1907e+1 (1.31e+0) –	9.6832e+0 (4.03e-1) +	2.1029e+1 (1.40e+0) –	1.9593e+1 (1.44e+0) –	1.1009e+1 (1.57e+0)
WFG5	3	2.3247e-1 (9.04e-4) +	2.2904e-1 (2.96e-3) +	2.2042e-1 (1.88e-3) +	3.3206e-1 (1.17e-2) –	3.0876e-1 (1.68e-2) –	3.6713e-1 (1.51e-2) –	2.9944e-1 (1.23e-2)
	5	1.2090e+0 (1.67e-3) +	1.2016e+0 (1.34e-2) +	1.1389e+0 (9.49e-3) +	1.4137e+0 (3.27e-2) +	1.3755e+0 (2.62e-2) +	1.5523e+0 (3.54e-2) –	1.4319e+0 (3.22e-2)
	10	5.8700e+0 (6.43e-2) –	5.2306e+0 (4.96e-2) +	5.2326e+0 (7.60e-1) +	5.3938e+0 (9.12e-2) +	1.0700e+1 (1.14e+0) –	6.7119e+0 (6.22e-1) –	5.5183e+0 (7.26e-2)
	15	1.1321e+1 (2.18e-1) –	9.2077e+0 (1.22e-1) +	9.9995e+0 (1.28e+0) ≈	1.1002e+1 (5.55e-1) –	2.1535e+1 (6.58e-1) –	1.5402e+1 (9.10e-1) –	9.7555e+0 (6.07e-1)
WFG6	3	2.4443e-1 (3.63e-3) +	2.3958e-1 (6.32e-3) +	2.2236e-1 (4.26e-3) +	3.4183e-1 (1.51e-2) –	3.3395e-1 (1.68e-2) –	3.6634e-1 (1.65e-2) –	3.0322e-1 (8.53e-3)
	5	1.2211e+0 (2.50e-3) +	1.2275e+0 (1.65e-2) +	1.1587e+0 (9.94e-3) +	1.4574e+0 (2.31e-2) ≈	1.4193e+0 (3.79e-2) +	1.5525e+0 (3.34e-2) –	1.4553e+0 (3.28e-2)
	10	6.1357e+0 (1.60e-1) –	5.3472e+0 (4.91e-2) +	5.3626e+0 (9.22e-1) +	5.4901e+0 (7.16e-2) +	7.8373e+0 (1.02e+0) –	7.5548e+0 (8.13e-1) –	5.5791e+0 (1.05e-1)
	15	1.1596e+1 (3.60e-1) –	9.4203e+0 (1.84e-1) +	1.1848e+1 (1.23e+0) –	1.0039e+1 (4.65e-1) –	1.8531e+1 (1.56e+0) –	1.6686e+1 (1.38e+0) –	9.8506e+0 (6.52e-1)
WFG7	3	2.2865e-1 (2.30e-3) +	2.2602e-1 (4.84e-3) +	2.1084e-1 (2.00e-3) +	3.2641e-1 (1.16e-2) –	3.4703e-1 (2.06e-2) –	3.8270e-1 (1.97e-2) –	2.9397e-1 (8.92e-3)
	5	1.2332e+0 (5.23e-3) +	1.2342e+0 (1.44e-2) +	1.1649e+0 (1.11e-2) +	1.4498e+0 (2.46e-2) ≈	1.4482e+0 (3.12e-2) ≈	1.5813e+0 (4.15e-2) –	1.4648e+0 (2.94e-2)
	10	6.0392e+0 (1.39e-1) –	5.3823e+0 (1.47e-1) +	5.7873e+0 (1.15e+0) ≈	5.4548e+0 (7.63e-2) +	8.3765e+0 (1.04e+0) –	8.1365e+0 (9.09e-1) –	5.6247e+0 (9.58e-2)
	15	1.1587e+1 (3.38e-1) –	9.4543e+0 (2.30e-1) +	1.1867e+1 (1.30e+0) –	9.9106e+0 (3.59e-1) ≈	1.9754e+1 (2.07e+0) –	1.7738e+1 (1.64e+0) –	1.0113e+1 (7.04e-1)
WFG8	3	2.5897e-1 (3.61e-3) +	2.6483e-1 (5.52e-3) +	2.4343e-1 (3.99e-3) +	3.4594e-1 (1.26e-2) –	3.6428e-1 (1.79e-2) –	3.7153e-1 (1.26e-2) –	2.9376e-1 (8.23e-3)
	5	1.2212e+0 (2.86e-3) +	1.2382e+0 (1.25e-2) +	1.1560e+0 (5.59e-3) +	1.4203e+0 (2.51e-2) –	1.3874e+0 (3.29e-2) ≈	1.5179e+0 (2.66e-2) –	1.3775e+0 (3.56e-2)
	10	5.9610e+0 (9.23e-2) –	5.2628e+0 (5.82e-2) +	5.0271e+0 (3.12e-1) +	5.4175e+0 (6.60e-2) ≈	8.2559e+0 (1.09e+0) –	7.5353e+0 (6.03e-1) –	5.4848e+0 (1.95e-1)
	15	1.1779e+1 (4.39e-1) –	9.2034e+0 (1.65e-1) +	1.1182e+1 (1.30e+0) ≈	9.5087e+0 (1.47e-1) +	1.7933e+1 (1.93e+0) –	1.6665e+1 (1.60e+0) –	1.1029e+1 (9.16e-1)
WFG9	3	2.4642e-1 (9.75e-3) +	2.2233e-1 (1.25e-2) +	2.0980e-1 (2.37e-3) +	3.1964e-1 (1.33e-2) –	3.0613e-1 (2.33e-2) –	3.5928e-1 (1.90e-2) –	2.8495e-1 (1.31e-2)
	5	1.2426e+0 (1.54e-2) +	1.1662e+0 (1.38e-2) +	1.1276e+0 (9.13e-3) +	1.3624e+0 (4.03e-2) +	1.3728e+0 (2.91e-2) +	1.4890e+0 (6.88e-2) –	1.4137e+0 (9.01e-2)
	10	5.6930e+0 (5.44e-2) –	4.8862e+0 (5.60e-2) +	4.9436e+0 (2.40e-1) +	5.3048e+0 (7.91e-2) +	1.1223e+1 (9.84e-1) –	7.6574e+0 (8.28e-1) –	5.4327e+0 (1.09e-1)
	15	1.1282e+1 (2.23e-1) –	8.8267e+0 (9.18e-2) +	9.4930e+0 (1.29e+0) +	9.7044e+0 (6.23e-1) ≈	2.1988e+1 (8.69e-1) –	1.6289e+1 (1.22e+0) –	9.8250e+0 (4.19e-1)
+/–/≈		15/28/2	37/8/0	27/14/4	13/22/10	4/36/5	5/40/0	

Table 3. Performance Compared on DTLZ Problems with Respect to the Average Hypervolume Values. The Best Average Hypervolume Values Among the Algorithms on Each Instance is Highlighted in Gray

Problem	M	ARMOEA	BCEIBEA	GFMMOEA	SPEA2SDE	SMSEMOA	HypE	Proposed Method
DTLZ1	3	8.3288e-1 (5.89e-3) −	8.3670e-1 (3.42e-3) −	8.3908e-1 (3.58e-3) ≈	8.2135e-1 (5.82e-2) −	7.9562e-1 (1.07e-1) −	6.3027e-1 (1.32e-1) −	8.4007e-1 (2.03e-3)
	5	9.0779e-1 (2.02e-1) −	9.6392e-1 (3.64e-3) +	9.5493e-1 (3.79e-2) +	9.4748e-1 (7.63e-3) +	2.8770e-1 (3.66e-1) −	5.9893e-1 (1.10e-1) −	9.3750e-1 (1.26e-1)
	10	7.3662e-1 (3.80e-1) −	1.6133e-1 (2.78e-1) −	7.0097e-1 (3.78e-1) −	9.7294e-1 (1.64e-2) −	9.5102e-3 (5.21e-2) −	5.3905e-1 (1.71e-1) −	9.9796e-1 (5.39e-4)
	15	5.0204e-1 (3.69e-1) −	7.9106e-2 (1.73e-1) −	6.4453e-1 (3.69e-1) −	9.4359e-1 (6.45e-2) −	3.1215e-4 (1.71e-3) −	3.8578e-1 (2.70e-1) −	9.6779e-1 (1.67e-1)
DTLZ2	3	5.5893e-1 (1.39e-4) −	5.5908e-1 (9.26e-4) −	5.6049e-1 (8.03e-4) −	5.6083e-1 (1.21e-3) −	5.4818e-1 (1.79e-3) −	5.3506e-1 (3.11e-3) −	5.6696e-1 (5.55e-4)
	5	7.7169e-1 (6.87e-4) −	7.7398e-1 (2.38e-3) −	7.7293e-1 (2.66e-3) −	7.8522e-1 (2.66e-3) −	7.7687e-1 (2.19e-3) −	7.0419e-1 (2.06e-2) −	7.9734e-1 (1.13e-3)
	10	9.3661e-1 (1.43e-3) −	9.4005e-1 (2.96e-3) −	9.2293e-1 (4.88e-3) −	9.4623e-1 (2.16e-3) −	7.9334e-1 (4.49e-2) −	7.1329e-1 (3.89e-2) −	9.5368e-1 (9.86e-4)
	15	7.0241e-1 (2.96e-2) −	9.7704e-1 (3.34e-3) +	9.4794e-1 (9.28e-3) −	9.8018e-1 (2.99e-3) ≈	7.7071e-1 (5.68e-2) −	7.0187e-1 (3.96e-2) −	9.7079e-1 (2.85e-2)
DTLZ3	3	9.4786e-2 (1.64e-1) ≈	1.3342e-1 (1.70e-1) +	1.1882e-1 (1.93e-1) +	1.6748e-1 (2.03e-1) +	4.5295e-2 (1.13e-1) ≈	8.3893e-2 (1.35e-1) +	1.9512e-2 (7.14e-2)
	5	0.0000e+0 (0.00e+0) ≈	1.5722e-2 (6.06e-2) ≈	5.1279e-3 (2.48e-2) ≈	1.2496e-1 (2.28e-1) +	0.0000e+0 (0.00e+0) ≈	3.5351e-2 (9.66e-2) ≈	1.0537e-2 (5.77e-2)
	10	0.0000e+0 (0.00e+0) −	0.0000e+0 (0.00e+0) −	0.0000e+0 (0.00e+0) −	3.1584e-1 (3.72e-1) +	0.0000e+0 (0.00e+0) −	6.0696e-4 (3.32e-3) ≈	6.8358e-2 (2.03e-1)
	15	1.6948e-2 (9.28e-2) ≈	0.0000e+0 (0.00e+0) −	0.0000e+0 (0.00e+0) −	5.2957e-2 (1.75e-1) ≈	0.0000e+0 (0.00e+0) −	0.0000e+0 (0.00e+0) −	1.1455e-2 (6.09e-2)
DTLZ4	3	3.9610e-1 (1.70e-1) −	5.5846e-1 (1.58e-3) ≈	4.7422e-1 (1.25e-1) +	5.1828e-1 (8.70e-2) ≈	4.4737e-1 (1.51e-1) −	3.9171e-1 (1.42e-1) −	4.6195e-1 (1.44e-1)
	5	7.5189e-1 (5.00e-2) −	7.7252e-1 (5.41e-3) ≈	7.3647e-1 (6.13e-2) −	7.6893e-1 (3.70e-2) +	7.0386e-1 (7.82e-2) −	5.3559e-1 (1.43e-1) −	7.6251e-1 (4.75e-2)
	10	9.4061e-1 (8.33e-4) ≈	9.4219e-1 (5.05e-3) ≈	9.3409e-1 (6.01e-3) −	9.3670e-1 (1.87e-2) ≈	8.2233e-1 (1.57e-1) −	7.5927e-1 (5.33e-2) −	9.3830e-1 (2.36e-2)
	15	8.5968e-1 (1.15e-2) −	9.7866e-1 (2.88e-3) ≈	8.6078e-1 (2.85e-1) −	9.7983e-1 (4.34e-3) +	5.5821e-1 (3.46e-1) −	7.5597e-1 (4.90e-2) −	9.6792e-1 (2.32e-2)
DTLZ5	3	1.9901e-1 (1.36e-4) −	1.9978e-1 (9.87e-5) −	1.9992e-1 (4.60e-5) −	1.9942e-1 (2.95e-4) −	1.9464e-1 (8.34e-4) −	1.9609e-1 (7.23e-4) −	1.9996e-1 (2.66e-4)
	5	9.9098e-2 (6.33e-3) −	1.0241e-1 (1.61e-2) −	5.2714e-2 (3.36e-2) −	1.1125e-1 (2.67e-3) ≈	1.1520e-1 (1.58e-3) +	1.2344e-1 (8.48e-4) +	1.1142e-1 (6.81e-3)
	10	9.1436e-2 (9.77e-4) +	3.3298e-2 (3.94e-2) ≈	4.2237e-2 (2.74e-2) ≈	8.9513e-2 (1.33e-3) +	8.2300e-2 (6.42e-3) +	9.6844e-2 (7.23e-4) +	4.8259e-2 (1.87e-2)
	15	9.1516e-2 (3.83e-4) +	3.7298e-2 (3.41e-2) ≈	3.9130e-2 (3.18e-2) ≈	8.9375e-2 (9.51e-4) +	7.4179e-2 (1.11e-2) +	9.2553e-2 (5.25e-4) +	3.1642e-2 (2.59e-2)
DTLZ6	3	1.9957e-1 (6.98e-5) +	2.0017e-1 (1.26e-4) +	2.0006e-1 (5.26e-5) +	1.9957e-1 (1.89e-4) +	1.9522e-1 (1.00e-3) +	1.1420e-1 (1.13e-2) +	7.4380e-2 (3.36e-2)
	5	9.4310e-2 (2.62e-2) +	6.0641e-2 (3.78e-2) ≈	3.4021e-2 (4.42e-2) −	1.0503e-1 (4.82e-3) +	1.9889e-2 (3.67e-2) −	1.0068e-1 (6.25e-3) +	7.4636e-2 (3.82e-2)
	10	6.6939e-2 (4.11e-2) ≈	5.7471e-2 (4.30e-2) ≈	3.0861e-2 (3.60e-2) −	8.8499e-2 (8.26e-3) ≈	0.0000e+0 (0.00e+0) −	9.1565e-2 (9.07e-4) +	6.2715e-2 (3.77e-2)
	15	9.0977e-2 (3.00e-4) +	5.6713e-2 (4.41e-2) ≈	3.2003e-2 (2.86e-2) ≈	8.7117e-2 (1.68e-2) +	0.0000e+0 (0.00e+0) −	9.0975e-2 (2.83e-4) +	4.6320e-2 (4.45e-2)
DTLZ7	3	2.6537e-1 (1.39e-2) −	2.6417e-1 (2.08e-2) −	2.5779e-1 (2.78e-2) −	2.7757e-1 (6.58e-4) −	2.5979e-1 (1.84e-2) −	1.9517e-1 (1.22e-3) −	2.7995e-1 (7.90e-3)
	5	2.0720e-1 (5.06e-3) −	2.5728e-1 (5.53e-3) −	2.3878e-1 (2.99e-2) −	2.3763e-1 (1.17e-2) −	2.0713e-1 (2.05e-2) −	1.7539e-1 (1.11e-2) −	2.6310e-1 (1.51e-2)
	10	3.1632e-2 (2.75e-2) −	1.6146e-1 (2.01e-2) −	1.1515e-1 (3.19e-2) −	3.3809e-2 (1.85e-2) −	1.2509e-1 (7.12e-3) −	1.2298e-1 (4.02e-3) −	1.7746e-1 (7.42e-3)
	15	2.4103e-4 (9.07e-4) −	1.1434e-1 (2.97e-2) −	7.0172e-2 (3.05e-2) −	1.7568e-3 (1.66e-3) −	5.2803e-2 (3.65e-2) −	1.0646e-1 (4.00e-3) −	1.4015e-1 (1.27e-2)
IDTLZ1	3	2.1353e-1 (4.70e-3) ≈	2.1776e-1 (2.83e-3) +	2.2186e-1 (8.03e-4) +	2.1924e-1 (1.22e-3) +	1.8858e-1 (3.75e-2) −	1.1345e-1 (1.33e-2) −	2.0787e-1 (2.00e-2)
	5	7.4289e-3 (9.15e-4) +	8.0141e-3 (1.22e-3) +	9.2017e-3 (3.95e-4) +	9.6742e-3 (3.14e-4) +	7.6095e-3 (1.53e-3) +	1.3037e-3 (4.30e-4) −	5.2507e-3 (1.30e-3)
	10	5.1989e-8 (4.95e-8) +	1.1236e-7 (8.44e-8) +	1.2442e-7 (1.24e-7) +	1.0470e-7 (5.72e-8) +	1.9757e-7 (5.59e-8) +	1.1224e-8 (1.51e-8) −	3.0921e-8 (1.17e-7)
	15	7.916e-14 (6.08e-14) +	2.984e-13 (3.53e-13) +	5.302e-13 (4.39e-13) +	4.708e-13 (2.81e-13) +	1.319e-12 (6.04e-13) +	3.237e-14 (7.24e-14) −	7.144e-14 (2.48e-13)
IDTLZ2	3	5.2376e-1 (3.10e-4) −	5.3607e-1 (1.12e-3) +	5.2521e-1 (7.28e-2) −	5.3106e-1 (2.64e-3) ≈	5.2525e-1 (2.19e-3) −	5.1489e-1 (3.76e-3) −	5.3144e-1 (3.17e-3)
	5	8.5271e-2 (2.42e-3) −	1.0998e-1 (2.51e-3) −	1.1231e-1 (2.75e-3) −	1.1722e-1 (1.43e-3) −	1.1009e-1 (1.61e-3) −	8.9511e-2 (3.26e-3) −	1.2281e-1 (1.01e-3)
	10	1.8579e-4 (1.86e-5) −	2.1005e-4 (2.68e-5) −	2.1682e-4 (2.11e-5) −	3.1430e-4 (2.06e-5) −	2.4155e-4 (2.41e-5) −	1.1612e-4 (1.60e-5) −	4.2827e-4 (1.07e-5)
	15	5.8427e-8 (4.38e-8) −	9.7276e-8 (2.63e-8) −	1.9399e-7 (3.07e-8) −	1.6329e-7 (2.09e-8) −	5.4151e-8 (2.89e-8) −	5.5574e-8 (6.11e-8) −	2.8548e-7 (3.33e-8)
+/−/≈		10/28/7	10/23/12	9/29/7	18/19/8	9/34/2	10/33/2	

It is worth noting that in WFG1 problem, the proposed method's performance was initially mediocre with only three objectives in comparison to BCEIBEA, GFMMOEA, and SPEA2SDE. However, as the number of objectives increased, starting from eight objectives, the performance of these Pareto-based EMOAs began to weaken, and by the time the number of objectives reached 15, the proposed method had significantly surpassed them, demonstrating its powerful ability to deal with higher-dimensional problems.

Overall, these results emphasize the superior performance of the proposed method in solving complex optimization problems with a higher number of objectives, further demonstrating its potential as a promising algorithm for solving real-world problems.

Table 2 reports the IGD test results of the proposed method on WFG1-9. The results demonstrated that, in most cases, BCEIBEA and GFMMOEA outperformed the proposed method and even surpassed all hypervolume-based EMOAs under the IGD indicator. This finding aligns with our intuition since BCEIBEA and GFMMOEA make environmental selection based on simulated PF and IGD, while the proposed method selects based on hypervolume, causing the proposed method's performance under IGD indicator to be comparatively weak.

Moreover, when the hypervolume-based EMOAs encountered concave PF, the solution set tended to concentrate in the center of the PF due to the hypervolume operation mechanism, which also influenced the IGD indicator measurement of the solution set generated by the hypervolume-based EMOAs to a certain extent.

5.2 Performance Comparisons on DTLZ Test Suite

Table 3 presents the hypervolume indicator test results of the proposed method on DTLZ{1-7} and IDTLZ{1-2}. The proposed method was the best at solving 18 out of 45 problems, while SPEA2SDE, BCEIBEA, HypE, and SMSEMOA achieved the best results in 9, 7, 6, and 3 problems, respectively. These results demonstrate the competitive performance of the proposed method compared to other state-of-the-art algorithms in solving multi-objective optimization problems with different levels of difficulty.

Overall, the proposed method does not perform as well on the DTLZ test set as it does on the WFG test set. This can be attributed to the fact that DTLZ mainly evaluates an algorithm's convergence, while the hypervolume metric measures both convergence and diversity but focuses more on diversity. This can be observed in the DTLZ1, DTLZ2, and DTLZ7 problems, where the main emphasis is on the diversity of EMOAs.

6 Conclusion

This paper proposes an algorithm for solving MaOPs. The algorithm extends the direction vector of the original R_2^{HCA} indicator to determine the intersection with each remaining solution in the solution set, and calculates the length of the line segment. This allows for the simultaneous calculation of the overall hypervolume contribution of solutions. Moreover, a data array storing R2 information is designed to simplify the computational complexity of the algorithm. In the experimental study, we compared our algorithm with two hypervolume-based EMOAs and four advanced EMOAs. We tested five cases - of WFG{1-9}, DTLZ{1-7}, and IDTLZ{1-2} - where the objective numbers were 3, 5, 10, and 15. The results show that our algorithm outperforms comparison algorithms.

In the future, we plan to integrate this algorithm into practical problems for experimental testing, such as multi-robot system task planning and UAV swarm decision-making. We also aim to expand the algorithm to computationally expensive MOPs, large-scale MOPs, and multi-modal MOPs.

References

1. Asafuddoula, M., Singh, H.K., Ray, T.: An enhanced decomposition-based evolutionary algorithm with adaptive reference vectors. IEEE Trans. Cybernet. **48**(8), 2321–2334 (2017)
2. Bader, J., Zitzler, E.: HypE: an algorithm for fast hypervolume-based many-objective optimization. Evol. Comput. **19**(1), 45–76 (2011)

3. Beume, N., Naujoks, B., Emmerich, M.: SMS-EMOA: multiobjective selection based on dominated hypervolume. Eur. J. Oper. Res. **181**(3), 1653–1669 (2007)
4. Coello Coello, C.A., Reyes Sierra, M.: A study of the parallelization of a coevolutionary multi-objective evolutionary algorithm. In: Monroy, R., Arroyo-Figueroa, G., Sucar, L.E., Sossa, H. (eds.) MICAI 2004. LNCS (LNAI), vol. 2972, pp. 688–697. Springer, Heidelberg (2004). https://doi.org/10.1007/978-3-540-24694-7_71
5. Deb, K., Pratap, A., Agarwal, S., Meyarivan, T.: A fast and elitist multiobjective genetic algorithm: NSGA-II. IEEE Trans. Evol. Comput. **6**(2), 182–197 (2002)
6. Deb, K., Thiele, L., Laumanns, M., Zitzler, E.: Scalable test problems for evolutionary multiobjective optimization. In: Evolutionary Multiobjective Optimization: Theoretical Advances and Applications, pp. 105–145. Springer, London (2005). https://doi.org/10.1007/1-84628-137-7_6
7. Emmerich, M., Beume, N., Naujoks, B.: An EMO algorithm using the hypervolume measure as selection criterion. In: Coello Coello, C.A., Hernández Aguirre, A., Zitzler, E. (eds.) EMO 2005. LNCS, vol. 3410, pp. 62–76. Springer, Heidelberg (2005). https://doi.org/10.1007/978-3-540-31880-4_5
8. Farina, M., Amato, P.: On the optimal solution definition for many-criteria optimization problems. In: 2002 Annual Meeting of the North American Fuzzy Information Processing Society Proceedings. NAFIPS-FLINT 2002 (Cat. No. 02TH8622), pp. 233–238. IEEE (2002)
9. García, I.C., Coello, C.A.C., Arias-Montano, A.: MOPSOhv: a new hypervolume-based multi-objective particle swarm optimizer. In: 2014 IEEE Congress on Evolutionary Computation (CEC), pp. 266–273. IEEE (2014)
10. Huband, S., Hingston, P., Barone, L., While, L.: A review of multiobjective test problems and a scalable test problem toolkit. IEEE Trans. Evol. Comput. **10**(5), 477–506 (2006)
11. Ishibuchi, H., Setoguchi, Y., Masuda, H., Nojima, Y.: Performance of decomposition-based many-objective algorithms strongly depends on pareto front shapes. IEEE Trans. Evol. Comput. **21**(2), 169–190 (2016)
12. Li, B., Li, J., Tang, K., Yao, X.: Many-objective evolutionary algorithms: a survey. ACM Comput. Surv. (CSUR) **48**(1), 1–35 (2015)
13. Li, M., Yang, S., Liu, X.: Shift-based density estimation for pareto-based algorithms in many-objective optimization. IEEE Trans. Evol. Comput. **18**(3), 348–365 (2013)
14. Li, M., Yang, S., Liu, X.: Pareto or Non-Pareto: Bi-Criterion evolution in multiobjective optimization. IEEE Trans. Evol. Comput. **20**(5), 645–665 (2015)
15. Qi, Y., Ma, X., Liu, F., Jiao, L., Sun, J., Wu, J.: MOEA/D with adaptive weight adjustment. Evol. Comput. **22**(2), 231–264 (2014)
16. Shang, K., Ishibuchi, H., Ni, X.: R2-based hypervolume contribution approximation. IEEE Trans. Evol. Comput. **24**(1), 185–192 (2019)
17. Tian, Y., Cheng, R., Zhang, X., Cheng, F., Jin, Y.: An indicator-based multiobjective evolutionary algorithm with reference point adaptation for better versatility. IEEE Trans. Evol. Comput. **22**(4), 609–622 (2017)
18. Tian, Y., Zhang, X., Cheng, R., He, C., Jin, Y.: Guiding evolutionary multiobjective optimization with generic front modeling. IEEE Trans. Cybernet. **50**(3), 1106–1119 (2018)
19. Zhang, Q., Li, H.: MOEA/D: a multiobjective evolutionary algorithm based on decomposition. IEEE Trans. Evol. Comput. **11**(6), 712–731 (2007)
20. Zhang, X., Tian, Y., Cheng, R., Jin, Y.: A decision variable clustering-based evolutionary algorithm for large-scale many-objective optimization. IEEE Trans. Evol. Comput. **22**(1), 97–112 (2016)

21. Zitzler, E., Laumanns, M., Thiele, L.: SPEA 2: improving the strength pareto evolutionary algorithm. TIK report 103 (2001)
22. Zitzler, E., Thiele, L.: Multiobjective optimization using evolutionary algorithms — a comparative case study. In: Eiben, A.E., Bäck, T., Schoenauer, M., Schwefel, H.-P. (eds.) PPSN 1998. LNCS, vol. 1498, pp. 292–301. Springer, Heidelberg (1998). https://doi.org/10.1007/BFb0056872

Reinforcement Learning-Based Policy Selection of Multi-sensor Cyber Physical Systems Under DoS Attacks

Zengwang Jin[1,2,3], Qian Li[1,3], Huixiang Zhang[1](✉), and Changyin Sun[4]

[1] School of Cybersecurity, Northwestern Polytechnical University, Xi'an 710072, China
zhanghuixiang@nwpu.edu.cn
[2] Ningbo Institute of Northwestern Polytechnical University, 218 Qingyi Road, Ningbo 315103, China
[3] Yangtze River Delta Research Institute of NPU, No. 27 Zigang Road, Taicang 215400, Jiangsu, China
[4] School of Artificial Intelligence, Anhui University, Hefei 230039, Anhui, China

Abstract. This paper focuses on the problem of optimal policy selection for sensors and attackers in cyber-physical system (CPS) with multiple sensors under denial-of-service (DoS) attacks. DoS attacks have caused tremendous disruption to the normal operation of CPS and it is necessary to assess this damage. The state estimation can reflect the real-time operation status of the CPS and provide effective prediction and assessment in terms of the security of the CPS. For a multi-sensor CPS, different that robust control method is utilized to depict the state of the system against DoS attacks, the optimal policy selection of sensors and attackers is positively analyzed by dynamic programming ideology. To optimize the strategies of both sides, game theory is introduced to study the interaction process between the sensors and the attackers. During the policy iterative optimization process, the sensors and attackers dynamically learn and adjust strategies by incorporating reinforcement learning. To explore more state information, the restriction of state set is loosened, that is the transfer of states are not limited compulsorily. Meanwhile, the complexity of the proposed algorithm is decreased by introducing a penalty in the reward function. Finally, simulation results of the CPS containing three sensors show that the proposed algorithm can effectively optimize the policy selection of sensors and attackers in CPS.

Keywords: cyber-physical system · DoS attacks · multi-sensor · state estimation

This work was supported in part by the National Key Research and Development Project with Grant 2022YFB3104005, the National Natural Science Foundation of China under Grant 62003275, Basic Research Programs (2022) of Taicang with Grant TC2022JC17, Ningbo Natural Science Foundation with Grant 2021J046.

B. Xin et al. (Eds.): IWACIII 2023, CCIS 1931, pp. 298–309, 2024.
https://doi.org/10.1007/978-981-99-7590-7_24

1 Introduction

With the rapid development of information technology, the integration of cyber system and physical system has become an inevitable trend in recent years, thus the cyber-physical system (CPS) has emerged. With the characteristics of high flexibility and easy scalability, CPS enables the aggregation of system information and real-time data sharing [1]. Their deployment in critical infrastructures has shown the potential to revolutionize the world, such as smart grids [2], digital manufacturing [3], healthcare [4] and so on. In most of these applications, the information is delivered over a wireless channel. However, the attacks become easier to implement in the process of transmitting information through the wireless channel [5,6].

There are many types of common cyber attacks in CPS, such as deception attacks [7], replay attacks [8], denial-of-service (DoS) attacks [9–11] and so on. Among them, DoS attacks are easier and less costly to execute. In order to analyze the damage caused by DoS attacks on CPS, many scholars employ state estimation based on Kalman filter to evaluate the operation of CPS [12,13]. In [12], distributed Kalman filter is designed to address event-triggered distributed state estimation problems. In [13], two Kalman filter-based algorithms are presented for detection of attacked sensors. Thus, in this paper, the state estimation algorithm based on Kalman filter is conceived to assess the state of CPS under DoS attacks.

In CPS, the defender and the attacker can be considered as a two-player game. The interactive decision process between a system with countermeasures and an attacker is studied under the framework of game theory in [14,15]. In the game, the Nash equilibrium is used to find the point of convergence so that an optimal strategy can be determined. In [16], the Nash equilibrium algorithm is investigated as a means to enable each player to dictate their individual strategies and attain maximum benefit. Owing to the game theory is excellent at solving complex problems, a system model under the DoS attacks is constructed based on the game theory to solve for the optimal strategy.

Nowadays, reinforcement learning is rapidly spreading to a variety of domains. The long-term vision of the reinforcement learning algorithm allows it to be ideal for gaming between sensors and attackers in a CPS. For example, the literature [17] from the perspective of two reinforcement learning algorithms analyzes the security problem for the state estimation of CPS, both of which obtain the corresponding optimal policies. In [18], the distributed reinforcement learning algorithms for local information based sensors and attackers are proposed to find their Nash equilibrium policy, respectively. In this paper, we introduce reinforcement learning algorithms into the secure state estimation to solve the game problem between sensors and attackers.

Based on the above discussion, this paper presents a novel state estimation method based on reinforcement learning for a multi-sensor CPS under DoS attacks. Different from other papers, the main contributions of the paper are as follows: (i) The existing achievements of the single-sensor CPS are not guaranteed to meet with the needs in realistic scenarios, thus the CPS secure issue is extended to the multi-sensor CPS to explore the optimal policy selection problem of sensors and attackers under DoS attacks. (ii) Different from other works that passively describe the state of a system after DoS attacks, we positively analyze the optimal policy selection of sensors and attackers in a multi-sensor CPS. (iii) To further release the restriction on the set of states, the state space is unrestricted in order to comprehensively describe the state transition of the constructed markov chain in this paper. Besides, the complexity of the algorithm is decreased by introducing a penalty in the reward function.

The remainder of the paper is organized as follows. Section 2 portrays the system model for multi-sensor CPS under DoS attacks as well as the state estimation based on Kalman filter, and illustrates the state estimation processes. In Sect. 3, the secure state estimation algorithm based on reinforcement learning for multi-sensor CPS in confronting DoS attacks is proposed. The simulation results for a 3-sensor CPS in Sect. 4 demonstrate the effectiveness of the algorithm, and conclusions are drawn in Sect. 5.

2 Problem Formulation and Preliminaries

2.1 System Model

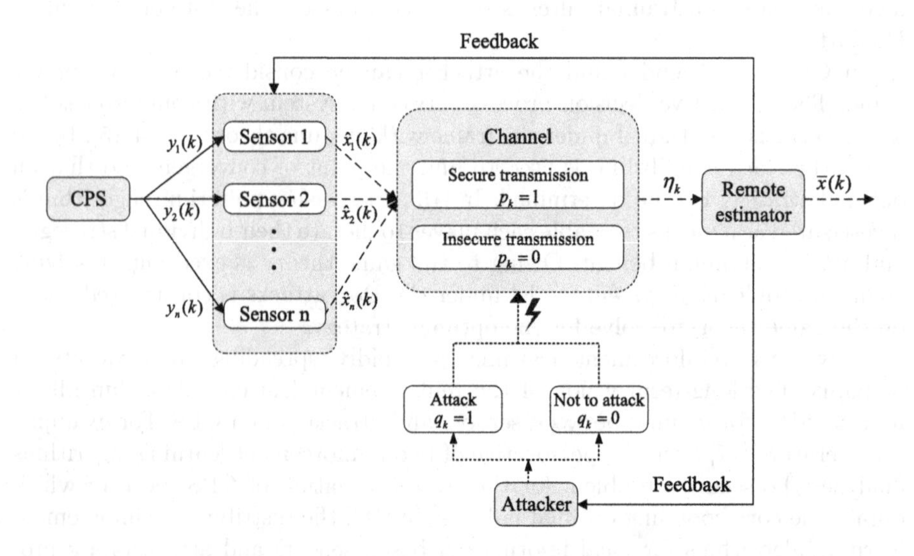

Fig. 1. The single-target multi-sensor system model under DoS attacks

Consider a CPS with n sensors and a remote estimator as shown in Fig. 1, where different sensors work together to monitor a specified CPS. At time k, the expression of sensor m under DoS attacks can be given by:

$$\begin{cases} x(k+1) = Ax(k) + Bw(k) \\ y_m(k) = Cx(k) + v_m(k), \end{cases} \tag{1}$$

where $k \in \mathbb{Z}$ indicates the discrete time step. $x(k) \in \mathbb{R}^{d_x}$ refers to the state vector of the system and $y_m(k) \in \mathbb{R}^{d_{y_m}}$ is the sensor measurement vector by sensor m at time k. $w(k) \in \mathbb{R}^{d_w}$ and $v_m(k) \in \mathbb{R}^{d_{v_m}}$ represent the process and measurement noises with zero mean, and their covariance matrices are $Q(k)$ and $R_m(k)$, respectively. A, B, and C are coefficient matrices with corresponding dimensions.

As seen the system model in Fig. 1, data measured by n sensors is transmitted to the remote estimator via a wireless channel. Each sensor $m \in \{1, 2, ..., n\}$ has the option of secure transmission $p_k = 1$ or insecure transmission $p_k = 0$ in the channel. In the channel between the sensor and the remote estimator, the attackers has two actions that can be chosen respectively denoted as $q_k = 1$ and $q_k = 0$. The former indicates that the attackers launch DoS attacks on the communication channel, while the latter on the contrary. At time k, the state estimation based on the packet from sensor m denoted by $\bar{x}(k)$. The symbol η_k indicates whether the packet is successfully received by the remote estimator. We denote η_k to indicate whether packet is lost at time k, which can be expressed as

$$\eta_k = \begin{cases} 1, p_k = 0, q_k = 1 \\ 0, p_k = 0, q_k = 0 \\ 0, p_k = 1, q_k = 1 \\ 0, p_k = 1, q_k = 0. \end{cases} \tag{2}$$

2.2 State Estimation Based on Kalman Filter

State estimation is performed employing a local Kalman filter to recursively update the system state. For each sensor m, the initial state $x(0)$ is a zero-mean Gaussian random vector with non-negative covariance. At each time k, the Kalman filter is run to obtain the minimum mean-squared error (MMSE) $\hat{x}(k)$ of the state vector $x(k)$ based on the measured data. The MMSE estimate of sensor m is denoted by:

$$\hat{x}_m(k) = \mathbf{E}\left[x(k) \mid y_m(1), \ldots, y_m(k)\right], \tag{3}$$

with its corresponding estimation error covariance

$$P_m(k) = \mathbf{E}\big[\left(x(k) - \hat{x}_m(k)\right)\left(x(k) - \hat{x}_m(k)\right)^T \mid y_m(1), \ldots, y_m(k)\big]. \tag{4}$$

According to the Kalman filter equations, $\hat{x}_m(k)$ and $P_m(k)$ are updated recursively. For simplicity, the Lyapunov and Riccati operators h and $\tilde{g}_m$ are defined as

$$\begin{aligned} h(X) &\triangleq AXA^T + Q \\ \tilde{g}_m(X) &\triangleq X - XC^T\left[CXC^T + R_m\right]^{-1} CX. \end{aligned} \tag{5}$$

Then the recursive updating equation of Kalman filter can be expressed as follows:

$$\begin{aligned}
&\hat{x}_m(k \mid k-1) = A\hat{x}_m(k-1) \\
&P_m(k \mid k-1) = h\left(P_m(k-1)\right) \\
&K_m(k) = P_m(k \mid k-1)C^T\left[CP_m(k \mid k-1)C^T + R_m\right]^{-1} \\
&\hat{x}_m(k) = \hat{x}_m(k \mid k-1) + K_m(k)\left(y_m(k) - C\hat{x}_m(k \mid k-1)\right) \\
&P_m(k) = \tilde{g}_m\left(P_m(k \mid k-1)\right).
\end{aligned} \tag{6}$$

2.3 State Estimation Process

The remote estimator performs state estimation based on the packet from the sensor m. In this paper, we define $\bar{x}(k)$ and $P(k)$ to denote the state estimation of remote estimator and the corresponding error covariance respectively. To simplify the game as well as the reinforcement learning algorithm, we assume that the error covariance matrix has converged to the steady state, i.e., $P(k) = \bar{P}_m$.

The estimation process can be formulated as follows: if the local estimation arrives, the estimator synchronizes its own estimate with it; otherwise, the estimator predicts $\bar{x}(k)$ according to the optimal estimate from the previous time step, i.e.,

$$\bar{x}(k) = \begin{cases} A\bar{x}(k-1), & \eta_k = 1 \\ \hat{x}_m(k), & \eta_k = 0, \end{cases} \tag{7}$$

with the corresponding estimation error covariance

$$P(k) = \begin{cases} h(P(k-1)), & \eta_k = 1 \\ \bar{P}_m, & \eta_k = 0. \end{cases} \tag{8}$$

In order to elaborate the changing process of the error covariance $P(k)$, an interval is defined as $\tau_k \triangleq k - \max_{0 \leq l \leq k}\{l : \eta_l = 1\}$, which is obtained by the time interval between the current time k and the time l when the packet is last received. When no packet loss occurs, τ_k recursively increases by 1, otherwise τ_k is updated to 0, that is,

$$\tau_k = \begin{cases} 0, & \eta_k = 1 \\ \tau_{k-1} + 1, & \eta_k = 0. \end{cases} \tag{9}$$

The estimation error covariance based on the time interval τ_k can be derived from (8) and (9) as

$$P(k) = h^{\tau_k}(\bar{P}_m). \tag{10}$$

Here, we assume that the packet successfully arrives at the remote estimator at the beginning of transmission, so $\tau_0 = 0$. Therefore, the initial value of the estimation error covariance is $P(0) = \bar{P}_m$.

In this paper, a Markov chain is introduced to represent the stochastic process of transition among states. On the basis of Markov property, the probability distribution of the next state can only be determined by the current state, independent of the state of the previous time series. Figure 2 represents the Markov chain of state transition with two sensors as an example.

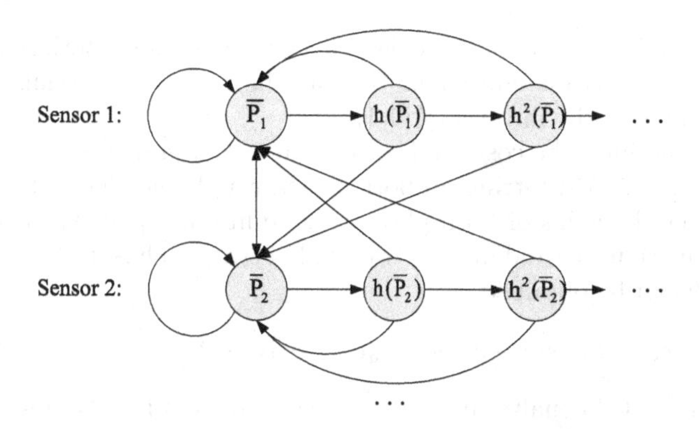

Fig. 2. The transition of Markov chain P(k)

3 Secure State Estimation Based on Reinforcement Learning

Reinforcement learning is an important research branch of machine learning algorithms, which focuses on learning through the interaction process between agents and their environment to achieve their own goals. As a value-based algorithm in reinforcement learning methods, the Q-learning algorithm constructs a Q-table of states and actions to store the expectation of gain as Q-value at each time. The Q-value is continuously updated during the learning process, and the action that obtains the greatest gain is selected based on the Q-table.

The goal of reinforcement learning is to find the optimal policy for a given Markov decision process (MDP). The MDP considers agents interacting with the environment by actions to obtain rewards. To more briefly describe the interaction, the MDP can be represented by a five-tuple: $MDP ::= \langle \mathcal{S}, \mathcal{A}, \mathcal{P}_{sa}, \gamma, \mathcal{R} \rangle$ [19], where $\mathcal{S}$ and $\mathcal{A}$ refer to the finite set of states and actions, respectively. $\mathcal{P}_{sa}$ denotes the set of state transition probabilities, indicating the probability of taking a certain action at a state and transferring to the next state. γ represents a discount factor for the decision process and $\mathcal{R}$ refers to the reward obtained at that moment given the current state and action. In this paper, an MDP is established to describe the interaction decision-making of the system defender and attacker, as depicted in detail below.

a) State: In a CPS, there are n sensors, corresponding to n initial states. Whenever a packet is lost, a new state is generated.
b) Action: The action combinations are defined as $a_k = (n_k, p_k, q_k)$, where n_k represents the selected sensor serial number, p_k represents whether to spend cost to transmit, and q_k represents whether the attacker initiates DoS attacks.
c) State transition: Consider the state of system as $s_k = P(k)$. Since packet loss may or may not occur, a corresponding transition will take place between the states, which is detailed described in the Sect. 2.3.

d) Discount factor: In order to achieve better algorithmic performance, a discount factor γ is introduced in the calculation of the cumulative reward, which locates in the interval of $[0, 1]$.
e) Reward function: The costs and actions are taken into account in the reward function, as the cost settings of both attacker and defender and the actions of them affect the gains of both players. In addition, a penalty is added to the reward function to avoid infinite traversal of states. Thus, the reward function at time k can be obtained as

$$r_k = Tr\left(s_k\right) + c_m p_k - c_a q_k + heaviside(s_k - h^i(P)) * t, \tag{11}$$

where t is the added penalty, and i is set as the number of packet loss occurrences as the condition for adding the penalty.

Remark 1. The system defender and the attacker have opposite objectives of minimizing the reward and maximizing the reward, respectively. For the attacker, it theoretically contributes to its reward maximization objective and makes it easier to launch DoS attacks. However, this is not the case in the actual operation of the algorithm. This is because the inhibitory effect of the cost of spending on the attacks partially offsets the promotional effect of a larger reward on the attacks, and in general the attacks are not facilitated.

For an MDP involving system defender and attacker, a Q-learning based algorithm is proposed to solve the optimal decision-making problem for the two players. The steps of the algorithm are as follows.

a) Initialization
Based on the input set of steady-state error covariance matrix and the number of sensors n, a set of states $S = \{\bar{P}_1, \bar{P}_2, \cdots, \bar{P}_n\}$ can be obtained.

Lemma 1. *For a system with finite actions and finite states, reinforcement learning uses the optimal action-value function to guide the agent to make decisions. Suppose the state of the system is s_k and action a_k is taken, the optimal action-value function can be expressed as*

$$Q_\star\left(s_k, a_k\right) = \max_\pi Q_\pi\left(s_k, a_k\right), \quad \forall s_k \in \mathcal{S}, \quad a_k \in \mathcal{A}.$$

According to Lemma 1, the size of the Q-value table is determined by the number of states and actions, so a Q-value table with n rows and $4n$ columns is initialized, where the initial value of the table is set to u.

b) Sensor and action selection
At each moment, the system randomly selects actions with the probability of ε or selects the optimal action with the probability of $1 - \varepsilon$ according to the minimum-maximum principle.
c) Observation of rewards and states
After the actions of the sensor and the attacker are determined, the reward obtained at the current moment is also available. The reward is observed at each moment according to (11).

d) Updating the Q-value table
Before updating the Q-value table at each moment, we determine whether there is a corresponding row in the Q-value table of the next state. Then, the corresponding value in the Q-value table is updated according to the following formula.

$$\begin{aligned}\widetilde{Q}_{k+1}(s,n,p,q) &= (1-\alpha_k)\,\tilde{Q}_k(s,n,p,q) \\ &+ \alpha_k\left(r_k + \rho \max_{q_{k+1}} \min_{p_{k+1}} \widetilde{Q}_k(s,n,p,q)\right).\end{aligned} \tag{12}$$

e) Obtaining the Nash equilibrium strategy
When the loop satisfies the termination condition, it means that the Q-value table has reached convergence. That is, the optimum Q-value table $\widetilde{Q}_\star(s,n,p,q)$ can be obtained.

4 Simulations and Experiments

Consider a CPS with three sensors and a remote state estimator. The system parameters are given as follows:

$$\mathbf{A} = \begin{bmatrix} 1 & 0.5 \\ 0 & 1 \end{bmatrix}, \mathbf{C} = \begin{bmatrix} 1 & 0 \end{bmatrix}, \mathbf{Q} = \begin{bmatrix} 0.8 & 0 \\ 0 & 0.8 \end{bmatrix}.$$

The three sensors have different measurement accuracy and their noise measurement covariance matrices are respectively $R_1 = 0.08$, $R_2 = 0.4$ and $R_3 = 0.8$. Running the Kalman filter, the steady state error covariance matrices of the three sensors are obtained as

$$\bar{\mathbf{P}}_1 = \begin{bmatrix} 0.0758 & 0.0577 \\ 0.0577 & 2.1043 \end{bmatrix}, \bar{\mathbf{P}}_2 = \begin{bmatrix} 0.3314 & 0.2343 \\ 0.2343 & 2.2627 \end{bmatrix}, \bar{\mathbf{P}}_3 = \begin{bmatrix} 0.6 & 0.4 \\ 0.4 & 2.4 \end{bmatrix},$$

where traces $Tr(\bar{P}_1)$, $Tr(\bar{P}_2)$ and $Tr(\bar{P}_3)$ respectively are 2.1801, 2.5941 and 3.

In the game between defender and attacker in a multi-sensor CPS, the defender can choose whether to spend a certain cost on defense depending on the situation. The defense cost of the three sensors decreases sequentially with cost values of $c_1 = 10.7$, $c_2 = 9.2$ and $c_3 = 6.6$. The cost for attackers to launch DoS attacks in the system model is set to $c_a = 1.5$.

In the MDP corresponding to this simulation experiment, there are three initial states $\bar{P}_1$, $\bar{P}_2$ and $\bar{P}_3$. A new state is generated only when a new packet loss condition occurs. According to the reward setting in (11), when two consecutive packet losses occur, a penalty of $t = 10$ is added to the reward function. When the number of sensors in a multi-sensor system is determined, the number of action combinations $a_k = (n_k, p_k, q_k)$ is also determined.

After the secure state estimation algorithm based on reinforcement learning is executed for 5000 iterations, the converged system contains 11 states, which are $\bar{P}_1, \bar{P}_2, \bar{P}_3, h(\bar{P}_1), h(\bar{P}_2), h(\bar{P}_3), h^2(\bar{P}_1), h^2(\bar{P}_2), h^2(\bar{P}_3), h^3(\bar{P}_2), h^3(\bar{P}_3)$. To facilitate the presentation, some of the states such as $\bar{P}_1, \bar{P}_2, \bar{P}_3, h(\bar{P}_2), h(\bar{P}_3), h(\bar{P}_1)$

Table 1. $\widetilde{Q}^*(s, n, p, q)$ matrix for convergence of multi-sensor system

Q-value \ State / Action	$\bar{P}_1$	$\bar{P}_2$	$\bar{P}_3$	$h(\bar{P}_2)$	$h(\bar{P}_3)$	$h(\bar{P}_1)$
(0,0,0)	8.143	8.166	8.038	20.708	9.270	6.091
(0,0,1)	9.764	41.345	27.854	44.299	22.923	16.101
(0,1,0)	18.749	18.749	18.738	28.044	20.616	18.901
(0,1,1)	17.475	17.343	17.238	26.512	100.000	33.670
(1,0,0)	19.494	20.830	19.494	75.335	21.497	19.495
(1,0,1)	9.731	37.776	27.854	44.073	29.249	23.095
(1,1,0)	28.574	16.082	14.694	27.916	23.537	14.779
(1,1,1)	13.195	13.209	13.194	33.639	13.281	13.253
(2,0,0)	19.962	19.958	19.900	40.464	19.936	100.000
(2,0,1)	11.374	28.684	27.854	50.952	22.986	15.790
(2,1,0)	11.500	11.500	11.500	12.860	12.394	11.500
(2,1,1)	10.000	10.002	10.000	33.877	100.000	10.372

are extracted as shown in Table 1. The action combination in the Table 1 is $a = (n, p, q)$, denotes serial number of the selected sensor, the sensor and attacker action selection. Each value in Table 1 represents the convergent value of the corresponding state action pair $\widetilde{Q}(s, n, p, q)$.

Taking the initial state of the three sensors as an example, the learning process of $\widetilde{Q}(s, n, p, q)$ is plotted in Fig. 3. According to the Fig. 3, it can be concluded that with the continuous iterations of the reinforcement learning algorithm, the attacker and defender gradually converge to the Nash equilibrium solution $\widetilde{Q}^*(s, n, p, q)$ eventually. In the first 500 iterations, the algorithm follows the $\varepsilon - greedy$ strategy in the trial-and-error exploration phase, and the elements of the Q-table are monotonically non-increasing. Through iterations learning of 500 – 5000, the elements of the Q-table can converge to a stable value.

Table 2. Nash equilibrium strategy for multi-sensor systems (n, p) and q

State	Nash equilibrium strategy	
	Defender strategy (n, p)	Attacker strategy q
$\bar{P}_1$	(0,0)	1
$h(\bar{P}_1)$	(2,1)	0
$h^2(\bar{P}_1)$	(0,0)	0
$\bar{P}_2$	(2,1)	0
$h(\bar{P}_2)$	(2,1)	0
$h^2(\bar{P}_2)$	(2,0)	0
$\bar{P}_3$	(2,1)	0
$h(\bar{P}_3)$	(0,0)	1
$h^2(\bar{P}_3)$	(0,0)	0

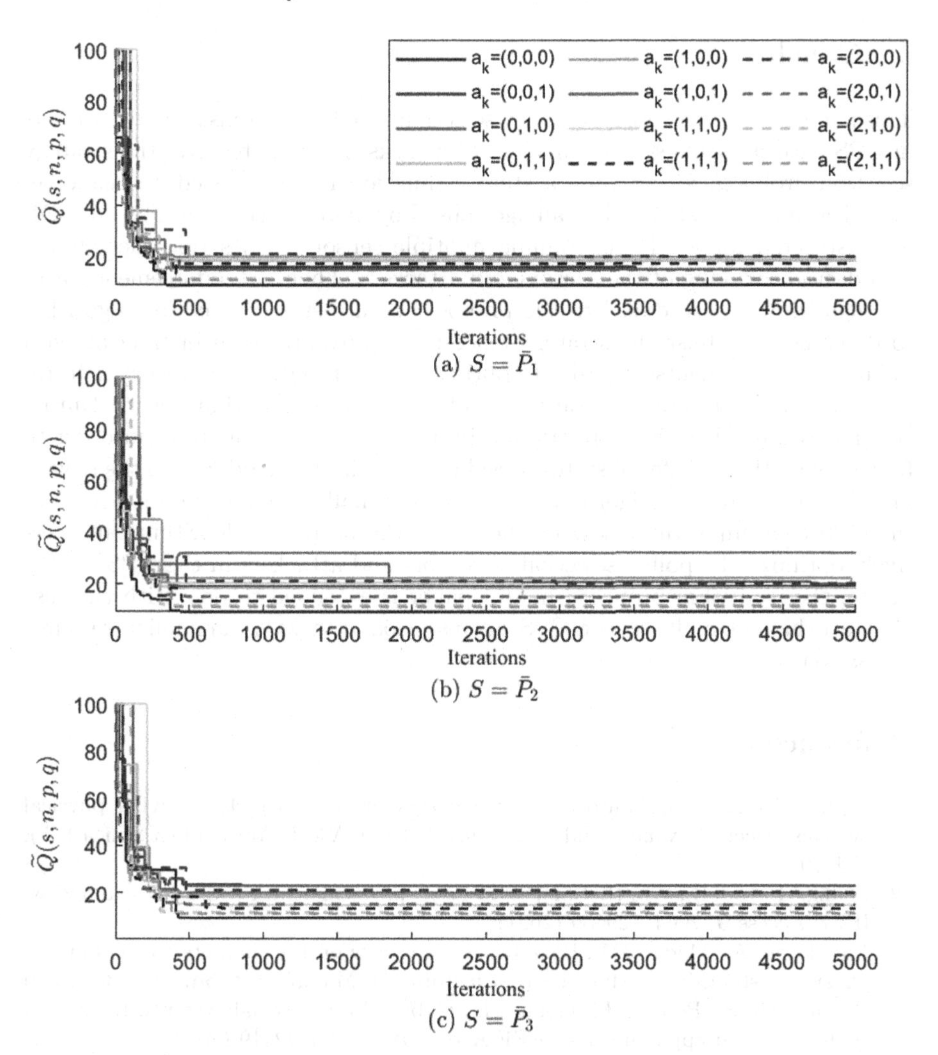

(a) $S = \bar{P}_1$

(b) $S = \bar{P}_2$

(c) $S = \bar{P}_3$

Fig. 3. Learning process $\widetilde{Q}(s, n, p, q)$ for multi-sensor system

By solving the convergent values in the Q-value table adopting a linear programming approach, the Nash equilibrium strategy of the game can be obtained, as shown in Table 2. The defender's strategy consists of choosing the serial number of the sensor and whether to defend, i.e., (n, p). The attacker's strategy is whether to launch DoS attacks, i.e., q. For example, in state $s = h(\bar{P}_1)$, the Nash equilibrium strategies of the defender and the attacker are $(1, 0)$ and 1 respectively. This means that sensor 1 is chosen in this state and no defense is taken, meanwhile the attacker chooses to launch DoS attacks in this case.

5 Conclusion

This paper studies the optimal policy selection problem for sensors and attackers in CPS with multiple sensors under DoS attacks. In order to solve the strategy selection problem, we propose a state estimation method based on reinforcement learning to evaluate the damage caused by DoS attacks. Initially, an MDP is constructed for a CPS containing multiple sensors to describe the interaction decision-making of the sensors and attackers. Then, a reinforcement learning algorithm is introduced to the proposed secure state estimation algorithm to dynamically adjust the strategy since it has advantages in interacting with unknown environments. In order to optimize the strategy, game theory is introduced to discuss the interaction process between sensors and attackers. During the interaction, there is no restriction imposed on the set of states in order to fully explore the transfer of states. Besides, a penalty is introduced to the reward function to ensure the algorithm's feasibility. Finally, the simulation results of the CPS containing three sensors show that the proposed algorithm can effectively optimize the policy selection of sensors and attackers in the CPS. In the future, the algorithm proposed in this paper is extended to deal with the case of multi-channel multi-sensor CPS against DoS attacks to maximize resource utilisation.

References

1. Duo, W., Zhou, M., Abusorrah, A.: A survey of cyber attacks on cyber physical systems: recent advances and challenges. IEEE/CAA J. Automatica Sinica **9**(5), 784–800 (2022)
2. Zhang, H., Liu, B., Wu, H.: Smart grid cyber-physical attack and defense: a review. IEEE Access **9**, 29641–29659 (2021)
3. Napoleone, A., Macchi, M., Pozzetti, A.: A review on the characteristics of cyber-physical systems for the future smart factories. J. Manuf. Syst. **54**, 305–335 (2020)
4. Pasandideh, S., Pereira, P., Gomes, L.: Cyber-physical-social systems: taxonomy, challenges, and opportunities. IEEE Access **10**, 42404–42419 (2022)
5. Amin, M., El-Sousy, F.F.M., Aziz, G.A.A., Gaber, K., Mohammed, O.A.: CPS attacks mitigation approaches on power electronic systems with security challenges for smart grid applications: a review. IEEE Access **9**, 38571–38601 (2021)
6. Burg, A., Chattopadhyay, A., Lam, K.Y.: Wireless communication and security issues for cyber-physical systems and the internet-of-things. Proc. IEEE **106**(1), 38–60 (2018)
7. Han, Z., Zhang, S., Jin, Z., Hu, Y.: Secure state estimation for event-triggered cyber-physical systems against deception attacks. J. Franklin Inst. **359**(18), 11155–11185 (2022)
8. Zhai, L., Vamvoudakis, K.G.: A data-based private learning framework for enhanced security against replay attacks in cyber-physical systems. Int. J. Robust Nonlinear Control **31**(6), 1817–1833 (2021)
9. Sun, Q., Zhang, K., Shi, Y.: Resilient model predictive control of cyber-physical systems under DoS attacks. IEEE Trans. Ind. Inf. **16**(7), 4920–4927 (2020)

10. Li, T., Chen, B., Yu, L., Zhang, W.A.: Active security control approach against DoS attacks in cyber-physical systems. IEEE Trans. Autom. Control **66**(9), 4303–4310 (2021)
11. Li, Z., Li, Q., Ding, D.W., Wang, H.: Event-based fixed-time secure cooperative control for nonlinear cyber-physical systems under denial-of-service attacks. IEEE Trans. Control Netw. Syst. 1–11 (2023)
12. Liu, Y., Yang, G.H.: Event-triggered distributed state estimation for cyber-physical systems under DoS attacks. IEEE Trans. Cybern. **52**(5), 3620–3631 (2022)
13. Basiri, M.H., Thistle, J.G., Simpson-Porco, J.W., Fischmeister, S.: Kalman filter based secure state estimation and individual attacked sensor detection in cyber-physical systems. In: 2019 American Control Conference (ACC), pp. 3841–3848 (2019)
14. Jin, Z., Zhang, S., Hu, Y., Zhang, Y., Sun, C.: Security state estimation for cyber-physical systems against dos attacks via reinforcement learning and game theory. In: Actuators, vol. 11, p. 192. MDPI (2022)
15. Li, Y., Yang, Y., Chai, T., Chen, T.: Stochastic detection against deception attacks in CPS: performance evaluation and game-theoretic analysis. Automatica **144**, 110461 (2022)
16. Wang, X.F., Sun, X.M., Ye, M., Liu, K.Z.: Robust distributed Nash equilibrium seeking for games under attacks and communication delays. IEEE Trans. Autom. Control **67**, 4892–4899 (2022)
17. Jin, Z., Ma, M., Zhang, S., Hu, Y., Zhang, Y., Sun, C.: Secure state estimation of cyber-physical system under cyber attacks: Q-learning vs. SARSA. Electronics **11**(19), 3161 (2022)
18. Dai, P., Yu, W., Wang, H., Wen, G., Lv, Y.: Distributed reinforcement learning for cyber-physical system with multiple remote state estimation under DoS attacker. IEEE Trans. Netw. Sci. Eng. **7**(4), 3212–3222 (2020)
19. Russell, S.J.: Artificial Intelligence a Modern Approach. Pearson Education Inc., London (2010)

A UAV Penetration Method Based on the Improved A* Algorithm

Shitong Zhang, Qing Wang(✉), Bin Xin, and Yujue Wang

School of Automation, Beijing Institute of Technology, Beijing, China
wangqing1020@bit.edu.cn

Abstract. This paper presents an algorithm for addressing the penetration problem between coverage regions. The algorithm combines the subregion coverage sequence determination and the A* algorithm-based drone obstacle avoidance methods to achieve path planning from the drone's starting point to the destination. The boustrophedon algorithm is employed to calculate optimal paths for independent subregions, while a genetic algorithm is utilized to determine an optimized coverage sequence, thereby minimizing the overall path length between regions. Additionally, the Laguerre graph construction algorithm is introduced, leveraging the Laguerre graph derived from the Voronoi graph to effectively describe the positional relationships of a given set of disjoint circles. To ensure obstacle avoidance, the A* algorithm is then applied to search for the shortest path within the Laguerre graph.

Keywords: UAVs · reconnaissance · area coverage · penetration path planning · region allocation

1 Introduction

In recent years, unmanned aerial vehicles (UAVs) have been widely utilized in various fields, including automated harvesting [1, 2], forest detection [3], geological surveying [4], nature conservation [5], and logistics transportation [6]. They have also gained popularity in military applications [7], gradually replacing traditional manned aircraft in specific mission domains. The application of UAVs in reconnaissance has garnered significant attention alongside their widespread adoption. One of the key challenges in accomplishing UAV reconnaissance missions is the area coverage problem [8], which involves finding an ordered list of waypoints and free paths between them in free space. The objective is to ensure complete coverage of the designated area as the aircraft moves along the planned path.

Some algorithms have been proposed to solve the area coverage problem. One classic area coverage algorithm is the boustrophedon coverage method [9]. The coverage path of the UAV can be calculated by this method when the shape of the area and the direction

This work was supported in part by the National Natural Science Fund of China under Grant 62003044 and the Basic Science Center Programs of NSFC under Grant 62088101.

B. Xin et al. (Eds.): IWACIII 2023, CCIS 1931, pp. 310–323, 2024.
https://doi.org/10.1007/978-981-99-7590-7_25

of entry are known. In practical reconnaissance missions, however, it is often necessary to consider reasonable planning of the coverage area order and flight paths between regions when one UAV is responsible for multiple coverage areas. The objective is to minimize flight path length and avoid obstacles. Some classical penetration path planning algorithms include A* algorithm [10] and its improved versions such as D* [11], graph-based traditional path planning algorithms, as well as intelligent algorithms like particle swarm optimization [12], genetic algorithm [13], and ant colony algorithm [14]. All of these algorithms have their own advantages and disadvantages: The A* algorithm seeks the shortest path by combining heuristic cost with actual cost. It can easily find the path with the minimum cost. However, in complex scenarios with multiple obstacles and a large search space, the presence of numerous irrelevant nodes in the search process can lead to increased search time. The Particle Swarm Optimization (PSO) algorithm simulates the survival behavior of birds to perform searches in complex spaces. However, this algorithm is prone to premature convergence and has poor global performance. The Genetic Algorithm (GA) simulates the theory of biological evolution to manipulate encodings, resulting in highly fit individuals that have broad coverage and are favorable for global search. However, due to its stochastic nature, the GA cannot consistently obtain the optimal solution. The Ant Colony Optimization (ACO) algorithm mimics the foraging behavior of ants in nature to find the optimal path. It enhances the optimization capability through positive feedback mechanisms and can be easily combined with other algorithms. However, this algorithm has slower search speed and may not efficiently find the optimal solution within a limited time.

This paper primarily employs the Genetic Algorithm to solve the traversal order problem in different regions, and combines it with an improved version of the A* algorithm to plan the breakthrough paths between regions.

The rest of this paper is organized as follows. Section 2 presents the problem description of inter-regional breakthrough and the models of the map and unmanned aerial vehicles (UAVs). Section 3 proposes a method using the Genetic Algorithm to calculate the optimal order for UAVs to visit different regions. Section 4 introduces the inter-regional breakthrough algorithm based on the improved A* algorithm. Section 5 provides the simulation experimental results and discussions. Finally, Sect. 6 presents the conclusion of this paper.

2 Problem Formulation

2.1 Problem Description

The problem of penetration path planning between coverage areas can be described in terms of the visiting order problem and the penetration problem. The visiting order problem involves determining the optimal sequence for a drone to conduct reconnaissance on multiple areas, with the objective of minimizing the straight-line distance outside the surveyed regions. This problem can be formulated as a Traveling Salesman Problem (TSP). However, since the drone does not always follow a straight-line path outside the surveyed regions due to considerations such as terrain and no-fly zones, it is necessary to plan penetration paths in order to meet safety requirements and navigate around these obstacles and threats.

2.2 Problem Modeling

UAV Model. To simplify the problem, the following assumptions are made: 1) The drone maintains a constant height during flight, allowing the problem to be solved in a two-dimensional area with terrain constraints represented as obstacles. 2) The drone model can be equivalently represented as an isosceles triangle, where the direction indicated by the vertex of the triangle represents the orientation of the drone. 3) The constraint imposed on the drone's turning radius, denoted as R, ensures that the minimum turning radius at minimum speed v_{min} is not violated. In other words, during a turn, the drone's turning radius should be greater than or equal to R. This constraint is essential to maintain safe maneuverability and prevent any potential risks or instability associated with sharp turns or small turning radii.

Due to the aforementioned design, the kinematic model of the drone can be represented by the following equation.

$$\begin{cases} \dot{x} = vcos\varphi \\ \dot{y} = vsin\varphi \\ \dot{\varphi} = \frac{gtan\theta}{v} \end{cases} \tag{1}$$

where:

- $\dot{x}$ and $\dot{y}$ represent the derivatives of the drone's position in the x and y directions, respectively:
- v represents the drone's linear velocity,
- φ represents the drone's orientation or heading angle,
- $\dot{\varphi}$ represents the derivative of the orientation angle,
- g represents the gravitational acceleration.
- θ represents the roll angle of the UAV.

These equations describe the relationship between the drone's position, velocity, and orientation, and allow for the prediction and control of its movement within the given constraints.

Modeling of Map Obstacles and Target Areas. The information of k target areas to be covered is known. These target areas do not overlap and have a certain distance between them. There are m threats and terrain obstacles between the target areas. The shapes of the target areas and obstacles are irregular and may have varying concavities and convexities.

Figure 1 shows an example of the model of map obstacles and target areas. The map obstacles and target areas are randomly distributed sets of vertices on the two-dimensional map plane. Since these vertices can form non-unique polygons, they are sorted in a certain order: for a set of vertices $A \sim L$ within a collection, the centroid M is taken as the pole, and the vertices are sorted counterclockwise based on the polar angles α formed by each vertex and the pole. This forms a fixed polygon representing the shape of the obstacles and target areas. The target areas to be covered and the map obstacles do not overlap and are known input information before path planning.

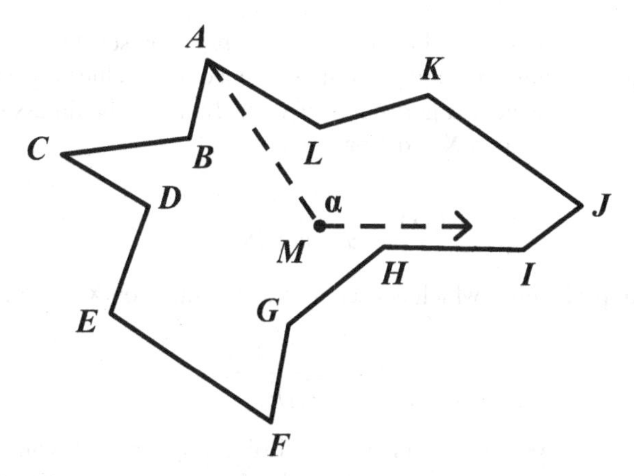

Fig. 1. Example of map obstacles and target areas model

3 Subregion Coverage Sequence Determination

The boustrophedon algorithm have been used to calculate the coverage path for the target region, as shown in Fig. 2. By applying the boustrophedon algorithm, we can determine two different directional entry and exit points for the coverage region, as well as a coverage path. After calculating the paths for each independent subregion using the boustrophedon algorithm, we obtain the path points for each subregion. Each subregion can be covered in four different paths, depending on the entry point. It is evident that regardless of the entry point chosen, the path length and the number of turns within the region almost remain the same. However, when considering the paths between different subregions, the choice of entry and exit points significantly affects the inter-region paths. Therefore, it is necessary to make decisions about the subregion coverage sequence to minimize the path length between regions. GA can be employed to solve this problem.

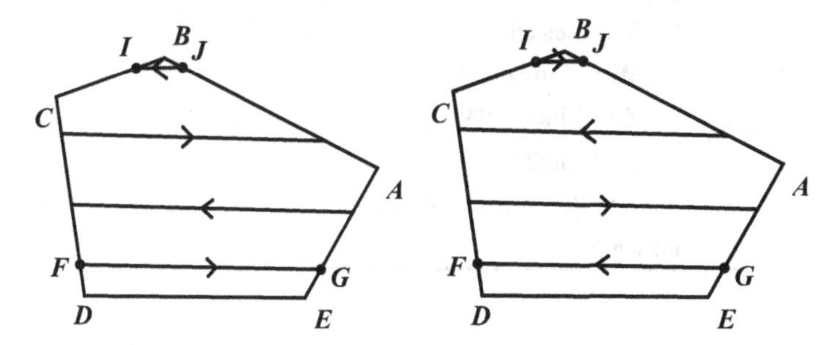

Fig. 2. Selection of Entry Points for Coverage Area

For N subregions that have undergone convex decomposition and removal of concave points resulting in M subregions, we have $4M$ different entry/exit combinations. These combinations can be encoded using a coding length of $2M$. The first M bits represent the coverage sequence of the different subregions, using a non-repeating sorting coding method with values ranging from 1 to M. The last M bits represent the entry/exit methods for each subregion, using a repeating coding sequence with values ranging from 1 to 4. The pseudo code for a genetic algorithm is given in Table 1. The fitness function which is used to evaluate population X is defined as:

$$f_i(x) = \frac{M}{distance(S, E)}. \tag{2}$$

The survival probability which is used to select population X is defined as:

$$g(x) = \frac{f_i(x)}{\sum_i f_i(x)}. \tag{3}$$

Randomly select individuals which is survival in population X with probability P_r, perform crossover operations between these individuals with probability P_c, and perform mutation operations with probability P_m. Solve gene conflicts in the coding using the partial crossover mapping method. The partial crossover mapping method is illustrated in Fig. 3. By exchanging the gene sequences indicated in the figure, conflicts occur. The mapping relationship is recorded using the hash table on the right side of the figure. Then, the conflict segments (excluding the exchanged segments) are replaced according to the mapping relationship to obtain the lower part of the figure, representing the gene sequence after the exchange, thereby resolving the conflict.

Table 1. Pseudo code of Genetic Algorithm(GA)

Algorithm1 Genetic Algorithm(GA)	
1	**X**←Generate_Initital_Population
2	Evaluate(**X**)
3	**while** (Stopping criterion is not met or optimal solution is not reached) **do**
4	**Y** ← Selection(**X**)
5	**W** ← Crossover(**Y**)
6	**Z** ← Mutation(**W**)
7	Evaluate(**Z**)
8	**X**← Replacement(**Z**, **X**)
9	**end while**

Using this algorithm, a complete path from the initial position S of the drone to the end position after covering all the subregions can be obtained. However, it does not consider obstacles or path planning between the subregions. Further refinement is required to determine the actual path between the marked points representing the subregions.

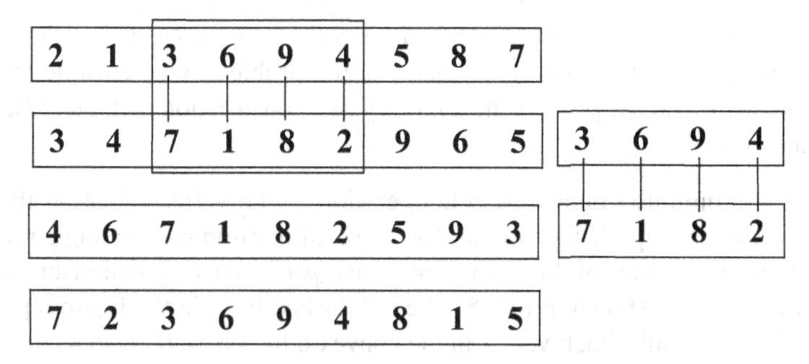

Fig. 3. Partial Crossover Mapping for Resolving Gene Conflicts

4 A* Algorithm-Based Drone Obstacle Avoidance

4.1 Laguerre Graph Construction Algorithm

The Laguerre graph is used to describe the positional relationships of a point set under a given set of disjoint circles. As shown in Fig. 4, a set *Z* of n circles on the plane constitutes a collection. For a circle *S* in the collection *Z*, there exists a region *ABCDEF* such that the length of any tangent from any point in region *ABCDEF* to circle *S* is smaller than the length of the tangent to any other circle in set *Z*. This region is referred to as the Laguerre region associated with circle *S*. The vertices and edges formed by the Laguerre regions of all circles in set *Z* constitute the Laguerre graph (highlighted in red in the figure).

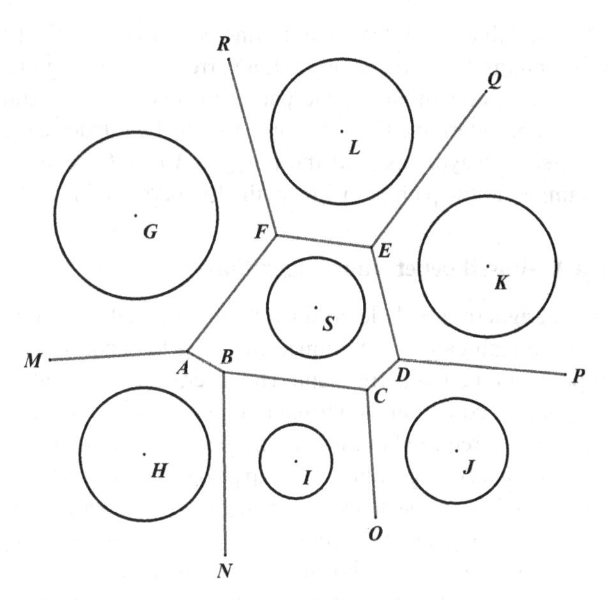

Fig. 4. Laguerre Graph in a 2D Plane (Color figure online)

The vertices of the Laguerre graph can be obtained by considering the minimum bounding circle of the threatening obstacles in the problem. One feasible method for constructing the Laguerre graph is the auxiliary line construction method [17], which is outlined as follows:

1) Apply the minimum bounding circle operation to the vertex set of all threatening obstacles on the map plane to obtain the set of all minimum bounding circles.
2) Extract the coordinates of the circle centers and perform triangulation and determine the convex hull, as shown in Fig. 5. The red dashed lines in the figure represent the obtained convex hull. Each vertex in the convex hull corresponds to a circle center.
3) Traverse each triangle in the triangulated convex hull and determine its circumcircle, as shown in Fig. 6. Circle *E* and circle *F* in the figure correspond to the circumcircles of triangle *ABD* and triangle *BCD* in the convex hull, respectively.
4) Calculate the intersection points between the circumcircle of the convex hull and the circles determined by the triangle vertices. From these intersection points, obtain the pairwise lines.
5) Find the intersection points of the two sets of non-parallel lines. Draw perpendicular lines from each intersection point to the corresponding edges of the triangles. [15] has already proven that the three perpendicular lines of a non-right triangle must intersect at one point. This intersection point represents a vertex of the Laguerre graph, and the perpendicular lines drawn from the intersection point to the three edges of the triangle form the edges of the Laguerre graph. As shown in Fig. 7, point L represents a waypoint of the Laguerre graph, and the three blue rays extending from L represent the edges of the Laguerre graph. The pseudo code for a minimum bounding circle algorithm is given in Table 2.

Subsequently, a feasible path L from the starting point to the destination point can be constructed by combining the vertices in the Laguerre graph. Using this feasible path, a minimum convex hull containing all the path points from the starting point to the destination point can be determined. As shown in Fig. 8, the Laguerre route *TVAFEWU* represents the Laguerre waypoints, and the polygon *VTEWU* represents the desired convex hull containing all the path points from the Laguerre graph.

4.2 Local Grid A*-Based penetration Algorithm

After obtaining the Laguerre graph determined by the original set of threatening obstacles, a rough feasible path can be determined by using the coordinates of the starting and destination points in the Laguerre graph. However, since this path is a feasible but suboptimal path determined within the virtual bounding circle obstacles, further search using the A* algorithm is required to find the shortest path within the range defined by this path. In the traditional A* algorithm, the map grid is divided into pre-defined global map grids. However, in this approach, by integrating with the Laguerre graph method for search, local grid partitioning can be used to reduce the number of grid points, thereby improving computational speed. As shown in Fig. 9, using this method in a Laguerre graph pre-searched map ensures that obstacles can be avoided by searching through fewer grid points. And because these grids can be more refined, the paths will also be more precise.

Table 2. Pseudo code of **Minimum Bounding Circle Algorithm**

	Algorithm2 Minimum Bounding Circle Algorithm
1	**A** ← Circle_Circle(**O**)
2	**C** ← Circle_Center(**Circle**)
3	**H** ← Convex_Hull(**C**)
4	**for** neighbor point P_{i-1} P_i P_{i+1} in **H**
5	**T**[i] ← Triangle(P_{i-1} P_i P_{i+1})
6	**end for**
7	**for** all the triangle **T**[i]
8	**B**[i] ← Circle_Circle(**T**[i])
9	**end for**
10	**for** all circle **A**
11	**for** all circle **B**[i]
12	point **I**[i] ← Calculate_Intersection(**B**[i],**A**)
13	**end for**
14	**end for**
15	**for** point pair $\mathbf{P1_i}$ $\mathbf{P2_i}$ in **I**[i]
16	**L**[i] ← Calculate_Line($\mathbf{P1_i}$,$\mathbf{P2_i}$)
17	**end for**
18	**for** all **L**[i-1] **L**[i] **L**[i+1] in **L**
19	point **D**[i-1] ← Calculate_Intersection(**L**[i-1], **L**[i])
20	point **D**[i] ← Calculate_Intersection(**L**[i], **L**[i+1])
21	point **D**[i+1] ← Calculate_Intersection(**L**[i+1], **L**[i-1])
22	**Laguerre**[i] ← Calculate_Perpendicular_point(**T**[i],D[i-1,i,i+1])
23	**end for**

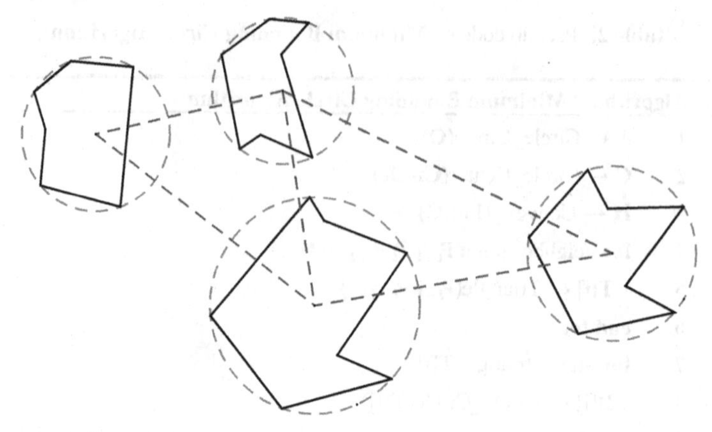

Fig. 5. Convex hull of minimum bounding circles for the threatening obstacle vertices. (Color figure online)

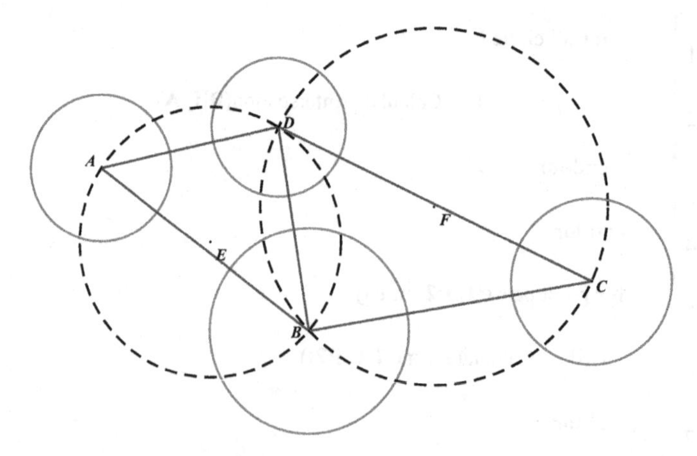

Fig. 6. Circumcircles of triangles in the convex hull.

By performing a pre-search using the Laguerre graph and further search using the A* algorithm, the rapid identification of the breakthrough paths between the target coverage areas can be achieved. Combining the paths obtained from the boustrophedon algorithm, genetic algorithm, and A* algorithm, a discrete path from the drone's starting point to the mission's endpoint can be obtained. However, these paths may not be smooth. Therefore, this paper utilizes Dubins curves to further process the drone's turning paths and satisfy the constraints of the drone's turning radius.

4.3 Path Smoothing Algorithm Based On Dubins

Dubins curves can be planned based on the starting and ending positions, directions, and the curvature constraints of the drone to obtain the shortest path that satisfies the curvature constraint. [16] has proven the existence of a path between two points, and

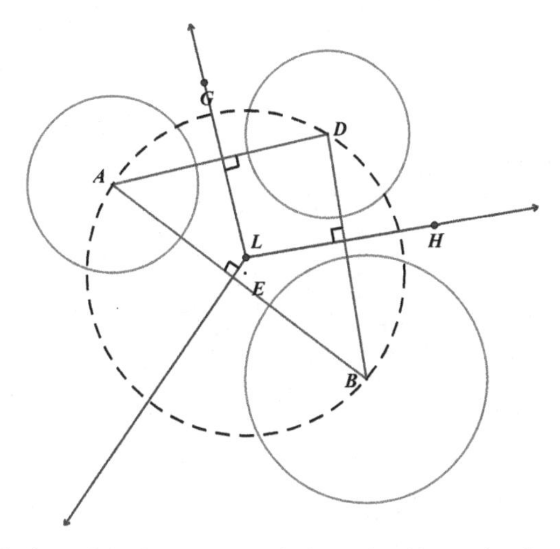

Fig. 7. Vertices and edges of the Laguerre graph determined by a triangle. (Color figure online)

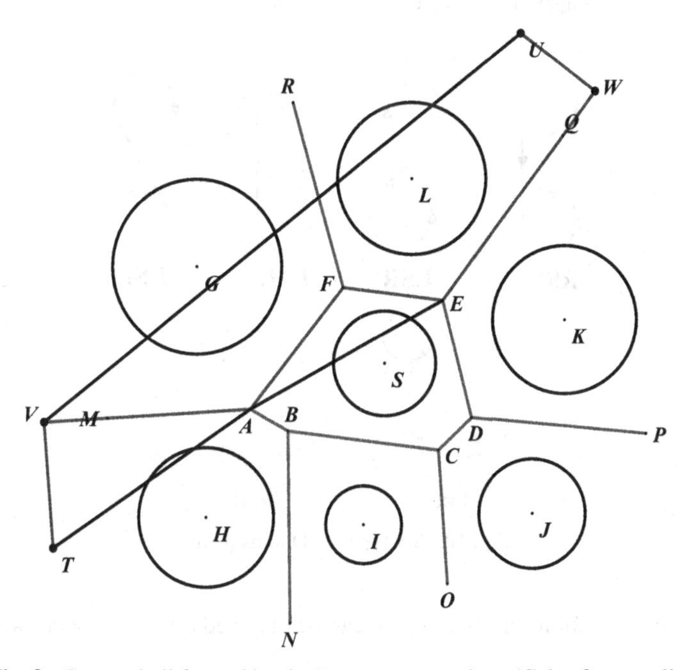

Fig. 8. Convex hull formed by the Laguerre waypoints. (Color figure online)

any path can be composed of two segments of circular arc paths and one straight line. As shown in Fig. 10, there are six types of Dubins paths: LSL, RSR, LSR, RSL, LRL, and RLR. Here, L represents counterclockwise circular arc motion, R represents clockwise circular arc motion, and S represents straight-line motion. By continuously inputting

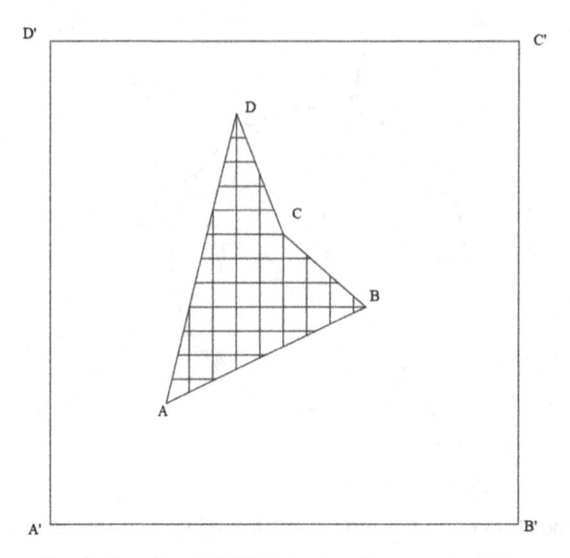

Fig. 9. Localized Grid Method on Pre-searched Map

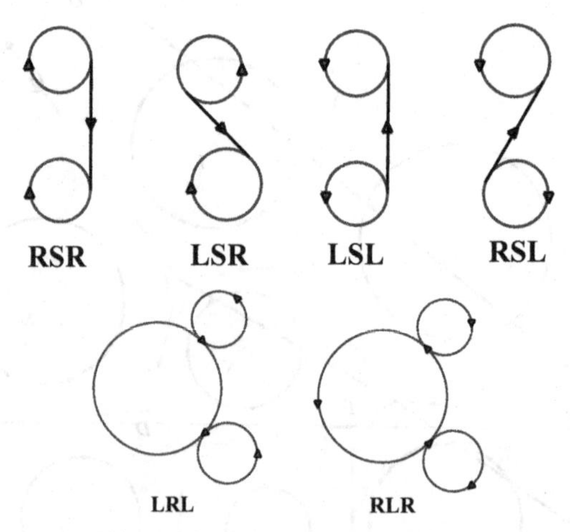

Fig. 10. Six types of Dubins paths.

path points and directions, a smooth path can be obtained after processing with Dubins curves.

5 Simulation Example

The enhanced A* algorithm was simulated by setting the map obstacles and UAV parameters in a simulation software. The specific settings are as follows.

On a 600 × 600 grid map, a simulation is conducted with the improved A* algorithm, starting from (0, 200) and ending at (600, 500). The simulation results are shown in

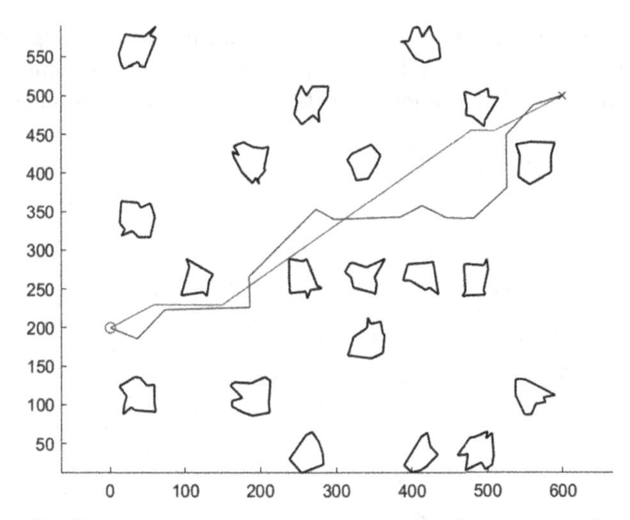

Fig. 11. Improved A* algorithm penetration simulation results

Fig. 11. The black enclosed area represents the map obstacles, the blue routes are the paths formed by the Laguerre graph algorithm's pre-search, and the red routes are formed by further using the A* algorithm within the convex hull formed by the blue route points. The Dubins algorithm is used to set the turning radius of the unmanned aerial vehicle, and the local effect is shown in Fig. 12. The green line in the figure represents the original flight route, and the purple line represents the Dubins path.

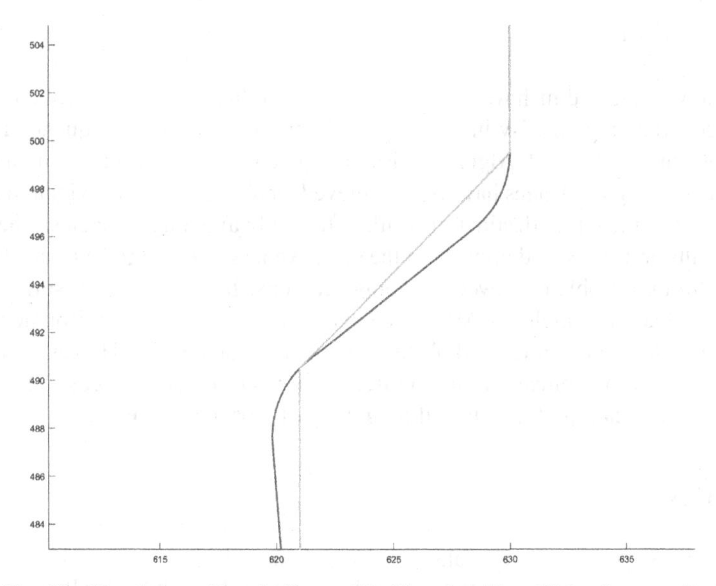

Fig. 12. Simulation of local effects of Dubins algorithm (Color figure online)

The runtime tests of the improved A* algorithm and the traditional algorithm on some cases are shown in Fig. 13. It can be seen that the improved algorithm can accelerate the search speed of the A* algorithm and better achieve breakthroughs in scenarios with larger ranges.

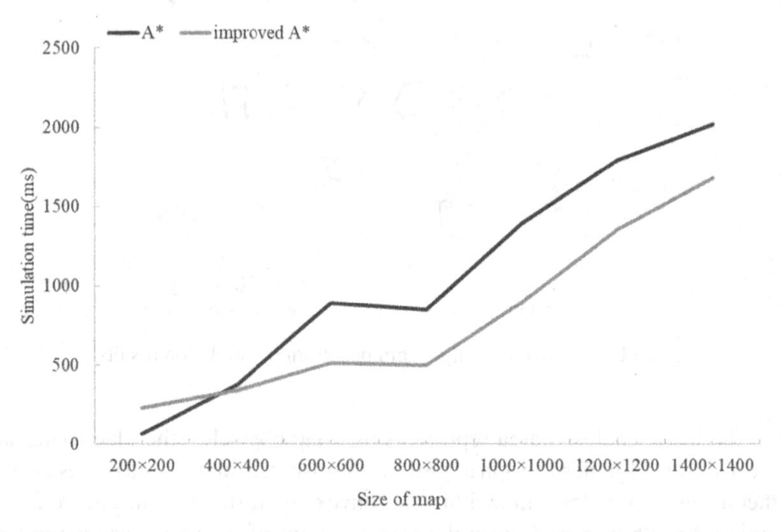

Fig. 13. The comparison results between the A* algorithm and the improved algorithm

6 Conclusion

In this study, a algorithm have been proposed to address the breakthrough problem between coverage regions. By integrating the Subregion Coverage Sequence Determination algorithm and the A* Algorithm-Based Drone Obstacle Avoidance method, efficient path planning for drones have been achieved, enabling them to navigate from their starting point to the target destination while effectively avoiding obstacles. The experimental results and analysis demonstrate the effectiveness of our algorithm in addressing the breakthrough problem between coverage regions. It offers a promising approach for unmanned aerial vehicles (UAVs) to navigate through complex environments while maintaining efficient coverage and obstacle avoidance capabilities. However, this algorithm is too ideal to applicate. In the future, further consideration needs to be given to the practical application of the algorithm under real-world constraints.

References

1. Ollis, M., Stentz, A.: First results in vision-based crop line tracking. In: Proceedings Conference IEEE International Robotics and Automation **1**, pp. 951–956 (1996)
2. Ollis, M., Stentz, A.: Vision-based perception for an automated harvester. In: Proceedings IEEE/RSJ Int. Intelligent Robots and Systems IROS'97. Conference. **3**, pp. 1838–1844 (1997)

3. Muid, A., et al.: Potential of UAV application for forest fire detection. J. Physics: Conference Series **2243**(1) (2022)
4. Mercuri, M., Conforti, M., Ciurleo, M., Borrelli, L.: UAV application for short-time evolution detection of the vomice landslide (South Italy). Geosciences, **13**(2) (2023)
5. Woellner, R., Wagner, T.C.: Saving species, time and money: application of unmanned aerial vehicles (UAVs) for monitoring of an endangered alpine river specialist in a small nature reserve. Biological Conservation, **233** (2019)
6. Ghelichi, Z., Gentili, M., Mirchandani, P.B.: Drone logistics for uncertain demand of disaster-impacted populations. Transportation Research Part C: Emerging Technologies **141**, 103735 (2022)
7. Yang, S., Cheng, H., Li, T., Sun, W.: Application research of UAV reconnaissance image target localization in military. Infrared Technol. **38**(06), 467–471 (2016)
8. Galceran, E., Carreras, M.: A survey on coverage path planning for robotics. Robotics and Autonomous Systems **61**(12) (2013)
9. Juan, C.R., Juris, V., Rolf, R.: Adaptive path planning for autonomous UAV oceanic search missions. In: AIAA 1st Intelligent Systems Technical Conference, Chicago (2004)
10. Zhou, X.: Principles and Applications of Artificial Intelligence. Aviation Industry Press, Beijing (1997)
11. Stentz, A.: The focused D* algorithm for real-time replanning. In: Proceedings of the 14th International Joint Conference on Artificial Intelligence, pp. 1652–1659 (1995)
12. Xu, H., Shan, J.: Global path planning of mobile robots based on improved particle swarm algorithm. Mechanical Engineer, **233**(11), 24–26 (2010)
13. Wu, Y., Li, Z., Sun, C., Wang, Z., Wang, D., Yu, Z.: Measurement and control of system resilience recovery by path planning based on improved genetic algorithm. Measurement and Control **54**(7–8) (2021)
14. Zhao, J., Gao, X., Fu, X., et al.: Improved ant colony optimization algorithm for mobile robot path planning. Control Theory Appl. **28**(04), 457–461 (2011)
15. Wang, S., Wei, R., Shen, D., Qi, X., Luo, P.: Construction algorithm of laguerre diagram for route planning. Systems Eng. Electronics **35**(03), 552–556 (2013)
16. Dubins, L.E.: On curves of minimal length with a constraint on average curvature, and with prescribed initial and terminal positions and tangents. Am. J. Math. **79**(3), 497–516 (1957)

Hybrid D-DEPSO for Multi-objective Task Assignment in Hospital Inspection

Chun Mei Zhang[1,2](✉), Xin Yao Ma[1,2], and Bin Zhai[1,2]

[1] Taiyuan University of Science and Technology, No.66, Waliu Road, Wanbailin District, Taiyuan, Shanxi, China
zhangchunmei@tyust.edu.cn, {S20201503011,S202115110213}@stu.tyust.edu.cn
[2] Shanxi Key Laboratory of Advanced Control and Equipment Intelligence, Wanbailin District, Taiyuan, Shanxi, China

Abstract. Hospital inspection tasks include temperature measurement, disinfection, emergency treatment, etc. Inspection robots can assist people in carrying out autonomous inspections and reduce the pressure on hospital staff. In this paper, we focus on the task assignment of hospital inspection robots, i.e., assigning multiple tasks to different robots to achieve the highest level of task completion. For the task assignment model of the inspection robots, a multi-objective mathematical model of task assignment is established, considering the benefit, cost, and execution time of task assignment. For the optimization scheme, a hybrid algorithm of discrete differential evolution and particle swarm optimization (D-DEPSO) algorithm is designed, applying differential mutation operation to the population initialization process of the particle swarm optimization (PSO) algorithm to expand the diversity of the population and improve the optimization-seeking ability of the algorithm. For coordination among objectives, the adjustment method of objective weights is proposed so as to achieve balance among objectives. The experimental results show that the designed method can improve the utility of task assignment in the hospital inspection process and thus efficiently complete the hospital inspection tasks.

Keywords: Task allocation · Multi-objective · Hybrid D-DEPSO · Hospital inspection

1 Introduction

During the hospital inspection process, the use of robotic inspection offers significant advantages. In the complex scene of hospital inspection, tasks such as temperature measurement, disinfection and emergency treatment need to be reasonably assigned to the robots. The feasibility of applying intelligent robots in hospitals is analyzed [1], G. Z. Yang et al. [2] introduce the development and

R. B. G. thanks Natural Science Fund of Shanxi Province (202203021211198).

B. Xin et al. (Eds.): IWACIII 2023, CCIS 1931, pp. 324–337, 2024.
https://doi.org/10.1007/978-981-99-7590-7_26

application of intelligent robotic systems in hospitals and provides a reasonable outlook on their future direction. The task allocation problem is a popular research topic in the fields of artificial intelligence and robotics, and scholars have conducted extensive research on the mathematical models and algorithms for this problem. Literature [3] proposes a task scheduling algorithm that provides reasonable scheduling for tasks with request migration. The algorithm also allocates a model and proposes a corresponding algorithm, which better guarantees the task's latency requirements. However, the model's establishment does not take into account the interrelationship between multiple tasks. The utility-driven allocation model is proposed, which only takes into account the system's tasks and disregards their resource demands [4]. An adaptive heuristic resource allocation algorithm [5] is presented to prioritize resources based on their utility per unit of allocation, resulting in more efficient utilization of resources. Rajkumar et al. [6] proposed a task allocation model that maximizes the overall system utility The model and the proposed algorithm lack consideration of contingencies, which hinders their ability to guarantee real-time allocation. Literature [7] investigates the optimal task allocation for a multi-UAS cooperative strategy, optimizing four objectives in a real-world application scenario. The problem of multi-intelligent task allocation and proposes corresponding algorithms [8,9] is analyzed for distributed multi-objective optimization. Jeon et al. [10] propose a fleet optimization method that assigns multiple tasks to a single robot and introduce an algorithm aimed at reducing the computation required for finding paths. Lee et al. [11] propose a resource allocation model based on discrete business metrics and provide an optimal solution algorithm using dynamic programming as well as an approximation algorithm utilizing local search. Literature [12] has modelled resource allocation as a mixed integer programming problem, however, the method requires linearizing the objective function during application, which increases the algorithm's complexity.

For above analyzation, existing models generally consider the system utility as the optimization objective and establish a single-objective optimization model. However, they overlook the cost of resource as well as the time required for assignment, which results in a lack of comprehensive consideration for the optimization objective. Additionally, the existing algorithms are not efficient in handling task assignment. This paper analyzes the relationship between tasks by studying the actual hospital inspection environment of robots. Multi-objective task assignment mathematical model is established fully considers the constraints of tasks in time and space. A hybrid discrete optimization algorithm has been designed to solve the task allocation problem. The adjustment method of objective weights is presented to achieve the purpose of balancing between multiple objectives for coordinating the task allocation problem.

The remainder of this paper is arranged as follows: In Sect. 2, the multi-objective task assignment problem is defined. In Sect. 3, the procedure of the hybrid discrete differential evolution and particle swarm optimization (D-DEPSO) is sufficiently described. Section 4 describes the adjustment method of objective weights. Section 5 shows a lot of experimental data and analysis. Conclusions are discussed in Sect. 6.

2 Multi-objective Task Assignment in Hospital Inspection

2.1 Task Allocation Revenue Function

When the robot A_i fulfil the task T_j, it generates the corresponding value. The size of revenue value depends on the importance of the task T_j which the robot A_i fulfilled. By parity of reasoning, the total revenue function formula after all tasks were allocated is as follows:

$$\mathrm{Re}\left(A_i, T_j\right) = \sum_{k=1}^{K} v_k \cdot \left\{ \prod_{i=1}^{A} \prod_{j=1}^{T} p_{ij} \cdot y_{ij} \right\} \tag{1}$$

$$y_{ij} = \begin{cases} 1, \mathrm{robot A_i executes the task T_j} \\ 0, \mathrm{others} \end{cases} \tag{2}$$

In the formula (1), v_k represents the corresponding value which the robot A_i finished the task T_j, p_{ij} represents the probability of robot i fulfilling the task j on time. y_{ij} represents whether the task T_j is executed by the robot A_i, given by the formula (2).

2.2 Task Allocation Revenue Function

The respective execution of the hospital inspection tasks by robots will consume the medical resources, energy and so on which the inspection tasks demand. The total consumption is the costs of robots, closely linked to task paths of the robots and the type of tasks. Define the costs of the robot A_i executing the task T_j as the below formula (3):

$$\mathrm{Cost}\left(A_i, T_j\right) = \mathrm{Resource}_{ij} + Path_{ij} \tag{3}$$

In this formula, $Resource_{ij}$ represents the medical resource consumption after the robot A_i fulfills the task T_j, $Path_{ij}$ represents the path length after the robot A_i fulfills the task T_j, and the value size is in direct proportion to the energy consumption.

2.3 Task Allocation Executing Time Function

To the multi-objective task allocation problem, not only the revenue maximization and the costs minimization, the most important thing is the time minimization. Every shorten seconds mean a more patient out of danger. So, it's needed that ensure that the time of all robots fulfilling tasks is the shortest, which means the minimum difference between the time that the robot A_i start to execute the task T_j and the time that the task T_j is fulfilled. The formula is as follows:

$$Time\left(A_i, T_j\right) = \max\left\{Tover_{ij} - Tstart_{ij}\right\} \tag{4}$$

In this formula, $Tover_{ij}$ and $Tstart_{ij}$ and represent respectively the time that the robot A_i start to execute the task T_j and the task T_j is fulfilled.

2.4 Task Allocation Distance Constrain

For a single robot, its voyage is limited. So, the range of motion is limited. Define the state radius under the epidemic environment as R_i, the constraint of the distance is as follows:

$$Path_{ij} \cdot x_{ij} \leq R_i \tag{5}$$

x_{ij} represents the matching matrix, where the elements are 1 (the robot A_i performs the task T_j) or 0 (the robot A_i does not perform the task T_j).

Constraints ensure every robot can only execute one task, as follows:

$$\sum_{i=1}^{A} y_{ij}\left(t\right) = 1, \left(i = 1, 2, \cdots, A\right) \tag{6}$$

$$\sum_{j=1}^{T} y_{ij}\left(t\right) = 1, \left(j = 1, 2, \cdots, T\right) \tag{7}$$

2.5 Multi-objective Task Allocation Mathematical Model

Task assignment problem in hospital inspection is a multi-objective optimization decision problem, which is translated into one-objective problem by combining the above-mentioned function with constraints, which forms the multi-objective task allocation mathematical model finally. This mathematical model and constraint condition is as follows:

$$\max f\left(Y\right) = \sum_{i=1}^{N}\sum_{j=1}^{N}\left(\begin{array}{c}\omega_1 \cdot \mathrm{Re}\left(A_i, T_j\right) - \omega_2 \cdot \mathrm{Cost}\left(A_i, T_j\right) \cdot y_{ij} \\ -\omega_3 \cdot Time\left(A_i, T_j\right) \cdot y_{ij}\end{array}\right) \tag{8}$$

s.t.

$$\sum_{i=1}^{A} y_{ij}\left(t\right) = 1, \left(i = 1, 2, \cdots, A\right) \tag{9}$$

$$\sum_{j=1}^{T} y_{ij}\left(t\right) = 1, \left(j = 1, 2, \cdots, T\right) \tag{10}$$

$$Path_{ij} \cdot x_{ij} \leq R_i \tag{11}$$

$$\omega_1 + \omega_2 + \omega_3 = 1 \tag{12}$$

$$A = T = N \tag{13}$$

Equation (8) is the objective function and refers to the overall utility that can be achieved in the task assignment process. The constraint (9) ensures that every task can only be executed by a robot. The formula (10) states that every robot can only fulfil one task. The constraint (11) stipulates the largest range that robot motions can't surpass. The formula (12) indicates that the sum of the weight coefficient w_1, w_2 and w_3 is 1. The formula (13) shows the number of machines and tasks is equal.

3 Hybrid D-DEPSO Algorithm

3.1 Discrete Difference Mutant Operator of de

The idea of the mutation operation of the classical DE is that the weighted difference vectors of two vectors are added to the third vector [13,14], as shown in Eq. (14).

$$V_{i,g} = \psi_{r0,g} + (F * (\psi_{r1,g} - \psi_{r2,g})) \tag{14}$$

where, g represents the evolutionary generation; $r_1 \neq r_2 \neq r_3 \neq j$, $j = 1, 2, \cdots, NP$, make sure the three variables are different variables, as is shown F is the mutation rate. Under the standard DE framework, for initial individuals encoded with positive integers, redefines addition, subtraction, and multiplication operations are redefined based on the replacement method.

The given n element displaces $[n] = \{1, 2, \cdots, n\}$, $S(n)$ denotes the n element displacing the group, $\psi(\varphi) = \phi$(the position of the element ϕ in the displacement ψ is φ).

Definition 1: $\psi_1, \psi_2 \in S(n)$,then the discrete addition of the substitutions ψ_1 and ψ_2 is defined as:

$$\psi_1 \oplus \psi_2 = \psi_1 * \psi_2 \tag{15}$$

That is, $\psi_1, \psi_2 \in S(n)$,for all $\varphi \in [n]$, there exists $(\psi_1 * \psi_2)(x) = \psi_1(\psi_2(x))$.

Definition 2: For setting $\psi_1, \psi_2 \in S(n)$, the discrete subtraction of the substitutions ψ_1 and ψ_2 is defined as:

$$\psi_1 \Theta \psi_2 = \psi_2^{-1} * \psi_1 \tag{16}$$

In the formula (16), for all φ, $\phi \in [n]$, $\psi(\varphi) = \phi$, $\psi^{-1}(\phi) = \varphi$ and the formula (12) is satisfied.

$$\psi_1 \oplus (\psi_2 \Theta \psi_3) = \psi_1 * \left(\psi_3^{-1} * \psi_2\right) \tag{17}$$

Set the value of mutation $F \in [0,1]$, the permutations $\psi \in S(n)$, $H \in \psi$, H is defined as generators of the permutation group ψ, then the discrete multiplication of F with the permutation ψ is defined as:

$$F \otimes \psi = F * \psi = F * (h_1 * h_2 * \cdots * h_\lambda) = h_1 * h_2 * \cdots * h_L \tag{18}$$

In the formula (18), $\psi = h_1 * h_2 * \cdots * h_\lambda, (h_1, h_2, \cdots, h_\lambda \in H), L = [F * \lambda]$, means that the result should not exceed its smallest integer value. $F * \psi$ is defined as the truncation operation of the substitution, $F = 0.5$. We redefine the difference-variation operator in discrete DE as follows:

$$V_{i,g} = \psi_{r1,g} \oplus (F \otimes (\psi_{r2,g} \Theta \psi_{r3,g})) = \psi_{r1,g} * \left(F * \left(\psi_{r3,g}^{-1} * \psi_{r2,g}\right)\right) \tag{19}$$

A demonstration of a mutation operation strategy is shown as Table 1.

Table 1. A demonstration of a mutation operation strategy

variables	integer values for variables			
$\boldsymbol{\Phi}_{r1}$	1	3	5	4
$\boldsymbol{\Phi}_{r2}$	2	4	3	5
$\boldsymbol{\Phi}_{r3}$	5	3	1	4
$\boldsymbol{\Phi}_3^{-1} * \boldsymbol{\Phi}_2$	5	4	2	1
$\boldsymbol{F} * \boldsymbol{\Phi}$	2	1	4	3
$\boldsymbol{V}$	3	1	4	5

3.2 Discrete D-DEPSO Algorithm

In this paper, D-DEPSO algorithm was presented to solve the above multi-objective task assignment problem in hospital inspection. Differential evolution mutation operations are used for mutating personal optimal position, and partial swarm optimization (PSO) [15,16] is mainly used to record personal optimal positions and update speed values. The formula for updating velocity is presented in the following equation:

$$v(t+1) = v(t) + c_1 (p_{best} - x) + c_2 (g_{best} - x) \tag{20}$$

v is velocity vector and x is position vector, c_1 and c_2 are the coefficients. p_{bset} is individual optimal value, and g_{bset} is the global optimal value . As it is based on a binary system, a binary-based discrete PSO is proposed [17], which follows the elementary PSO. The update formula of v in the continuous particle swarm remains unchanged. Its position update formula is as follows:

$$s(v) = \frac{1}{[1 + \exp(-v)]} \tag{21}$$

$$x = \begin{cases} 1, r < s(v) \\ 0, otherwise \end{cases} \tag{22}$$

The value of r is derived from the $u(0,1)$ random number generated by the distribution. The D-DEPSO algorithm flow is as follows. Step1: Initialization involves setting up the position matrix x, vanity matrix v, and algorithm parameters: NP, F, c_1, c_2. Additionally, the individual optimal value p_{best} is initialized.

Step2: calculate the global optimal value g_{best} and the objective function, If the condition is satisfied, the optimal solution will be output. Otherwise, go to the next step.

Step3: Discrete differential mutation is applied to the individual position matrix x.

Step4: Update the individual optimal value p_{best}, and the global optimal value g_{best}, and record both the global optimal individual and its corresponding optimal value.

Step5: Update the velocity matrix v and position matrix x, and discretize them.

Step6: Record the velocity matrix v and position matrix x after discretizing.

Step7: Proceed to the next iteration and return to Step2.

4 The Adjustment Method of Objective Weights

To balance the objective weights ω_1, ω_2 and ω_3, each ω_i $(i = 1, 2, 3)$ consists of the attribute weight $\omega_i{}'$ and $\omega_i{}''$. The attribute weight vector $\omega_i{}'$ is calculated by the subjective assignment method AHP the attribute weight vector $\omega_i{}''$ is calculated by the objective entropy value method [18]. Denote α, β for the coefficients of $\omega_i{}'$ and $\omega_i{}''$ respectively, and combine the subjective weight vector and the objective weight vector such that:

$$\begin{cases} \omega_i = \alpha\omega_i{}' + \beta\omega_i{}'' \\ 0 \leq \omega_i \leq 1, \sum\limits_{i=1}^{3} \omega_i = 1 \end{cases} \tag{23}$$

The evaluation objective value of each alternative is obtained as:

$$\begin{cases} g_k = \sum\limits_{i=1}^{3} a_i^k \omega_i = \sum\limits_{i=1}^{n} a_i^k \left(\alpha W' + \beta W''\right) \\ \quad s.t. \begin{cases} \alpha^2 + \beta^2 = 1 \\ \alpha, \beta \geq 0 \end{cases} \end{cases} \tag{24}$$

a_i^k is the expert score for the kth alternative solution of w_i. A linear weighted method single-objective optimization model with equal weights can be constructed as follows:

$$\begin{cases} MaxZ = \sum\limits_{k=1}^{p} g_k = \sum\limits_{k=1}^{p} \sum\limits_{i=1}^{n} a_i^k \left(\alpha W' + \beta W''\right) \\ \qquad s.t. \begin{cases} \alpha^2 + \beta^2 = 1 \\ \alpha, \beta \geq 0 \end{cases} \end{cases} \tag{25}$$

where, p is the number of alternative solutions of w_i. The model can be solved by constructing the Lagrange function:

$$L = \sum_{k=1}^{p} \sum_{i=1}^{n} a_i^k \left(\alpha W' + \beta W''\right) + \frac{\omega}{2} \left(\alpha^2 + \beta^2 - 1\right) \tag{26}$$

ω is the Lagrangian multiplier, find the partial derivatives of ∂, β and make $\partial L / \partial\alpha = 0, \partial L / \partial\beta = 0$, obtain the value of the optimal solution of the optimization model ∂, β:

$$\begin{cases} \alpha = \dfrac{\sum_{k=1}^{p} \sum_{i=1}^{n} a_{ij}^k W_i'}{\sqrt{\left(\sum_{k=1}^{p} \sum_{i=1}^{n} a_{ij}^k \omega_i'\right)^2 + \left(\sum_{k=1}^{p} \sum_{i=1}^{n} a_{ij}^k \omega_i''\right)^2}} \\ \beta = \dfrac{\sum_{k=1}^{p} \sum_{i=1}^{n} a_{ij}^k W_i''}{\sqrt{\left(\sum_{k=1}^{p} \sum_{i=1}^{n} a_{ij}^k \omega_i'\right)^2 + \left(\sum_{k=1}^{p} \sum_{i=1}^{n} a_{ij}^k \omega_i''\right)^2}} \end{cases} \tag{27}$$

Normalize ∂, β:

$$\begin{cases} \overline{\alpha} = \frac{\alpha}{\alpha+\beta} \\ \overline{\beta} = \frac{\beta}{\alpha+\beta} \end{cases} \tag{28}$$

Equation (28) is substituted into Eq. (23) and the optimized weight can be obtained. An example of optimized weights is shown as Table 2. After 10 iterations of adjustment, the optimization weights tend to be stable, as shown in Table 3.

Table 2. An example of optimized weights

Indicator	AHP	entropy weight	Optimization(w)
w_1	0.1	0.1	0.1
w_2	1.35	1.35	1.35
w_3	50	50	50

Table 3. Iterative weight adjustment values

Iteration	w_1	w_2	w_3
1	0.509	0.317	0.173
2	0.502	0.331	0.166
3	0.597	0.342	0.161
4	0.493	0.352	0.154
5	0.491	0.360	0.148
6	0.484	0.363	0.153
7	0.467	0.365	0.168
8	0.489	0.348	0.163
9	0.494	0.325	0.181
10	0.503	0.351	0.146

5 Experiments and Analysis

5.1 Comparative Experiments

The performance of D-DEPSO for the multi-objective assignment in hospital inspection is examined using comparative experiments. We compare the D-DEPSO with improved discrete DE (IDE) [19], improved discrete PSO (IPSO) [20].

Based on simulated data using stochastic functions, simulation results for three sets of 10-dimensional trials with various populations are obtained. According to 10-dimensional simulation trials, ten robots must be linked with ten task requirements. The jobs in this example include 1 remote body temperature monitoring task, 4 periodic disinfection tasks, 3 medication dispensing tasks, 2 medical material handling tasks, and 3 drug dispensing tasks. Any of these jobs can be accomplished by any of the general-purpose robots. Based on the various benefits that different robots receive from completing each task, the various medical resources used, the various amounts of energy needed, and the various amounts of time needed to complete the tasks, or the various benefits obtained in the final aggregate, the distribution scheme with the highest benefit is determined.

The Fig. 1 compares the experimental simulation results of IDE, IPSO, and D-DEPSO in 10 dimensions for 50, 100, and 200 populations, respectively. Comparing D-DEPSO to the other two algorithms, it is evident that it not only achieves the best result when the population size is high, but also does so when the population size is low. Moreover, the final optimal solution of D-DEPSO has a clear advantage of the population number setting. The analysis indicates that D-DEPSO outperforms IPSO in achieving the optimal solution for the multi-objective task assignment problem particularly in the 10-dimensional scenario.

The simulation experiments conducted in 20 dimensions are presented in Fig. 2. Notably, these results exhibit distinct trends from those obtained in the 10-dimensional experiments. Specifically, when the population number is set to

50, D-DEPSO outperforms the others by identifying a solution with a larger gain value. As the population size increases to 100 and 200, IPSO exhibits a significantly faster convergence rate in comparison to the other two algorithms. From the standpoint of optimal solutions, the D-DEPSO algorithm consistently identifies the optimal solution assignment scheme that maximizes the gain.

Coming to the multi-objective task assignment problem in a 50-dimensional epidemic environment, the simulation of fifty tasks assigned to fifty robots is performed in this part. The complexity of this problem increases significantly due to the increase in the number of tasks, which makes the algorithm require a larger population size. Thus, the population size is set to 100, 200, and 300 for the three simulation experiments. The simulation results are shown in Fig. 3. The comparison shows that when NP is 100, the convergence trends of the three algorithms are not much different, the final optimal solutions D-DEPSO and IPSO are also much better than IDE, but the optimal solution of D-DEPSO is still slightly better than IPSO. When NP is 200, the D-DEPSO convergence rate clearly catches up with IPSO's convergence rate, and at the same time, it is guaranteed to obtain the assignment scheme with greater gains.

As NP increases to 300, the advantages of the D-DEPSO algorithm in dealing with the 20-dimensional mufti-objective task assignment problem are fully demonstrated, both in terms of convergence speed and optimal solution.

Since the mufti-objective task assignment problem in a 100-dimensional epidemic environment has a more complex computation, the population number NP is also increased to 100, 200, and 300. Comparative plots of simulation simulations are shown in Fig. 4. It can be intuitively seen that D-DEPSO has an absolute advantage in searching for the global optimal solution of the task assignment with the maximum gain for dealing with the mufti-objective task assignment problem in a high-dimensional environment. When the population size is 100, D-DEPSO has a better optimal solution compared with other algorithms, and the convergence speed is only less obvious at this point. However, when the population size is set to 200 and 300, D-DEPSO has faster convergence speed and more profitable assignment solution.

According to the comparison of 10, 20, 50 and 100 dimensions in Figs. 1, Figs. 2, Figs. 3 and Figs. 4, we can learn that when the population size of the three algorithms is set relatively small, the convergence speed does not differ much in the early stage of the algorithm operation, and D-DEPSO can generally find the optimal solution at this time. For low-dimensional simulations, IPSO can have good convergence speed, but its disadvantage is that it is optimum situation.

Correspondingly, D-DEPSO is much more effective than IDE and IPSO in searching for global optimal solutions. For a high-dimensional simulation experiment, as long as the population size is set enough, then D-DEPSO will outperform the other two discrete algorithms in terms of convergence speed and optimal solutions across the board.

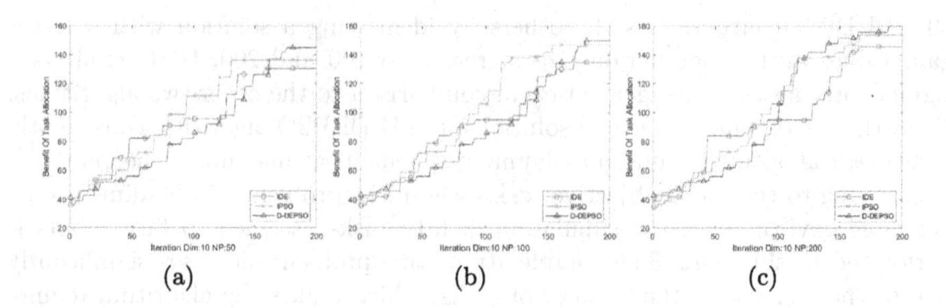

Fig. 1. Comparison of the benefits of three algorithms in 10 dimensions

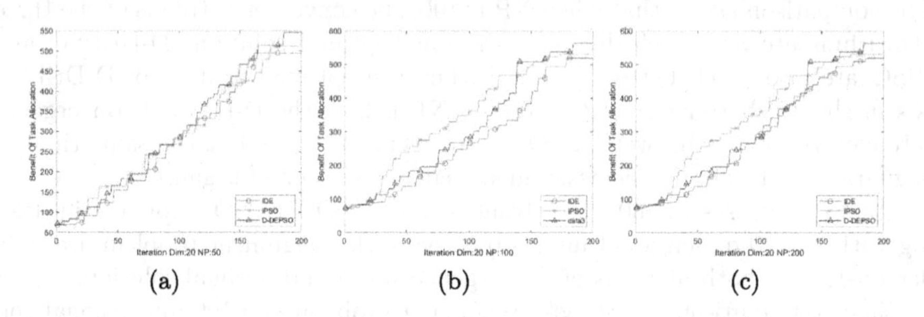

Fig. 2. Comparison of the benefits of three algorithms in 20 dimensions

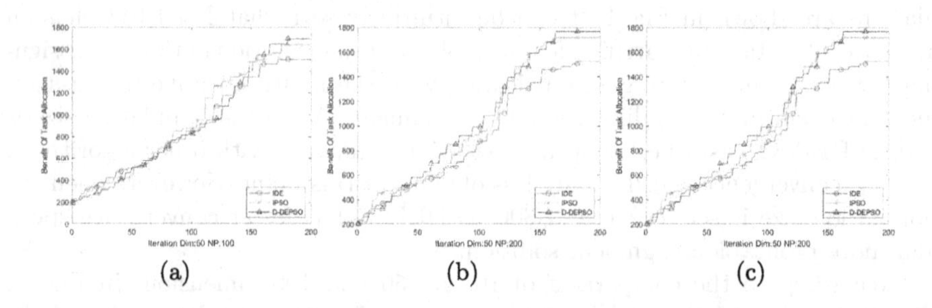

Fig. 3. Comparison of the benefits of three algorithms in 50 dimensions

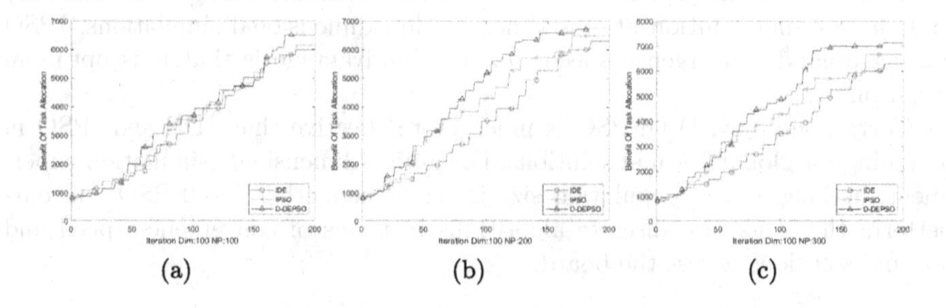

Fig. 4. Comparison of the benefits of three algorithms in 100 dimensions

5.2 Analysis of Experimental Statistical Data

The simulation statistics of IDE, IPSO and D-DEPSO in dealing with multi-objective task assignment problems in hospital inspection are summarized in Tab. 4. The following conditions need to be stated in advance. The parameter setting of NP is set to 200 for 10- dimensional and 20-dimensional experiments. NP is set to 300 for 50-dimensional and 100-dimensional experiments. For the experimental data of each dimension, the mean, standard deviation and mean time are obtained from 100 sets of data statistics. The maximum gain (Max), mean value, standard deviatio (Mean±Std), and average time (Timeavg) corresponding to each algorithm, respectively. And 1k means the equivalent of a thousand RMB gain. The unit of average time is second (s). The bolded font is the optimal parameter for each dimension of the simulation experiment.

The IDE algorithm has the fastest running speed for both 10-dimensional and 20-dimensional experimental data. However, D-DEPSO has the largest gain value and mean value, and the smallest variance. D-DEPSO has smaller standard deviation, which means the results of D-DEPSO algorithm are less volatile and more stable than IDE and IPSO algorithms. Analysis of the 50-dimensional and 100-dimensional simulation experimental data shows that D-DEPSO obtains excellent results in all aspects, including maximum gain value, mean value, standard deviation, and running time. So, D-DEPSO can generate the optimal assignment scheme for the multi-objective task assignment problem.

Table 4. Iterative weight adjustment values

Algorithm	Dim	10	20	50	100
IDE	$Max\,(k)$	146.64	520.65	1506.03	6296.18
	$Mean \pm Std\,(k)$	142.76 ± 0.83	506.65 ± 3.52	1458.51 ± 5.05	6185.52 ± 5.62
	$Time_{avg}\,(s)$	**0.75**	**1.75**	4.46	14.56
IPSO	$Max\,(k)$	154.64	531.45	1768.13	6684.15
	$Mean \pm Std\,(k)$	151.67 ± 0.82	520.52 ± 4.21	1685.22 ± 6.25	6547.548 ± 6.14
	$Time_{avg}\,(s)$	0.89	1.81	4.15	15.36
D-DEPSO	$Max\,(k)$	**160.26**	**564.60**	**1806.17**	**7150.02**
	$Mean \pm Std\,(k)$	**158.12 ± 0.63**	**556.06 ± 3.38**	**1729.26 ± 4.13**	**6989.01 ± 4.56**
	$Time_{avg}\,(s)$	1.78	1.89	**3.88**	**13.12**

6 Conclusion

This paper discusses the problem of multi-objective task assignment for hospital inspection by assigning multiple tasks to different robots to achieve the highest level of task completion. The multi-objective model is built by considering revenue as well as the cost of resource and time. Hybrid D-DEPSO algorithm is presented to solve the task assignment problem. Moreover, in order to balance the objectives, an adjustment method of objective weights is proposed. The experimental results show that the proposed method can achieve more satisfactory solution.

References

1. Zhang, H., et al.: Research on intelligent robot systems for emergency prevention and control of major pandemics (in Chinese). Sci. Sin Inf. **50**, 1069–1090 (2020)
2. Yang, G.Z., et al.: Combating COVID-19-the role of robotics in managing public health and infectious diseases. Sci. Robot. **5**, 5589 (2020)
3. Li, J., Ni, H., Wang, L.F., Ling, F., Chun, J.: Request migration based task scheduling algorithm in VoD system. J. Jilin Univ. (Eng. Tech. Ed.) **3**, 938–945 (2015)
4. Wang, X.G., Ni, H., Zhu, M., Liu, L.: A two-level adaptive scheduler model and algorithm for DVB-C2 system toward QOS. J. Univ. Sci. Tech. China **4**, 300–305 (2013)
5. Wan, L.-J., Yao, P.Y., Sun, P.: Distributed task allocation method of manned/unmanned combat agents. Syst. Eng. Electron. **35**(2), 310–316 (2013)
6. Pradas, D., Vazquez-Castro, M.A.: NUM-based fair rate-delay balancing for layered video multicasting over adaptive satellite networks. IEEE J. Selected Areas Commun. **29**, 969–978 (2011)
7. Zhou, J., Zhao, X.Z., Xu, Z., Lin, Z., Zhao, X.P.: Many-objective task allocation method based on D-NSGA-III algorithm for multi-UAVs. Syst. Eng. Electron. **43**, 1240–1247 (2021)
8. Zhou, J., Zhao, X.Z., Zhong, X.P., Zhao, D.D., Li, H.H.: Task allocation for multi-agent systems based on distributed many-objective evolutionary algorithm and greedy algorithm. IEEE Access **8**, 19306–19318 (2020)
9. Zhou, J., Zhao, X., Zhao, D., Lin, Z.: Task allocation in multi-agent systems using many-objective evolutionary algorithm NSGA-III. In: Zhai, X.B., Chen, B., Zhu, K. (eds.) MLICOM 2019. LNICST, vol. 294, pp. 677–692. Springer, Cham (2019). https://doi.org/10.1007/978-3-030-32388-2_56
10. Jeon, S., Lee, J., Kim, J.: Multi-robot task allocation for real-time hospital logistics. In: Proceedings of IEEE International Conference on System, Man, Cybernetics (SMC), Banff, AB, Canada, pp. 2465–2470. IEEE ()2017
11. Lee, C., Lehoczky, J., Siewiorek, D., Rajkumar, R., Hansen, J.: A scalable solution to the multi-resource QoS problem. In: Proceedings of 20th IEEE Real-Time System Symposium, pp. 315–326 (2002)
12. Jiang, Y., et al.: Research on intelligent robot systems for emergency prevention and control of major pandemics. Scientia Sinica Inf. **50**, 1069–1090 (2020)
13. Storn, R., Price, K.: Differential evolution-a simple and efficient heuristic for global optimization over continuous spaces. J. Global Optim. **11**, 341–359 (1997)
14. Zhang, C.M., Chen, J., Xin, B.: Distributed memetic differential evolution with the synergy of Lamarckian and Baldwinian learning. Appl. Soft Comput. **13**, 2947–2959 (2013)
15. Kennedy, J., Eberhart, R.: Particle swarm optimization. In: Proceedings of International Conference Neural Network, vol. 4, pp. 1942–1948 (1995)
16. Xin, B., Chen, J., Zhang, J., Fang, H., Peng, Z.H.: Hybridizing differential evolution and particle swarm optimization to design powerful optimizers: a review and taxonomy. IEEE Trans. Syst. Man Cybern. C **42**, 744–767 (2012)

17. Shi, Y., Eberhart, R.: A modified particle swarm optimizer. In: Proceedings of IEEE International Conference on Evolutionary Computer Proceedings. IEEE World Congress on Computational Intelligent, pp. 69–73 (1998)
18. Sun, Y., Huang, H.F., Ding, J.H.: Adaptive algorithm for adjusting weights in multiple attributes group decision making. Comput. Eng. Appl. **50**(2), 35–38 (2014)
19. Sallam, K.M., Elsayed, S.M., Chakrabortty, R.K., Ryan, M.J.: Improved multi-operator differential evolution algorithm for solving unconstrained problems. In: Proceedings of IEEE Congress on Evolutionary Computation (CEC), pp. 1–8 (2020)
20. Tong, L., Du, B., Liu, R., Zhang, L.: An improved multi-objective discrete particle swarm optimization for hyperspectral endmember extraction. IEEE Trans. Geosci. Remote Sens. **57**(10), 7872–7882 (2019)

An Analysis of the Generalized Tit-for-Tat Strategy Within the Framework of Memory-One Strategies

Yunhao Ding, Jianlei Zhang, and Chunyan Zhang(✉)

College of Artificial Intelligence, Nankai University, No. 38, Tongyan Road, Jinnan District, Tianjin, China
2120220508@mail.nankai.edu.cn, {jianleizhang,zhcy}@nankai.edu.cn

Abstract. The Tit-for-tat strategy is a traditional strategy in game theory. In the Prisoner's Dilemma, the TFT strategy has been proven to be strong. However, within a four-component Memory-One strategy framework, the experiment proves that when the TFT strategy faces a random strategy, the expected payoffs of both sides are equal in most cases, that is, the TFT strategy does not show an absolute advantage. In this paper, a Generalized TFT strategy is set up to study the impact of each variable on its strengths. The results show that: within the framework, in order to ensure a higher expected payoff, the user of the Generalized TFT strategy should appropriately increase the possibility of choosing a strategy opposite to the opponent's strategy in the previous round currently. These works expound the nature of the TFT strategy from a new perspective, enrich its core, and solve the problem of defining the TFT strategy in evolutionary games.

Keywords: Game Theory · the Prisoner's Dilemma · Memory-One Strategy · Generalized Tit-for-Tat Strategy

1 Introduction

Cooperation permeates the natural world, serving as a vital foundation for civilization and progress [1]. However, the decision of individuals to engage in cooperation rather than prioritize personal interests remains an intriguing enigma [2–4]. Despite the expectation that natural selection would compel individuals to seek their own optimal chances of survival and reproduction, the collective benefits derived from group cooperation are substantial. This paradoxical nature of cooperation renders its emergence a captivating realm of exploration.

In species with advanced social intelligence, cooperation and defection constitute the two basic states of dynamic individual relationships. Individuals can opt to cooperate, driven by shared interests and objectives, or they can defect, pursuing self-centered gains. This fluctuating nature of relationships imbues animal societies with complexity and dynamism. Nevertheless, these states are not fixed; they possess the ability to convert from one to the other upon specific circumstances [5]. To examine the interdependence

B. Xin et al. (Eds.): IWACIII 2023, CCIS 1931, pp. 338–347, 2024.
https://doi.org/10.1007/978-981-99-7590-7_27

of cooperation and defection, scholars have devised the most classic game model, known as the Prisoner's Dilemma (PD).

The Prisoner's Dilemma stands out as a quintessential illustration of non-zero-sum and non-cooperative games within the realm of game theory [6]. The PD game involves two completely rational players who independently make decisions and only pursue their own benefits. It presents two optional strategies: cooperation (C) and defection (D). In the game, if two players choose cooperation (C), they can both obtain the payoff $\boldsymbol{R}$ (reward). If they both choose defection (D), they will only receive a lesser payoff $\boldsymbol{P}$ (punish). If one player chooses cooperation (C) and the other chooses defection (D), the naive cooperator will receive the least payoff $\boldsymbol{S}$ (sucker), while the selfish defector will receive the best payoff $\boldsymbol{T}$ (temptation).

The game parameters satisfy $T > R > P > S$ and $2R > T + S$. So, in this game, each prisoner will find that whatever the opponent chooses, it is always better to choose defection. Therefore, the stable result of the game is that both prisoners choose to defect. This outcome is called Nash Equilibrium. It means that neither player can improve their payoff by changing their strategy, given that the other player's strategy remains the same.

Throughout the trajectory of biological and social evolution, individuals often partake in games that recur over time, thereby sparking researchers' interest in exploring the concept of repeated games [7]. In such scenarios, individuals have the chance to modify their strategies for ensuing rounds by referencing the strategies that they or their opponent utilized previously [8]. It is widely assumed that individuals possessing longer memories, capable of recalling earlier strategies and payoffs, hold an advantage. However, theoretical analyses have revealed that longer memory provides no advantage [9]. Hence, a significant class of strategies is proposed, known as the Memory-One strategies [10]. The Memory-One strategy exclusively considers the most recent move or outcome, thereby streamlining the decision-making process. It enables a simpler decision-making process, alleviates the computational complexity and cognitive burden for the player.

Specifically, in a PD game between Player $\boldsymbol{X}$ and Player $\boldsymbol{Y}$, C and D represent cooperation and defection respectively, then $\boldsymbol{XY} \in (CC, CD, DC, DD)$ can depict the four outcomes of last round. Player $\boldsymbol{X}$'s strategy is set as $\boldsymbol{p} = (p_1, p_2, p_3, p_4)$ to correspond to the game result. Similarly, Player $\boldsymbol{Y}$'s strategy is $\boldsymbol{q} = (q_1, q_2, q_3, q_4)$ to correspond to the game result $\boldsymbol{XY} \in (CC, DC, CD, DD)$ [11, 12].

Multiple classic strategies have been developed and studied in repeated games, among which the TFT strategy is a famous one. The TFT strategy is simple and effective. In the first round, the player cooperates with their opponent, and in subsequent rounds, they replicate the opponent's previous move. Within the framework of the Memory-One strategy, the TFT strategy can be represented by $\boldsymbol{p} = (1, 0, 1, 0)$.

Experimental studies have demonstrated the effectiveness of the TFT strategy. In the late 1970s, Axelrod organized two experiments in which "players" were computer programs submitted by participants. The winning strategies in both experiments were the TFT strategy [10]. These results highlight the power and utility of the TFT strategy in promoting cooperation and resolving conflicts in a wide range of contexts.

Since the emergence of the Memory-One strategy and the TFT strategy, a multitude of associated studies have rapidly evolved. Axelrod first put forth the Generous Tit-for-Tat (GTFT) strategy, which sporadically opts for cooperation even when met with betrayal from the opponent [13]. In 1992, Nowak and Sigmund introduced the Win-Stay, Lose-Shift (WSLS) strategy. This tactic persists with the current action if it leads to a favorable outcome and converts to the opposite action when faced with a detrimental result [14]. More contemporarily, in 2012, Press and Dyson unveiled a hybrid strategy that empowers a player to unilaterally dictate its opponent's score and impose an extortive linear relationship between their payoffs. This is known as the Zero-Determinant (ZD) strategy [9]. This paper opens a new research direction. Pan et al. apply the ZD strategy to the public goods game, and found that the user can control the ratio of his own payoff to the total payoff of other individuals within a certain range [15, 16]. Adami finds that the ZD strategy is not stable. But this situation can be changed if the ZD strategy has an informational advantage over other players [17]. Zhang et al. name a new class of strategies "self-bad, partner-worse strategies", which share some similarities with the ZD strategy. These strategies have survival advantages and robust fitness in terms of evolutionary processes [1].

The remainder of this paper is structured as follows: Sect. 2 delineates the game model and the computation of payoffs and introduces the methodologies employed in this study. Section 3 utilizes heat maps to illustrate the impact of four variables. Finally, Sect. 4 concludes the paper with a comprehensive discussion and conclusion.

2 Model and Methods

2.1 The Model Settings

In this paper, we set $R = 3$, $S = 0$, $T = 5$, $P = 1$. The payoff matrix of the PD game is shown in Table 1.

We denote the Memory-One strategy by $\boldsymbol{p} = (P_R, P_S, P_T, P_P)$. Each parameter of the quadruple corresponds to the probability that the individual will choose a cooperative strategy in current round, given the result of the previous round. Besides, $(P_R, P_S, P_T, P_P) \in [0, 1]$. For simplicity, let us consider two individuals $\boldsymbol{X}$ and $\boldsymbol{Y}$ playing a repeated prisoner's dilemma. We use $\boldsymbol{p} = (p_1, p_2, p_3, p_4)$ to denote individual $\boldsymbol{X}$'s strategy corresponding to the game result $\boldsymbol{XY} = (CC, CD, DC, DD)$. Similarly, we use $\boldsymbol{q} = (q_1, q_2, q_3, q_4)$ to denote individual $\boldsymbol{Y}$'s strategy corresponding to the game result $\boldsymbol{XY} = (CC, DC, CD, DD)$.

Table 1. Payoff matrix of player $\boldsymbol{X}$ in the PD game

X	Y	
	C	D
C	R (3)	S (0)
D	T (5)	P (1)

2.2 Payoff Calculation

In the PD game between individual $\boldsymbol{X}$ and $\boldsymbol{Y}$, the payoff matrix of individual $\boldsymbol{X}$ is defined as $\boldsymbol{R}_X = (3, 0, 5, 1)$, and correspondingly, the payoff matrix of $\boldsymbol{Y}$ is $\boldsymbol{R}_Y = (3, 5, 0, 1)$. After an interaction between the two competitors, the expected payoffs of individual $\boldsymbol{X}$ and $\boldsymbol{Y}$ can be calculated by the following formulas.

$$r_X = \frac{\boldsymbol{\mu} \cdot \boldsymbol{R}_X}{\boldsymbol{\mu} \cdot 1} = \frac{D(\boldsymbol{p}, \boldsymbol{q}, \boldsymbol{R}_X)}{D(\boldsymbol{p}, \boldsymbol{q}, 1)} \tag{1}$$

$$r_Y = \frac{\boldsymbol{\mu} \cdot \boldsymbol{R}_Y}{\boldsymbol{\mu} \cdot 1} = \frac{D(\boldsymbol{p}, \boldsymbol{q}, \boldsymbol{R}_Y)}{D(\boldsymbol{p}, \boldsymbol{q}, 1)} \tag{2}$$

in which $\boldsymbol{\mu}$ is the stationary vector of Markov matrices $\boldsymbol{p}$ and $\boldsymbol{q}$. Besides,

$$\begin{aligned} \boldsymbol{\mu} \cdot \boldsymbol{h} &\equiv D(\boldsymbol{p}, \boldsymbol{q}, \boldsymbol{h}) \\ &= \det \begin{bmatrix} -1 + p_1 q_1 & -1 + p_1 & -1 + q_1 & h_1 \\ p_2 q_3 & -1 + p_2 & q_3 & h_2 \\ p_3 q_2 & p_3 & -1 + q_2 & h_3 \\ p_4 q_4 & p_4 & q_4 & h_4 \end{bmatrix} \end{aligned} \tag{3}$$

2.3 Methods

Within the framework mentioned above, the TFT strategy is set to be $\boldsymbol{p} = (1, 0, 1, 0)$ [16, 18]. However, we found that if individual $\boldsymbol{X}$ adopts this TFT strategy, regardless of the strategy chosen by $\boldsymbol{Y}$, the payoffs for both $\boldsymbol{X}$ and $\boldsymbol{Y}$ are identical. This situation diverges from our initial expectations.

In fact, some literatures also mentioned the problems above. Zhang and Liu found that when $p_1 = 1$ and $q_1 = 1$ (corresponding to the situation in which the two participants both choose to cooperate when they both cooperated in the last round), the two participants will get the same payoffs [1]. Schmid proposed to define the ALLC strategy as $\boldsymbol{p} = (0.99, 0.99)$, the ALLD strategy as $\boldsymbol{p} = (0.01, 0.01)$, and the TFT strategy as $\boldsymbol{p} = (0.99, 0.01)$ within the framework of 2-tuple, the essence of which is to show universality and eliminate calculation errors [19].

To avoid calculation errors and gain a deeper understanding of the TFT strategy within the framework of the Memory-One strategy, four infinitesimals that act on the TFT strategy are introduced. We define them as x_1, x_2, x_3, x_4, and they are close to 0. Thus, the Generalized TFT strategy can be expressed as follows:

$$\boldsymbol{p} = (1 - x_1, x_2, 1 - x_3, x_4) \tag{4}$$

The question is how the four infinitesimals influence the character of the TFT strategy.

Let us assume that individual $\boldsymbol{X}$ employs this Generalized TFT strategy, and another individual $\boldsymbol{Y}$ adopts a random strategy denoted as follows:

$$\boldsymbol{q} = (q_1, q_2, q_3, q_4) \tag{5}$$

By applying the payoff calculation rules, we can obtain the respective payoffs of both sides, which are defined as r_x and r_y, respectively. The parameter r is stipulated to indicate the payoff gap.

$$r = r_x - r_y = \frac{r_{numerator}}{r_{denominator}} \tag{6}$$

$$\begin{aligned} r_{numerator} = &\begin{pmatrix} 5q_1x_4 - 5q_2q_4 - 5q_4x_4 - 5q_1q_2x_4 + 5q_1q_4x_2 - 5q_1q_3x_4 \\ -5q_1q_4x_3 + 5q_2q_4x_3 - 5q_3q_4x_2 + 5q_2q_4x_4 + 5q_3q_4x_4 \end{pmatrix} x_1 \\ &+ 5x_4 - 5q_1x_4 + 5q_4x_2 - 5q_3x_4 - 5q_4x_3 - 5q_1q_4x_2 + 5q_1q_3x_4 \\ &+ 5q_1q_4x_3 - 5q_2x_3x_4 + 5q_3x_2x_4 - 5q_4x_2x_4 + 5q_4x_3x_4 + 5q_1q_2x_3x_4 \\ &- 5q_1q_3x_2x_4 + 5q_1q_4x_2x_4 - 5q_1q_4x_3x_4 \end{aligned} \tag{7}$$

$$\begin{aligned} r_{denominator} = &\begin{pmatrix} q_1q_2 - q_1 + q_1q_3 - 2q_1q_4 - q_2q_3 + q_2q_4 + q_1x_2 - q_1x_4 \\ -q_3x_2 + q_4x_4 - q_1q_2x_2 + q_1q_2x_4 - q_1q_3x_3 + q_1q_4x_2 \\ +q_2q_3x_2 - q_1q_3x_4 + q_1q_4x_3 + q_2q_3x_3 - q_2q_4x_3 - q_3q_4x_2 \\ -q_2q_4x_4 + q_3q_4x_4 \end{pmatrix} x_1 \\ &+ q_1 + q_3 - 2q_4 + x_2 - x_4 - q_1q_3 + 2q_1q_4 - q_2q_4 - q_1x_2 + q_1x_4 \\ &+ q_2x_3 - q_3x_2 - q_3x_3 + q_4x_2 - q_3x_4 + q_4x_3 + q_4x_4 - q_1q_2x_3 \\ &+ q_1q_3x_2 + q_1q_3x_3 - q_1q_4x_2 + q_1q_3x_4 - q_1q_4x_3 + q_2q_4x_2 - 2q_1q_4x_4 \\ &- q_2q_3x_4 + q_2q_4x_3 - q_3q_4x_2 + q_2q_4x_4 + q_3q_4x_4 - q_2x_2x_3 + q_3x_2x_3 \\ &+ q_2x_3x_4 - q_4x_3x_4 + q_1q_2x_2x_3 - q_1q_3x_2x_3 - q_1q_2x_3x_4 - q_1q_3x_2x_4 \\ &+ q_1q_4x_2x_4 + q_2q_3x_2x_4 - q_2q_4x_2x_3 + q_1q_4x_3x_4 + q_2q_3x_3x_4 \\ &- q_2q_4x_2x_4 + q_3q_4x_2x_3 - q_3q_4x_3x_4 - 1 \end{aligned} \tag{8}$$

The computation of the payoff gap entails eight variables: four elements of the random strategy and four infinitesimals of the Generalized TFT strategy. In this study, we harness a control variable method. By systematically adjusting each variable while keeping the rest constant, we can discern the impact of each component on the payoff gap. This methodology allows us to ascertain the contribution of each variable to the aggregate outcome and to establish the significance of each component within the TFT strategy as per the framework of the Memory-One strategy.

We set $b = 10^{-4}$ as a base, and the logarithmic multiples change from -1 to 3 at intervals of 0.25 (corresponding to the actual multiples range from 0.1 to 1,000), so that the four infinitesimals range from 0.00001 to 0.1. We set such range and base. On one hand, this range ensures that the Generalized TFT strategies still comply with the basic form and requirements of the TFT strategies, and at the same time, the infinitesimals are not too small to lose the meaning and value of calculation. On the other hand, the positive and negative logarithmic multiples ensure that the results do not lose generality. With this setup, there are only a few permutations and combinations to consider. For each combination, we randomly generate sufficient number of random strategies. Then we calculate their payoff gaps according to the formulas and record the winning rates of the current TFT strategy.

We divide all experiments into four parts, which are "one infinitesimal", "two infinitesimals", "three infinitesimals" and "four infinitesimals". For each part, we draw heat maps for logarithmic multiples ranging from -1 to 3 respectively. The heat maps indicate the frequency at which the TFT strategy can defeat random strategies. Here gives an example in "two infinitesimals": when x_1, x_2 are studied, we treat x_3, x_4 as 0. Then we apply formula to figure out the expression of the payoff gap between the two participants. Take x_1 and x_2 at different sampling points and substitute them into this expression to figure out the payoff gap. In this way, the influence of x_1 and x_2 on the TFT strategy's advantage is obtained. This approach enables us to visualize the effect of each variable on the performance of the TFT strategy in defeating random strategies.

3 Results

In fact, the case of "four infinitesimals" is the most complete and complex one, covering the other three cases. Therefore, this section only explains the results of the case "four infinitesimals".

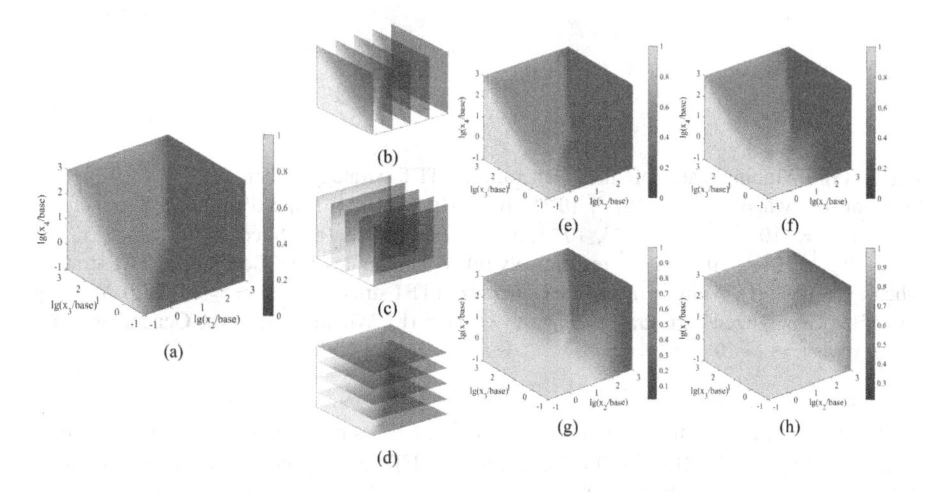

Fig. 1. (a). Winning rate of the Generalized TFT strategy when $x_1 = 10^{-5}$ **(b).** Slice of (a) when $x_2 = 10^{-5}, 10^{-4}, 10^{-3}, 10^{-2}, 10^{-1}$ respectively **(c).** Slice of (a) when $x_3 = 10^{-5}, 10^{-4}, 10^{-3}, 10^{-2}, 10^{-1}$ respectively **(d).** Slice of (a) when $x_4 = 10^{-5}, 10^{-4}, 10^{-3}, 10^{-2}, 10^{-1}$ respectively **(e).** Winning rate of the Generalized TFT strategy when $x_1 = 10^{-4}$ **(f).** Winning rate of the Generalized TFT strategy when $x_1 = 10^{-3}$ **(g).** Winning rate of the Generalized TFT strategy when $x_1 = 10^{-2}$ **(h).** Winning rate of the Generalized TFT strategy when $x_1 = 10^{-1}$

Figure 1 indicates that when $x_1 = 10^{-5}$, in the sampling set, the subset with a winning rate of over 50% for the Generalized TFT strategy comprises approximately 1/3 of the total. The distribution of this subset is primarily concentrated in instances where x_2, x_4 are relatively small and x_3 is relatively large. As x_2, x_4 increase, x_3 decreases, the advantage of the Generalized TFT strategy gradually diminishes. Besides, it indicates that as x_1 increases from small to large, the set of points with a winning rate of the Generalized TFT strategy over 50% continues to expand, and at the point where $x_1 =$

10^{-1}, the set of points reaches 95%. In other words, as x_1 increases, the overall winning rate of the Generalized TFT strategy increases significantly, and its advantages become more pronounced.

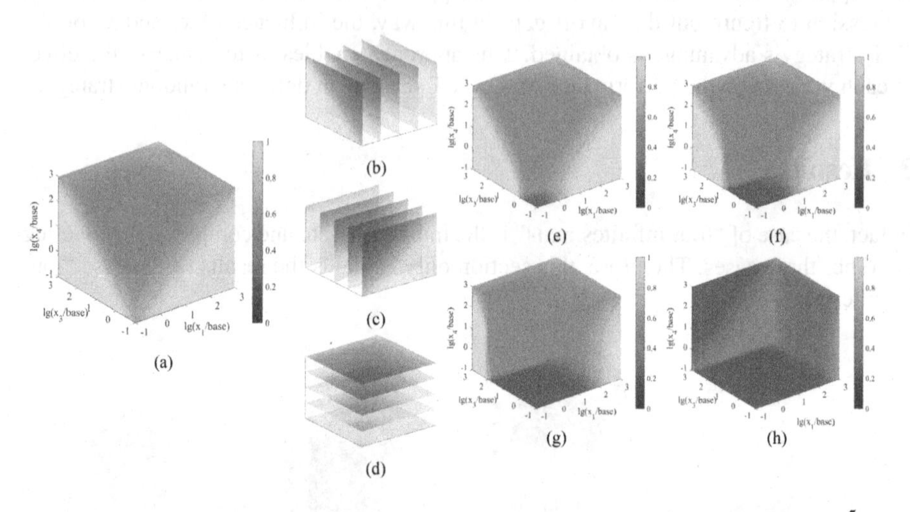

Fig. 2. (a). Winning rate of the Generalized TFT strategy when $x_2 = 10^{-5}$ **(b).** Slice of (a) when $x_1 = 10^{-5}, 10^{-4}, 10^{-3}, 10^{-2}, 10^{-1}$ respectively **(c).** Slice of (a) when $x_3 = 10^{-5}, 10^{-4}, 10^{-3}, 10^{-2}, 10^{-1}$ respectively **(d).** Slice of (a) when $x_4 = 10^{-5}, 10^{-4}, 10^{-3}, 10^{-2}, 10^{-1}$ respectively **(e).** Winning rate of the Generalized TFT strategy when $x_2 = 10^{-4}$ **(f).** Winning rate of the Generalized TFT strategy when $x_2 = 10^{-3}$ **(g).** Winning rate of the Generalized TFT strategy when $x_2 = 10^{-2}$ **(h).** Winning rate of the Generalized TFT strategy when $x_2 = 10^{-1}$

Figure 2 indicates that when $x_2 = 10^{-5}$, in the sampling set, the subset with a winning rate of over 50% for the Generalized TFT strategy comprises approximately 2/3 of the total. The distribution of this subset is primarily concentrated in instances where x_1, x_3 are relatively large and x_4 is relatively small. As x_1, x_3 decrease, x_4 increases, the advantage of the Generalized TFT strategy gradually diminishes. Besides, it indicates that as x_2 increases from small to large, the set of points with a winning rate of the Generalized TFT strategy over 50% continues to shrink, and at the point where $x_2 = 10^{-1}$, the set of points reaches 5%. In other words, as x_2 increases, the overall winning rate of the Generalized TFT strategy decreases significantly, and its advantages become less pronounced.

Figure 3 indicates that when $x_3 = 10^{-5}$, in the sampling set, the subset with a winning rate of over 50% for the Generalized TFT strategy comprises approximately 1/3 of the total. The distribution of this subset is primarily concentrated in instances where x_2, x_4 are relatively small and x_1 is relatively large. As x_2, x_4 increase, x_1 decreases, the advantage of the Generalized TFT strategy gradually diminishes. Besides, it indicates that as x_3 increases from small to large, the set of points with a winning rate of the Generalized TFT strategy over 50% continues to expand, and at the point where $x_1 = 10^{-1}$, the set of points reaches 95%. In other words, as x_3 increases, the overall winning

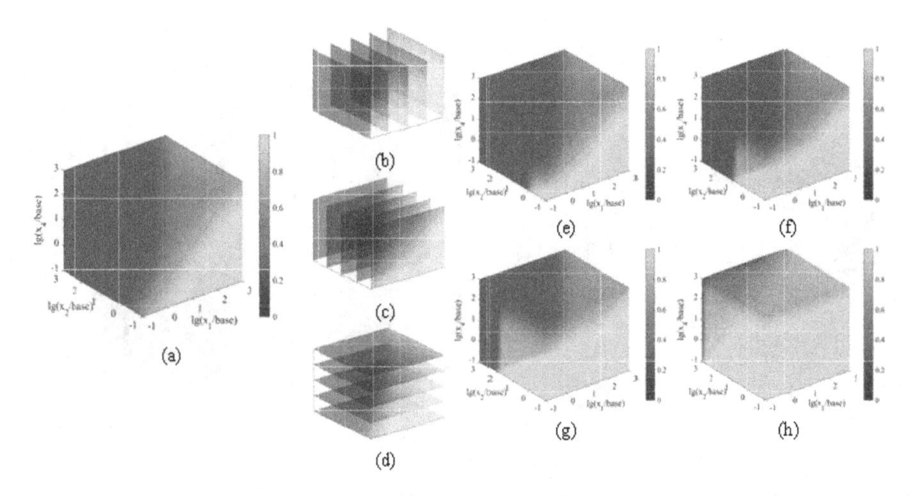

Fig. 3. (a). Winning rate of the Generalized TFT strategy when $x_3 = 10^{-5}$ **(b).** Slice of (a) when $x_1 = 10^{-5}, 10^{-4}, 10^{-3}, 10^{-2}, 10^{-1}$ respectively **(c).** Slice of (a) when $x_2 = 10^{-5}, 10^{-4}, 10^{-3}, 10^{-2}, 10^{-1}$ respectively **(d).** Slice of (a) when $x_4 = 10^{-5}, 10^{-4}, 10^{-3}, 10^{-2}, 10^{-1}$ respectively **(e).** Winning rate of the Generalized TFT strategy when $x_3 = 10^{-4}$ **(f).** Winning rate of the Generalized TFT strategy when $x_3 = 10^{-3}$ **(g).** Winning rate of the Generalized TFT strategy when $x_3 = 10^{-2}$ **(h).** Winning rate of the Generalized TFT strategy when $x_3 = 10^{-1}$

rate of the Generalized TFT strategy increases significantly, and its advantages become more pronounced.

Figure 4 indicates that when $x_4 = 10^{-5}$, in the sampling set, the subset with a winning rate of over 50% for the Generalized TFT strategy comprises approximately 2/3 of the total. The distribution of this subset is primarily concentrated in instances where x_1, x_3 are relatively large and x_2 is relatively small. As x_1, x_3 decrease, x_2 increases, the advantage of the Generalized TFT strategy gradually diminishes. Besides, it indicates that as x_4 increases from small to large, the set of points with a winning rate of the Generalized TFT strategy over 50% continues to shrink, and at the point where $x_4 = 10^{-1}$, the set of points reaches 5%. In other words, as x_4 increases, the overall winning rate of the Generalized TFT strategy decreases significantly, and its advantages become less pronounced.

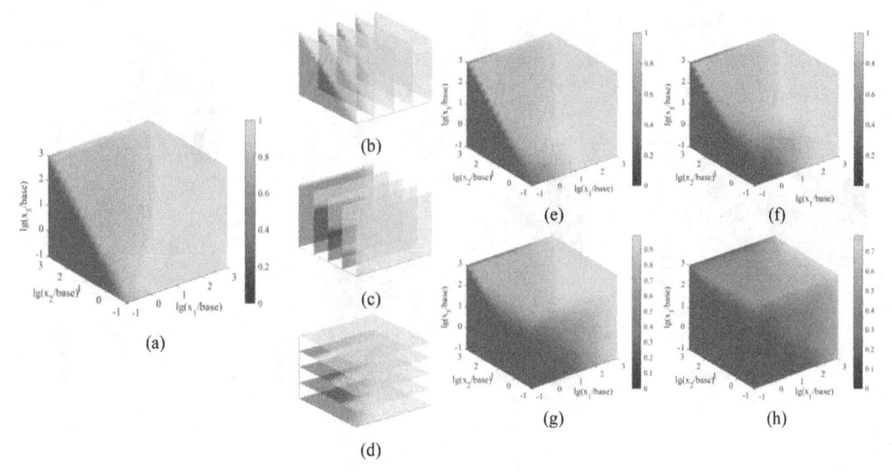

Fig. 4. (a). Winning rate of the Generalized TFT strategy when $x_4 = 10^{-5}$ **(b).** Slice of (a) when $x_1 = 10^{-5}, 10^{-4}, 10^{-3}, 10^{-1}$ respectively **(c).** Slice of (a) when $x_2 = 10^{-5}, 10^{-4}, 10^{-3}, 10^{-1}$ respectively **(d).** Slice of (a) when $x_3 = 10^{-5}, 10^{-4}, 10^{-3}, 10^{-1}$respectively **(e).** Winning rate of the Generalized TFT strategy when $x_4 = 10^{-4}$ **(f).** Winning rate of the Generalized TFT strategy when $x_4 = 10^{-3}$ **(g).** Winning rate of the Generalized TFT strategy when $x_4 = 10^{-2}$ **(h).** Winning rate of the Generalized TFT strategy when $x_4 = 10^{-4}$

4 Conclusion and Discussion

The nature of the Tit-for-Tat (TFT) strategy under the Memory-One strategy is investigated in this study using the control variable method. The results indicate that for the Generalized TFT strategy $p = (1 - x_1, x_2, 1 - x_3, x_4)$, an increase in variables x_1 and x_3 has a promoting effect on the advantage of it, while an increase in variables x_2 and x_4 has an inhibitory effect.

Consequently, users of the Generalized TFT strategy must enact specific counteractive measures during gameplay to optimize their payoffs. When the opponent showed cooperation in the prior round, the TFT user should strategically decrease their cooperation probability in the current round to attain a higher payoff than the adversary. In contrast, when the opponent committed a betrayal in the preceding round, the TFT user should augment their cooperation probability in the present round while diminishing the anticipated payoff to incentivize the opponent's cooperation in subsequent stages of the game.

These findings underscore the significance of the inherent nature of the Generalized TFT strategy in attaining optimal payoffs. Moreover, they delve into the intricate dynamics between the TFT strategy and other strategies within the Memory-One game, shedding light on the definition quandary surrounding the TFT strategy in the realm of evolutionary games. Through the deployment of well-planned counteractive measures during gameplay, individuals and organizations can bolster their decision-making capabilities and maximize their potential benefits.

References

1. Zhang, C., Liu, S., Wang, Z., et al.: The "self-bad, partner-worse" strategy inhibits cooperation in networked populations. Inf. Sci. **585**, 58–69 (2022)
2. Essam, E.L.S.: The effect of noise and average relatedness between players in iterated games. Appl. Math. Comput. **269**, 343–350 (2015)
3. Zhang, J., Chen, Z., Liu, Z.: Fostering cooperation of selfish agents through public goods in relation to the loners. Phys. Rev. E **93**(3), 032320 (2016)
4. Zeng, Z., Li, Y., Feng, M.: The spatial inheritance enhances cooperation in weak prisoner's dilemmas with agents' exponential lifespan. Physica A **593**, 126968 (2022)
5. Attari, S.Z., Krantz, D.H., Weber, E.U.: Reasons for cooperation and defection in real-world social dilemmas. Judgm. Decis. Mak. **9**(4), 316–334 (2014)
6. Fogel, D.B.: Evolving behaviors in the iterated prisoner's dilemma. Evol. Comput. **1**(1), 77–97 (1993)
7. Ramazi, P., Cao, M.: Global convergence for replicator dynamics of repeated snowdrift games. IEEE Trans. Autom. Control **66**(1), 291–298 (2020)
8. Xu, X., Rong, Z.: Extortion boosts cooperation through redistributing strategies in assortative networked systems. IFAC-PapersOnLine **52**(24), 267–271 (2019)
9. Press, W.H., Dyson, F.J.: Iterated prisoner's dilemma contains strategies that dominate any evolutionary opponent. Proc. Natl. Acad. Sci. **109**(26), 10409–10413 (2012)
10. Axelrod, R., Hamilton, W.D.: The evolution of cooperation. Science **211**(4489), 1390–1396 (1981)
11. Glynatsi, N.E., Knight, V.A.: Using a theory of mind to find best responses to memory-one strategies. Sci. Rep. **10**(1), 17287 (2020)
12. Mathieu, P., Delahaye, J.P.: Experimental criteria to identify efficient probabilistic memory-one strategies for the iterated prisoner's dilemma. Simul. Model. Pract. Theory **97**, 101946 (2019)
13. Wedekind, C., Milinski, M.: Human cooperation in the simultaneous and the alternating prisoner's dilemma: pavlov versus generous tit-for-tat. Proc. Natl. Acad. Sci. **93**(7), 2686–2689 (1996)
14. Nowak, M., Sigmund, K.: A strategy of win-stay, lose-shift that outperforms tit-for-tat in the prisoner's dilemma game. Nature **364**(6432), 56–58 (1993)
15. Pan, L., Hao, D., Rong, Z., et al.: Zero-determinant strategies in iterated public goods game. Sci. Rep. **5**(1), 13096 (2015)
16. Ichinose, G., Masuda, N.: Zero-determinant strategies in finitely repeated games. J. Theor. Biol. **438**, 61–77 (2018)
17. Adami, C., Hintze, A.: Evolutionary instability of zero-determinant strategies demonstrates that winning is not everything. Nat. Commun. **4**(1), 2193 (2013)
18. Nowak, M.A., Sigmund, K.: Tit for tat in heterogeneous populations. Nature **355**(6357), 250–253 (1992)
19. Schmid, L., Chatterjee, K., Hilbe, C., et al.: A unified framework of direct and indirect reciprocity. Nat. Hum. Behav.Behav. **5**(10), 1292–1302 (2021)

Stochastic Resource Allocation with Time Windows

Yang Li[1] and Bin Xin[2(✉)]

[1] Beijing Institute of Electronic System Engineering, Beijing 100854, China
[2] School of Automation, Beijing Institute of Technology, Beijing 100081, China
brucebin@bit.edu.cn

Abstract. The stochastic resource allocation problem with time windows (SRAPTW) refers to a class of combinatorial optimization problems which are aimed at finding the optimal scheme of assigning resources to given tasks within their time windows. In SRAPTW, the capability of resources to accomplish tasks is quantitatively characterized by probability. The expected allocation scheme should include not only the task-resource pairings but also their allocation time. This paper formulates SRAPTW as a nonlinear mixed 0–1 programming problem with the objective of maximizing the reward of completing specified tasks. Then, a general encoding/decoding method is proposed for the representation of solutions, and several different problem-solving methodologies are presented and compared. Results of computational experiments show that the utilization of SRAPTW-specific knowledge can bring in excellent performance, and a constructive heuristic combining maximal marginal return strategy and maximal probability strategy has remarkable advantages, especially in larger-scale cases.

Keywords: Stochastic Resource Allocation · Time Windows · Constructive Heuristic

1 Introduction

The stochastic resource allocation problem with time windows (SRAPTW) refers to a class of combinatorial optimization problems which are aimed at finding the optimal scheme of assigning resources to given tasks within certain time windows. As a complex variant of the stochastic resource allocation problems (SRAP) [1, 2], a prominent feature of SRAPTW is that the capability of resources to accomplish tasks is quantitatively characterized by probability within certain time windows.

SRAPTW widely exists in the control of complex systems which involve the allocation of multiple sensors and actuators for various control tasks. When sensors or actuators are used for moving targets especially those time-critical ones, the time windows of resource allocation are one of the key factors to check the feasibility of task scheduling and execution.

Previous studies on SRAP, such as sensor-target assignment problem and weapon-target assignment problem, were focused on the allocation of resources to tasks without time windows [1–8]. In fact, time windows widely exist in the resource allocation

B. Xin et al. (Eds.): IWACIII 2023, CCIS 1931, pp. 348–358, 2024.
https://doi.org/10.1007/978-981-99-7590-7_28

problems from diverse fields, e.g., emergency and rescue [9], communication channel management [10], traffic resources management [11], transportation and logistics [12–14], defense resources allocation [15]. Unlike common resource allocation problems involving time windows, the success probability of resources to accomplish tasks is time-dependent in SRAPTW. Therefore, the time windows in SRAPTW not only bring constraints, but also relate to the distribution of success probability, which means the task performance of resources is sensitive to the execution time of tasks. In this sense, solving SRAPTW involves the matching of resources and tasks as well as determining the execution time of tasks, so as to optimize the overall task performance.

This paper is aimed at formulating SRAPTW and proposing efficient strategies and algorithms for solving it. The contribution of the paper is summarized as follows:

1) Formulate the SRAPTW as a nonlinear mixed 0–1 programming problem.
2) Provide a general encoding/decoding method for representing solutions and designing search algorithms to solve SRAPTW.
3) Propose four different SRAPTW algorithms which combine knowledge-dependent constructive heuristic and random search. Conduct comparative computational experiments to identify their pros and cons.

The rest of the paper is structured as follows. Section 2 presents the problem formulation. Section 3 presents the proposed algorithms. Section 4 presents computational experiments. Section 5 concludes the paper.

2 Problem Formulation

2.1 Problem Description

Stochastic resource allocation (SRA) refers to the scenarios in which the capability of a resource (e.g., a person or a robot) to complete tasks is characterized by probabilities. An allocation scenario of m resources and n tasks is considered. The probability of the ith resource ($i = 1, 2, \cdots, m$) to complete the jth task ($j = 1, 2, \cdots, n$) also depends on the allocation time. So, if the probability, denoted by Pij(t), is lower than certain threshold or even becomes 0, the allocation will be abandoned. Generally, the allocation of resources to tasks needs to be handled within specified time windows. Thus, the problem is an SRA problem with time windows (SRAPTW). Assume that each task, once accomplished, will bring a reward. Denote the reward of the jth task by v_j. The objective of SRAPTW is to maximize the total reward of allocating resources to tasks within specified time windows. Besides, resources when assigned usually have to satisfy some constraints. For example, for one resource, there will be some time interval between one allocation and its next, as well as an allowable maximal number of tasks to be allocated. For each task, it is allowed that multiple resources can be used but each resource can be used only once.

2.2 Optimization Model

The SRAPTW can be formulated as an optimization problem. Assume that the execution of tasks about each resource is mutually independent. Then, the objective of the problem can be presented as follows:

$$\max J_1(X,T) = \sum_{j=1}^{n} v_j\left(1 - \prod_{i=1}^{m}(1 - p_{ij}(t_{ij})x_{ij})\right) \tag{1}$$

where $J_1(X,T)$ represents the objective function, X and T are the decision matrices, $X = [x_{ij}]_{m\times n}$ represents the resource-task allocation relation, $T = [t_{ij}]_{m\times n}$ represents the task execution time for each resource-task pair when allocated, and $P = [p_{ij}(t_{ij})]_{m\times n}$ represents the success probability that a resource can accomplish a task at specific time. A detailed description of other parameters is presented in Table 1.

The objective shown in Eq. (1) can be equivalently converted to the following

$$\min J_2(X,T) = \frac{\sum_{j=1}^{n} v_j \prod_{i=1}^{m}(1 - p_{ij}(t_{ij})x_{ij})}{\sum_{j=1}^{n} v_j} \tag{2}$$

where $J_2(X,T)$ is a normalized objective function.

Table 1. Explanation of Model Parameters.

Symbol	Explanation
v_j	Reward of accomplishing task j
x_{ij}	$x_{ij} = 1$ means that task j is allocated to resource i; $x_{ij} = 0$ otherwise
t_{ij}	The task execution time for resource i to execute task j
$p_{ij}(t_{ij})$	The success probability that resource i can accomplish task j at the time t_{ij}
TW_{ij}	The time interval for resource i to execute task j
t_{ij}^-	The lower bound of the time interval for resource i to execute task j
t_{ij}^+	The upper bound of the time interval for resource i to execute task j

The following constraints have to be satisfied.

Range of Decision Variables. It is obvious that the allocation variable is of 0–1 type and the allocation time should be determined within specified time window.

$$x_{ij} \in \{0,1\}, \forall i \in \{1,2,\cdots,m\}, \forall j \in \{1,2,\cdots,n\} \tag{3}$$

$$t_{ij} \in [t_{ij}^-, t_{ij}^+], \forall i \in \{1,2,\cdots,m\}, \forall j \in \{1,2,\cdots,n\} \tag{4}$$

Maximal Number of Tasks to be Allocated for Each Resource. For resource i, the maximal number of tasks which can be allocated to it should not exceed n_i.

$$\sum_{j=1}^{n} x_{ij} \le n_i, \forall i \in \{1,2,\cdots,m\} \tag{5}$$

Maximal Number of Resources to be Allocated for Each Task. For task j, the maximal number of resources which can be allocated to it should not exceed m_j.

$$\sum_{i=1}^{m} x_{ij} \le m_j, \forall j \in \{1, 2, \cdots, n\} \tag{6}$$

Time Interval for Each Resource to Execute Two Tasks. For resource i, the execution of any two tasks assigned to it should be separated at least τ_i.

$$|t_{ij} - t_{ik}| \ge \tau_i, \forall i \in \{1, 2, \cdots, m\}, \forall j, k \in \{1, 2, \cdots, n\}, j \ne k \tag{7}$$

To sum up, the optimization model for SRAPTW can be formulated as follow:

$$\min J_2(X, T), \text{s.t.}(3) \sim (7).$$

Obviously, SRAPTW is a nonlinear mixed-variable programming problem.

2.3 Problem Analysis

From the problem formulation shown above, the following properties can be derived:

Property 1: The more resources are allocated without violating any constraints, the better the objective value will be. Stated another way, for any X, if more x_{ij} can become 1, $J_2(X,T)$ can be improved.

Property 2: For any X, the larger $p_{ij}(t_{ij})$ is, the better the objective value.

Property 3: If all $t_{ij}^* = \text{argmin}(p_{ij}(t_{ij})), \forall i \in \{1, 2, \cdots, m\}, \forall j \in \{1, 2, \cdots, n\}$, do not have conflicts w.r.t. constraints (7), then $t_{ij} = t_{ij}^*$ will be a necessary condition for any $x_{ij} = 1$.

As demonstrated later, these properties are beneficial to design efficient problem-specific strategies or operators to solve SRAPTW.

3 Algorithm Design

3.1 Solution Representation

The decision variables X (binary valued) and T (real valued) are straightforward representation of solutions to SRAPTW. However, it will be more convenient to utilize the problem-domain knowledge in SRAPTW by use of permutation-based encoding and decoding schemes.

According to Property 1, for a feasible X which does not violate the constraints (3), (5) or (6), changing more x_{ij} from 0 to 1 will bring the improvement of the objective value. In this sense, for an empty X (all zero elements), the order of making each x_{ij} become one while ensuing constraint satisfaction can be a key factor to generate a feasible and even high-quality solution. In other words, the permutation of all resource-task pairs can be used as a main component of solution representation.

For the execution time regarding each pair, we can determine the value of each t_{ij} from its time window according to certain rules, e.g., unbiased random sampling or biased heuristic selection.

To sum up, the permutation of all resource-task pairs and corresponding execution time can be used to represent a solution to SRAPTW. The detailed encoding-decoding scheme is described as follows.

Encoding Scheme. For a resource-task pair (briefly called an RT pair) denoted by *i-j* ($i \in \{1, 2, \cdots, m\}, j \in \{1, 2, \cdots, n\}$), we use the number $n \times (i-1) + j$ to represent it. Then, the permutation of all RT pairs is formally a permutation of the integers from 1 to $m \times n$. The matrix of the execution time regarding each RT pair, together with the permutation, constitutes the encoding scheme.

For example, for an SRAPTW of 2 resources and 2 tasks with time windows $TW_{11} = [0, 0.1]$, $TW_{12} = [0.3, 0.4]$, $TW_{21} = [0.2, 0.3]$ and $TW_{22} = [0.5, 0.9]$, the permutation 4-1-3-2 and the matrix of execution time [0.05, 0.38; 0.25, 0.78] represent a solution.

Decoding Scheme. For a given permutation Pe and a matrix of execution time$T = \left[t'_{ij}\right]_{m \times n}$, $x_{ij} = 1$ for each RT pair will be checked, in the order indicated by Pe, to see if it violates constraints (5), (6) or (7). If no constraints violation occurs, then let $x_{ij} = 1$ and $t_{ij} = t'_{ij}$; otherwise, let $x_{ij} = 0$ and $t_{ij} = \infty$.

Based on the above representation scheme, a feasible solution can be generated by decoding after determining a permutation of all RT pairs and a execution time matrix. Both the permutation and the execution time matrix can be generated in different ways. For example, they can be generated by random sampling or constructive heuristics.

3.2 Constructive Heuristics

Constructive heuristics represent a philosophy of straightforwardly generating a solution to a complex problem by determining its components step by step, rather than implementing samplings in solution space as widely adopted in various search algorithms such as improvement heuristics and metaheuristics [16–18]. In comparison with search algorithms, constructive heuristics do not need function evaluations to improve solutions in an iterative way. To design efficient constructive heuristics, the permutation of all RT pairs and the execution time matrix can be generated by utilizing SRAPTW-specific knowledge, e.g., the three properties shown in Subsect. 2.3. To utilize Properties 2 and 3, the execution time will be set to t^*_{ij} if it does not violate any constraint. The permutation can be constructed by using the rule-based method based on maximal marginal return (MMR) [6, 19]. The marginal return means the improvement of the objective value with reference to its current value brought by choosing one RT pair. MMR is a rule for constructing a solution incrementally, and in each step, the RT pair with maximal marginal return without constraint violation will be added into the allocation scheme. In fact, since the MMR-based procedure can generate a complete solution directly, it is unnecessary to get the permutation as a preliminary which is de facto implicated in the procedure. To save space, the rationale of MMR will not be presented here in more detail, and interested readers may refer to the reference [6]. Instead, we provide the pseudo-code of the procedure in **Procedure 1**.

Procedure 2. MMR(T)

```
Input: The initial execution time matrix T
Output: allocation scheme X, final execution time matrix T, objective value F
X = O_mxn ;   % All zeros
CS1 = O_mx1 ; % The flag of constraint 5
CS2 = O_nx1 ; % The flag of constraint 6
Temp = inf * ones(m,n) ; % All infinities
cnt = 0 ;
For i = 1 to m
   For j = 1 to n
      r(i,j) = v_j * p_ij(t_ij) ; % Calculate the initial marginal return for each RT
pair
   End
End
For k = 1 to m*n
   [w,g] = max(r) ; % Get the maximum for each column of the matrix r
                    % w records the maximum values
                    % g records the row number of each maximum in each column
   [u,h] = max(w) ; % Get the maximum of the vector w
                    % u records the maximum value
                    % g records the column number of the maximum in w
   i = g(h) ; % the hth element of the vector g
   j = h ;    % i-j indicates the RT pair with the maximal marginal return
   If CS1(i) < n_i && CS2(j) < m_j && min(|t_ij-Temp(i,:)|) ≥ τ_i
      X(i,j) = 1; CS1(i) = CS1(i) + 1 ; CS2(j) = CS2(j) + 1 ;
       Temp(i,CS2(j)) = t_ij ;
       If CS1(i) == m_i  cnt = cnt +1; end
      If CS2(j) == n_j  cnt = cnt +1; end
      If cnt == m+n break ; end
      r(:,j) = r(:,j) * (1 - p_ij(t_ij)) ;
   End
   r(i,j) = - inf ;
End
F = 0 ; q = ones(n,1) ;
For j=1 to n
   For i = 1 to m
      q(j) = q(j) * ( 1 - X(i,j) * p_ij(t_ij) ) ;
   End
   F = F + v_j * q(j) ;
End
T = (1./X) * T ; F = F / sum(v) ;
Return X, T, F
```

Since the core of this constructive heuristic is the MMR-based procedure combined with the execution time setting based on maximal probability, we name it by **MMR-MP**. The time complexity of determining all the execution time with maximal probability is

$O(mnl)$ where l represents the number of time samplings in each time window. From the pseudo-code of MMR-MP, the worst-case time complexity of MMR procedure is $O(m^2n^2)$. So, the time complexity of MMR-MP is $O(m^2n^2 + mnl)$.

3.3 Random Search

Different from the constructive heuristic MMR-MP, the random search here refers to certain unbiased search process to find better solutions. Since any solution involves two parts, i.e., the permutation and the execution time matrix, we have three different strategies to implement random search:

Strategy 1: MMR for permutation (*implicit*) & Random sampling of execution time
Strategy 2: Random permutation & Execution time with maximal probability
Strategy 3: Random permutation & Random sampling of execution time

Random permutation can be achieved in MATLAB by the command randperm(n * m) while the random sampling of execution time can be conducted by implementing $t_{ij} = t_{ij}^- + (t_{ij}^+ - t_{ij}^-) \times rand$ where *rand* is a random number following the uniform distribution in (0,1). The three versions of random search resulting from these strategies are named **MMR-RS, RP-MP** and **RP-RS**, respectively. In fact, similar to MMR-MP, both MMR-RS and RP-MP utilize SRAPTW-specific knowledge while maintaining certain randomness. In contrast, RP-RS is a pure random search without search bias. Denote the allowable number of random samplings (function evaluations) by N. Then, the time complexity of MMR-RS, RP-MP and RP-RS is $O(Nm^2n^2)$, $O((N + l)mn)$ and $O(Nmn)$, respectively.

4 Computational Experiments

The goal of the computational experiments is to validate the performance of the proposed algorithms including MMR-MP, MMR-RS, RP-MP and RP-RS. All tests are conducted on a Lenovo Laptop X1 with Intel Core i7 CPU (1.8GHz) and 16GB RAM. For each SRAPTW instance, the random search algorithms including MMR-RS, RP-MP and RP-RS will run 25 times for making a statistical analysis of results.

4.1 Experiments Configuration

Test case generator: The task rewards are generated by following a uniform distribution $v_j \sim U(1,1 + \delta)$ where $U(a,b)$ represents a uniform distribution in the interval (a,b), and δ is a control parameter. All time windows TW_{ij} are randomly generated from the interval (0,1). The success probability $p_{ij}(t_{ij}) \sim N(\mu_{ij},\sigma_{ij}^2)$ where $N(\mu, \sigma^2)$ represents a Gauss distribution with mean μ and standard deviation σ. Let $\mu_{ij} \sim U(0,1)$ and $(\sigma_{ij} - \varepsilon) \sim U(0,t_{ij}^+ - t_{ij}^-)$ where ε is a positive number for regulating the distribution of success probability. A larger ε implies a smaller change of success probability within corresponding time window. For constraints (5) and (6), $m_i = \lceil \alpha \times rand_i \rceil$, $n_j = \lceil \beta \times rand_j \rceil$ where $\lceil \cdot \rceil$ is the ceiling operator, and α and β are control parameters, and $rand_i, rand_j \sim U(0,1)$. For constraint (7), $\tau_i \sim U(\tau_i^-, \tau_i^+)$ where τ_i^- and τ_i^+ are control parameters.

The following cases are generated.

Case 1: m=5, n=5, $\delta = 0, \varepsilon = 1, \alpha = 1, \beta = 1, \tau_i^- = 0.1, \tau_i^+ = 0.4$
Case 2: m=5, n=5, $\delta = 1, \varepsilon = 1, \alpha = 1, \beta = 1, \tau_i^- = 0.1, \tau_i^+ = 0.4$
Case 3: m=5, n=5, $\delta = 1, \varepsilon = 0.7, \alpha = 1, \beta = 1, \tau_i^- = 0.1, \tau_i^+ = 0.4$
Case 4: m=5, n=5, $\delta = 9, \varepsilon = 1, \alpha = 1, \beta = 1, \tau_i^- = 0.1, \tau_i^+ = 0.4$
Case 5: m=5, n=5, $\delta = 9, \varepsilon = 0.7, \alpha = 2, \beta = 2, \tau_i^- = 0.1, \tau_i^+ = 0.4$
Case 6: m=5, n=10, $\delta = 9, \varepsilon = 0.4, \alpha = 2, \beta = 2, \tau_i^- = 0.1, \tau_i^+ = 0.4$
Case 7: m=10, n=5, $\delta = 9, \varepsilon = 0.1, \alpha = 2, \beta = 2, \tau_i^- = 0.1, \tau_i^+ = 0.4$
Case 8: m=10, n=5, $\delta = 9, \varepsilon = 0.4, \alpha = 2, \beta = 2, \tau_i^- = 0.1, \tau_i^+ = 0.4$
Case 9: m=10, n=10, $\delta = 9, \varepsilon = 0.1, \alpha = 2, \beta = 2, \tau_i^- = 0.1, \tau_i^+ = 0.4$
Case 10: m=20, n=10, $\delta = 9, \varepsilon = 0.1, \alpha = 2, \beta = 2, \tau_i^- = 0.1, \tau_i^+ = 0.4$

In each case, ten different instances are generated by following the probability distribution preset in the test case generator. Detailed data about the total 100 instances and related results will be released online and can be acquired by contacting the authors.

Parameter Setting: For all random search algorithms, the maximal number of function evaluations (samplings) is set as $10 \times m \times n$. For determining the time for the maximal probability in each time window, even samplings within [0,1] are conducted and the minimal spacing is set to 0.001.

4.2 Performance Comparison

For each instance in each case, the results of all algorithms, i.e., obtained objective values in 25 runs, will be compared by Wilcoxon ranksum test with 0.05 significance level. For two algorithms A and B, the indicator of the ranksum test $h = 1$ means the performance of A is significantly different from that of B. In this case, if the mean of the results obtained by A is smaller (larger) than that by B, then A gets a score of $+1$ (-1) and B gets a score of -1 (+1); otherwise, both A and B get zero. In each case, the highest and lowest scores are 30 and –30, respectively. The scores of four algorithms in 10 cases (totally 100 instances) are summarized in Table 2. To save space, detailed results for each instance will not be presented. The average runtime of the four algorithms in different cases is shown in Fig. 1.

Table 2. Statistical results of algorithm comparison (scores)

Algorithm	Case 1	Case 2	Case 3	Case 4	Case 5	Case 6	Case 7	Case 8	Case 9	Case 10
RP-RS	−16	−18	−24	−26	−24	−29	−30	−30	−30	−30
RP-MP	**25**	**28**	**27**	**25**	**15**	−11	9	12	−4	−10
MMR-RS	8	4	−3	−5	9	**22**	−7	−3	12	10
MMR-MP	−17	−14	0	6	0	18	**28**	**21**	**22**	**30**

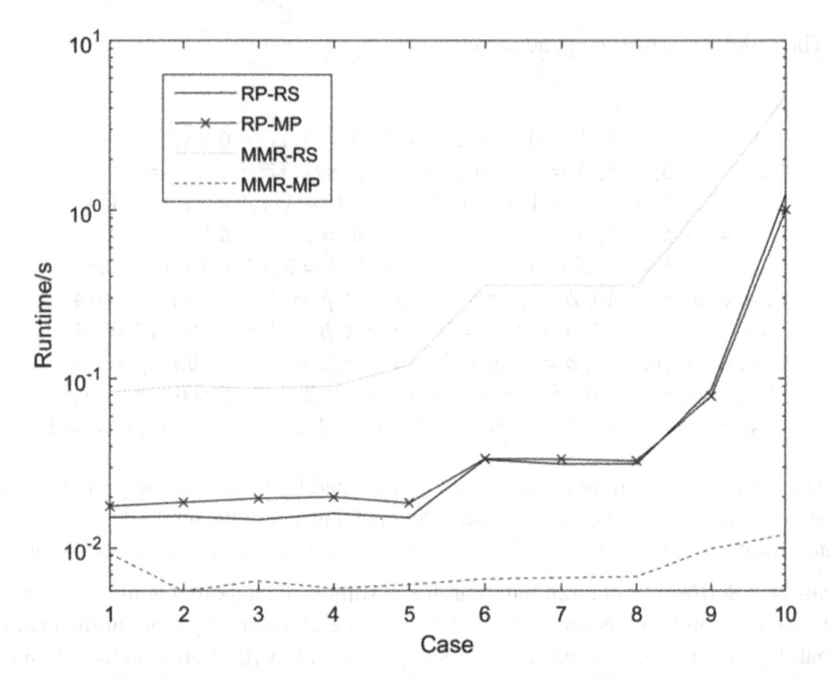

Fig. 1. The average runtime of four algorithms in different cases.

From Table 2, the following results can be found:

1) RP-MP has obvious advantages in Cases 1, 2, 3 and 4; in contrast, it is not a good solver in Cases 6 and 10;
2) MMR-MP takes the first place in Cases 7, 8, 9 and 10 (especially wins all in the larger-scale Case 10). However, it was defeated by RP-MP in Cases 1, 2, 3, 4 and 5 and performs poorly in Cases 1 and 2.
3) MMR-RS gets the highest score in Case 6, and shows its advantage over RP-MP and RP-RS but loses to MMR-MP in Cases 9 and 10.
4) RP-RS has no advantage in almost all cases. In Cases 7, 8, 9 and 10, it was completely defeated by the other algorithms. This implies that pure random search without use of problem-specific knowledge is inefficient, though it can theoretically cover the solution space and find the optimal solution with sufficient samplings.
5) It seems that the MMR strategy generally has dominant advantages in larger-scale cases, which is supported by (a) the superiority of MMR-MP over RP-MP in Cases 6 to 10, (b) the superiority of RP-MP over MMR-MP in Cases 1 to 5, and (c) the superiority of MMR-RS over RP-RS. The comparison implies that the permutation of RT pairs will play a more important role in determining the quality SRAPTW solutions as the scale of SRAPTW increases.
6) It seems that the MP strategy has advantages in smaller-scale cases, which is supported by (a) the superiority of MMR-MP over MMR-RS in Cases 7 to 10, (b) the superiority of MMR-RS over MMR-MP in Cases 1, 2 and 5, and (c) the superiority of RP-MP over RP-RS. Obviously, MP plays a crucial role in generating high-quality

solutions for smaller-scale cases, and it is also a necessary strategy for guaranteeing the performance of MMR-MP.

The average time cost of the four algorithms shown in Fig. 1 is consistent with the analysis about computational complexity in Sect. 3. Obviously, MMR-MP has the lowest time cost in all cases. The time costs of RP-RS and RP-MP are very close, which implies that the cost of obtaining the maximal probabilities (regulated by the parameter l) is negligible as compared to that of determining the permutation of RT pairs (regulated by the number of function evaluations N).

5 Conclusion

A mathematical programming model was built for the stochastic resource allocation problem with time windows. A general encoding/decoding scheme was proposed for representing SRAPTW solutions. Constructive heuristics and random search algorithms were proposed to solve SRAPTW. Comparative experiments show that MMR-MP and RP-MP have obvious advantages over MMR-RS and RP-RS in most cases. MMR-MP and RP-MP share the similarity in the use of preset time under maximal probability strategy, which implies the superiority and necessity of utilizing the problem-specific knowledge embodied in Properties 3 and 4. In some instances, MMR-RS is the single winner, which reflects that MMR is an effective strategy for determining the order of resource-task pairs in allocation process. In contrast, RP-RS did not take the first place in any instances, and also performs the worst in most cases. The failure of RP-RS as compared with its competitors confirms that pure random search without utilization of problem knowledge is not a satisfactory choice for algorithm design. Generally, misuse of inaccurate knowledge in problems may cause the loss of optimality or lead to local optima. On the other hand, no use of problem-specific knowledge usually results in slow convergence of iterative search process. A delicate balance between the two aspects may bring better strategies for solving complex SRAPTW. On the basis of the proposed general solution representation scheme, how to design advanced neighborhood search operator or meta-heuristics [20–22] also deserves further studies in the future.

Acknowledgement. We would like to thank the National Outstanding Youth Talents Support Program (Grant 61822304), the Basic Science Center Programs of NSFC (Grant 62088101), the Shanghai Municipal Science and Technology Major Project (2021SHZDZX0100), and the Shanghai Municipal Commission of Science and Technology Project (19511132101).

References

1. Castanon, D.A., Wohletz, J.M.: Model predictive control for stochastic resource allocation. IEEE Trans. Autom. Control **54**(8), 1739–1750 (2009)
2. Fan, G.M., Huang, H.J.: Scenario-based stochastic resource allocation with uncertain probability parameters. J. Syst. Sci. Complexity **30**(2), 357–377 (2017)
3. Li, J., Xin, B., Pardalos, P.M., Chen, J.: Solving bi-objective uncertain stochastic resource allocation problems by the s-based risk measure and decomposition-based multi-objective evolutionary algorithms. Ann. Oper. Res. **296**(1–2), 639–666 (2021)

4. Cassandras, C.G., Dai, L., Panayiotou, C.G.: Ordinal optimization for a class of deterministic and stochastic discrete resource allocation problems. IEEE Trans. Autom. Control **43**(7), 881–890 (1998)
5. Xin, B., Chen, J., Zhang, J., Dou, L.H., Peng, Z.H.: An efficient rule-based constructive heuristic to solve dynamic weapon-target assignment problem. IEEE Transactions on Systems Man and Cybernetics Part A - Systems and Humans **41**(3), 598–606 (2011)
6. Xin, B., Wang, Y.P., Chen, J.: An efficient marginal-return-based constructive heuristic to solve the sensor-weapon-target assignment problem. IEEE Transations on Systems, Man Cybernetics-Systems **49**(12), 2536–2547 (2019)
7. Asadpour, A., Wang, X., Zhang, J.: Online resource allocation with limited flexibility. Manage. Sci. **66**(2), 642–666 (2020)
8. Wang, Y.P., Xin, B., Chen, J.: An adaptive memetic algorithm for the joint allocation of heterogeneous stochastic resources. IEEE Transactions on Cybernetics **52**(11), 11526–11538 (2022)
9. Luan, D., Liu, A., Wang, X., Xie, Y., Wu, Z.: Robust two-stage location allocation for emergency temporary blood supply in postdisaster. Discrete Dynamics in Nature and Society **2022**, Article ID 6184170
10. Bankov, D., Khorov, E., Lyakhov, A., Famaey, J.: Resource allocation for machine-type communication of energy-harvesting devices in Wi-Fi HaLow networks. Sensors **20**, 2449 (2020)
11. Jiang, B., Fan, Z.P.: Optimal allocation of shared parking slots considering parking unpunctuality under a platform-based management approach. Transp. Res. Part E **142**, 102062 (2020)
12. Puglia Pugliesea, L.D., Ferone, D., Macrinac, G., Festa, P., Guerriero, F.: The crowd-shipping with penalty cost function and uncertain travel times. Omega **115**, 102776 (2023)
13. Wang, Y., Wang, X., Guan, X., Li, Q., Fan, J., Wang, H.: A combined intelligent and game theoretical methodology for collaborative multicenter pickup and delivery problems with time window assignment. Appl. Soft Comput. **113**, 107875 (2021)
14. Hoogeboom, M., Adulyasak, Y., Dullaert, W., Jaillet, P.: The robust vehicle routing problem with time window assignments. Transp. Sci. **55**(2), 395–413 (2021)
15. Almeida, R., Gaver, D.P., Jacobs, P.A.: Simple probability-models for assessing the value of information in defense against missile attack. Nav. Res. Logist. **42**(4), 535–547 (1995)
16. Meng, K., Chen, C., Xin, B.: MSSSA: a multi-strategy enhanced sparrow search algorithm for global optimization. Frontiers of Information Technol. Electronic Eng. **23**(12), 1828–1847 (2022)
17. Gao, G.Q., Mei, Y., Jia, Y.H., Browne, W.N., Xin, B.: Adaptive coordination ant colony optimization for multipoint dynamic aggregation. IEEE Trans. Cybernetics **52**(8), 7362–7376 (2022)
18. Guo, M., Xin, B., Chen, J., Wang, Y.P.: Multi-agent coalition formation by an efficient genetic algorithm with heuristic initialization and repair strategy. Swarm and Evolutonary Computation **55** (2020)
19. Gülpınar, N., Çanakoglu, E., Branke, J.: Heuristics for the stochastic dynamic task-resource allocation problem with retry opportunities. Eur. J. Oper. Res. **266**, 291–303 (2018)
20. Ding, Y.L., Xin, B., Zhang, H., Chen, J., Dou, L.H., Chen, B.M.: A memetic algorithm for curvature-constrained path planning of messenger UAV in air-ground coordination. IEEE Trans. Autom. Sci. Eng.Autom. Sci. Eng. **19**(4), 3735–3749 (2022)
21. Qi, M.F., Dou, L.H., Xin, B.: 3D smooth trajectory planning for UAVs under navigation relayed by multiple stations using Bezier curves. Electronics **12**(11), 2358 (2023)
22. Jiao, K.M., Chen, J., Xin, B., Li, L., Zhao, Z.X., Zheng, Y.F.: A framework for co-evolutionary algorithm using Q-learning with meme. Expert Systems With Applications **225** (2023)

Author Index

A
Alsumeri, Abdulrahman Abdo Ali I-3
Ariga, Harunobu I-189
Asaadi, Mohsen I-101
Aslansefat, Koorosh I-101

B
Bai, Mingsong II-57

C
Cao, Qi II-329
Chen, Bing II-250
Chen, Zhiwei II-223
Cheng, Lan I-15
Chugo, Daisuke II-343
Cui, Siyi II-329

D
Dai, Yaping II-262
Ding, Yunhao I-338
Dong, Chengyuan II-117
Dong, Hongcheng I-270
Dong, Lei I-89
Du, Zhaofeng I-255
Duan, Yiming II-197

E
Ebine, Kenta I-226

F
Fu, Si-Yue II-43

G
Gao, Hui I-116
Geng, Zhiqiang II-223
Grigorev, Maksim II-3

H
Han, Yongming II-223
Hashimoto, Daisuke II-304
Hashimoto, Hiroshi II-343
He, Hui I-167
He, Yan-Lin I-63
Ho, Yihsin II-117
Hong, Bowen II-223
Hoshino, Yukinobu I-241, II-141, II-150, II-304
Hu, Zhimin I-15
Huang, He I-255

I
Ishiguro, Keio II-343
Izadi, Iman I-101

J
Jiang, Pengcheng II-70
Jin, Yaning II-293
Jin, Ying II-57
Jin, Zengwang I-298
Jing, Yidi II-129
Ju, Shuang II-293

K
Kawamoto, Kazuhiko II-18
Ke, Wei I-63
Kera, Hiroshi II-18
Khan, Malak Abid Ali I-3, II-197
Kholodilin, Ivan II-3
Khriukin, Dmitry II-3
Kubota, Naoyuki II-81, II-105
Kuroda, Shigeki II-343
Kusunose, Shoya II-150

L
Li, Boyan I-89
Li, Lihua I-283
Li, Qian I-298
Li, Tianyi I-24
Li, Yang I-348
Li, Yuan II-70

B. Xin et al. (Eds.): IWACIII 2023, CCIS 1931, pp. 359–361, 2024.
https://doi.org/10.1007/978-981-99-7590-7

Liang, Xiao I-24
Lin, Chong II-250
Liu, Jiating II-11
Liu, Meng I-116
Liu, Zhen-Tao I-178
Lu, Yi I-50
Luo, Yi I-63

M
Ma, Hongbin I-3, I-283, II-57, II-197
Ma, Xin Yao I-324
Maeda, Nagamasa II-150
Maksimov, Nikita II-3
Mizanur, Rahman II-197
Moeurn, Si Kheang II-283
Mu, Yifen I-270

N
Na, Qin I-50
Nakajima, Taiga II-167

O
Obo, Takenori II-105
Ohno, Sumio II-11
Okamoto, Kazushi I-153
Okawa, Reiji II-150

P
Pei, Xiaoshuai I-167
Peng, Zexuan I-75, II-29
Peng, Zhihong I-167

Q
Qian, Lin I-215

R
Rathnayake, Namal I-241
Rehman, Zia Ur I-3, II-197

S
Sakai, Hiromi II-343
Sasagawa, Junichi II-141
Sato-Shimokawara, Eri II-129
Savosteenko, Nikita II-3
Shang, Bojin II-262
Shang, Peiqiao I-167
Shao, Shuai II-262
She, Jinhua II-343
Shibata, Hiroki I-215, I-226
Shinomiya, Yuki I-189, II-93
Song, Qingwei II-81, II-105
Su, Wei I-128
Sugahara, Kai I-153
Sugiyama, Takahiro II-156
Sun, Changyin I-298
Sun, Xiaojun I-24

T
Takama, Yasufumi I-215, I-226
Tian, Hongzhi II-319

U
Umeda, Hibiki II-93
Ushiwaka, Takashi II-150
Utami, Moegi I-241

W
Wang, Baoji I-89
Wang, Haoran II-209
Wang, Honghong II-250
Wang, Jing II-293
Wang, Jirong II-319
Wang, Lei I-37
Wang, Qing I-310, II-209
Wang, Qixuan I-89
Wang, Rennong II-343
Wang, Wenjie I-167
Wang, Xiaohan II-262
Wang, Xin I-3
Wang, Ya-fei I-116
Wang, Yujue I-310, II-209
Watamori, Michio II-141
Wei, Dong II-43
Wei, Hui I-139
Wen, Chengxin I-283
Wu, Nan II-18
Wu, Peng I-116
Wu, Qi I-75, II-29
Wu, Shuang II-57
Wu, Xianhao II-329

X
Xiao, Wei II-237
Xie, Qingyu I-128
Xin, Bin I-255, I-310, I-348, II-283
Xu, Gang II-250

Xu, Ning I-50
Xu, Xinying I-15
Xu, Yuan I-63

Y

Yamamoto, Shinpei II-150
Yan, Guangming I-24
Yang, Fan I-101
Yang, Xuanhao II-237
Yoshida, Shinichi II-156, II-167
Yuan, Yuan II-329

Z

Zeng, Xiang-Yu I-178
Zhai, Bin I-324
Zhang, Bo-Yang I-178
Zhang, Chun Mei I-324
Zhang, Chunyan I-139, I-338
Zhang, Hongzhan II-237
Zhang, Huanghui I-200
Zhang, Huixiang I-298
Zhang, Jia I-37
Zhang, Jianlei I-338
Zhang, Ming-Qing I-63
Zhang, Shitong I-310
Zhang, Tianyao I-75, II-29, II-182
Zhang, Wei I-63
Zhang, Weiqiang I-15
Zhang, Yang I-63
Zhang, Yuqi II-81, II-105
Zhang, Zhaohui I-75, II-29, II-182, II-329
Zhang, ZhenYu I-50
Zhang, Zimin II-182
Zhao, Xiaoyan I-75, II-29, II-182, II-329
Zhao, Yang I-63
Zheng, Shaowen II-329
Zheng, Zhi I-200
Zhou, Liu-Ying II-43
Zhu, Qun-Xiong I-63
Zhu, Yanfeng I-24
Zhu, Yan-ling I-116
Zuo, Zhe I-50

Printed in the United States
by Baker & Taylor Publisher Services